# Sliding Mode Control in Dynamic Systems

# Sliding Mode Control in Dynamic Systems

Editors

**Jamshed Iqbal**
**Ali Arshad Uppal**
**Muhammad Rizwan Azam**

MDPI • Basel • Beijing • Wuhan • Barcelona • Belgrade • Manchester • Tokyo • Cluj • Tianjin

*Editors*

Jamshed Iqbal
University of Hull
Hull, UK

Ali Arshad Uppal
COMSATS University
Islamabad
Islamabad, Pakistan

Muhammad Rizwan Azam
COMSATS University
Islamabad
Islamabad, Pakistan

*Editorial Office*
MDPI
St. Alban-Anlage 66
4052 Basel, Switzerland

This is a reprint of articles from the Special Issue published online in the open access journal *Electronics* (ISSN 2079-9292) (available at: https://www.mdpi.com/journal/electronics/special_issues/Sliding_Mode_Control_in_Dynamic_Systems).

For citation purposes, cite each article independently as indicated on the article page online and as indicated below:

LastName, A.A.; LastName, B.B.; LastName, C.C. Article Title. *Journal Name* **Year**, *Volume Number*, Page Range.

ISBN 978-3-0365-8356-3 (Hbk)
ISBN 978-3-0365-8357-0 (PDF)

# Contents

# About the Editors

**Jamshed Iqbal**

Jamshed Iqbal is currently working as a Senior Lecturer (Associate Professor) at the University of Hull, UK. Prior to joining Hull, he worked as a Professor at the University of Jeddah, KSA. He holds a PhD in Robotics from the Italian Institute of Technology (IIT) and three Master's degrees in various fields of Engineering from Finland, Sweden and Pakistan. With more than 20 years of multi-disciplinary experience in industry and academia, his research interests include robotics, mechatronics and control systems. He has published more than 120 journal papers and has an H-index of 40. He is a senior member of IEEE USA and a Senior Fellow of Advance HE, UK.

**Ali Arshad Uppal**

Dr. Ali Arshad Uppal received his PhD degree in Electrical Engineering from COMSATS University Islamabad (CUI), Islamabad, Pakistan, in 2016. He worked as a postdoctoral researcher at the department of Electrical and Computer Engineering, University of Porto from September 2020 to September 2021. He is currently working as a Tenured Associate Professor at the department of Electrical and Computer Engineering, CUI. His research interests include nonlinear control/estimation, sliding mode control, process control, optimal control, numerical methods, artificial intelligence in control systems and control of infinite-dimensional systems. He has mainly applied his research skills to energy conversion systems, e.g., underground coal gasification (UCG) processes, airborne wind energy systems (AWES), surface coal gasifiers, drum boiler turbines and fuel cell systems. He is currently working on the modeling estimation and control of battery management systems for electric vehicles. So far, he has authored/co-authored 28 peer-reviewed journal publications, 5 conference papers and 1 editorial, and as an h-index of 12. Some of his articles have been published in top-ranked journals of control systems and energy, such as *IEEE Transactions on Control Systems Technology, Journal of Process Control, International Journal of Control, Journal of the Franklin Institute* and *Energy*. He has also served as a reviewer for various journals and conferences related to control systems. Moreover, he has managed a Special Issue—Sliding Mode Control in Dynamic Systems—in the MDPI journal *Electronics* as a guest editor. He was elected to the grade of senior member, IEEE, USA as a recognition of his research contributions in the area of control systems in June 2022.

**Muhammad Rizwan Azam**

Muhammad Rizwan Azam has worked as an Assistant Professor in COMSATS University Islamabad (CUI), Islamabad, Pakistan, since 2014. He holds a PhD in control systems from Capital University of Science & Technology (CUST), Islamabad, Pakistan. The focus of his research was to apply sliding mode control-based techniques for the drug dosage design of p53, a cancer suppressor protein. His research interests include linear/nonlinear controller/observer design, sliding mode control, energy management systems and systems biology. So far, he has authored/co-authored a total of seven peer reviewed journal publications, five conference papers and one editorial, and has an h-index of 5. Currently, he is working on the modeling and estimation of Lithium-ion batteries for electric vehicle applications.

*electronics*

MDPI

*Editorial*

# Sliding Mode Control in Dynamic Systems

**Ali Arshad Uppal [1], Muhammad Rizwan Azam [1] and Jamshed Iqbal [2,***

[1] Department of Electrical and Computer Engineering, COMSATS University Islamabad, Islamabad 45550, Pakistan
[2] School of Computer Science, University of Hull, Hull HU6 7RX, UK
* Correspondence: j.iqbal@hull.ac.uk

## 1. Introduction

Due to its inherent robustness and finite time convergence, sliding mode control (SMC) is extensively used for the control of nonlinear uncertain systems. Apart from robustness against parametric variations and external disturbances, it also provides model order reduction. It has been employed in a large range of dynamic systems, including but not limited to; electric drives, power converters, power systems, aircrafts, autonomous vehicles, a wide range of mechanical systems, and industrial processes. The discontinuous control law in SMC is the reason behind its robustness. However, it also suffers from for the so-called chattering problem. Much of the research in SMC is carried out to address the chattering phenomenon, which gives rise to higher order SMC, adaptive SMC and chattering free SMC. Another important research topic pertaining to SMC is the finite-time convergence of the error dynamics, which leads to the advent of finite time convergent SMC (FTSMC). Contrary to conventional SMC, the sliding variable in FTSMC is nonlinear which results in finite time convergence of the sliding mode.

The objective of this special issue is to bring together an articulate set of papers that advance understanding of the theory and practice behind SMC and its variants. The special issue is timely since recent years have witnessed the notable experimental realisation of SMC-based control laws in applications of immense importance. It is anticipated that the wider dissemination of recent research trends in SMC and its variants will stimulate more exchanges and collaborations among the control community and contribute to further advancements from an applied perspective.

The topics of interest are as follows:

- Sliding mode control-Theory and practice
- Sliding mode observers
- Higher-order sliding mode controllers
- Fast integral terminal sliding mode controller
- Fixed-time nonlinear homogeneous sliding mode controller
- Adaptive or Neuro-adaptive global sliding mode controller
- Role of SMC in Industry 4.0 cyber physical systems

## 2. Review of Published Papers

Asif et al. in [1] propose a two-stage maximum power point tracking (MPPT) approach for photovoltaic (PV) systems with partial shading conditions. The proposed method combines machine learning (ML) and terminal sliding mode control (TSMC) to improve MPPT algorithm accuracy. The first stage generates the reference voltage for MPPT by employing a neuro fuzzy network (NFN), whereas, the second stage tracks the maximum power point (MPP) voltage by employing a TSMC. The proposed method is validated through simulations as well as experiments. The results indicate that it outperforms traditional MPPT algorithms in generating higher power and ensuring finite time convergence of the MPP voltage tracking. Moreover, it effectively handles shading-induced multiple peaks and

**Citation:** Uppal, A.A.; Azam, M.R.; Iqbal, J. Sliding Mode Control in Dynamic Systems. *Electronics* **2023**, *12*, 2970. https://doi.org/10.3390/electronics12132970

Received: 28 June 2023
Accepted: 4 July 2023
Published: 5 July 2023

performs well under irradiance, temperature and load variations. The proposed two-stage approach offers a promising and robust solution for enhancing PV system performance in PSCs.

In [2], Yao, G. et al. have designed a sliding mode based observer to reconstruct the q-axis current of the permanent magnet synchronous motor (PMSM), provided that the d-axis current is measurable. The authors have employed the 2-phase model of the PMSM in a static reference frame for observer synthesis. Apart from robustness, the proposed technique also eliminates the blind area of current reconstruction. The estimated current is used to control the speed of the PMSM using a cascaded control system. The inner loop controls the currents of d and q axes of synchronously rotating reference frame, whereas the outer loop is responsible for maintaining a desired speed. The performance of the control system is tested for nominal and perturbed systems, which shows that the proposed method is robust.

In [3], Vennemann, J. et al. have developed a cascaded control system for active magnetic bearings (AMBs) with large air gaps. The aim of the control strategy is to regulate the positions of x and y axes AMBs, which is attained by two independent position controllers in the outer loop. The inner current control loop employs the super-twisting algorithm (STA) for maintaining the desired trajectories of currents produced by the outer loop controllers. The use of STA is motivated due to the fact that it only requires the measurement of the output signal, yields reduced chattering and improved tracking performance. Furthermore, the proposed methodology is implemented on a physical system. The results show that the designed methodology with STA is superior as compared to conventional control techniques.

The authors in [4] have designed a model-based, chattering free sliding mode controller (CFSMC) to track the desired trajectory of the calorific value of the exit gas mixture for an underground coal gasification process. The design of the CFSMC employs a nonlinear mathematical model of the UCG process. In order to estimate the unmeasurable states of the system involved in the controller synthesis, a state dependent Kalman Filter (SDKF) is also designed. The structure and algorithm of the SDKF is similar to a discrete time Kalman Filter, however, it utilizes the quasi-linear model of the UCG process to estimate the states on the entire operating range. The advantage of using a quasi-linear model is that it decomposes the system in the form of state space representation, with state dependent matrices, retaining the original nonlinear dynamics of the process. The simulation results of the designed technique are compared with dynamic integral sliding mode control and a conventional sliding mode control, which show that the tracking performance of the CFSMC is superior as compared to its counterparts.

Riaz S. et al. in [5] propose a robust predefined time convergent sliding mode controller (PreDSMC) algorithm for precise position control of permanent magnet linear motor (PMLM) systems for industrial applications. The proposed algorithm eliminates the position tracking error within a predefined time. Moreover, PreDSMC effectively handles external disturbances and model parameter variations with bounded convergence error and constrained control input. Numerical simulations demonstrate current research work's effectiveness in reducing the impacts of friction and external disturbances compared to traditional control methods (i.e., PID and linear SMC). The proposed algorithm has potential applications in trajectory tracking control of a wide range of industrial nonlinear systems.

In [6], the authors have proposed a hybrid backstepping based super-twisting algorithm for speed control of a 3-phase induction machine (IM) with squirrel cage rotor. The discontinuities in the input are minimized using exact differentiator. The simulations are performed in MATLAB and the IM is operated in three different modes: start and stop mode, normal operation mode and disturbance rejection mode. A comprehensive simulation study is carried out in which the proposed technique is compared with conventional sliding mode control, super-twisting control and backstepping control. The quantitative comparison in terms of integral squared error, integral absolute error and integral time absolute error shows that the proposed methodology has superior disturbance rejection and tracking performances.

Alvaro J. P. et al. in [7] address the windup phenomenon by proposing and experimentally realizing two SMC-based control algorithms to improve the response of integral controllers, which are common in industrial setups. The algorithms, named as 'windup conditional reset' and 'windup instantaneous reset', rely on resetting the integral control action while maintaining the disturbance rejection capability of the system. An anti-saturation function is used for anticipating and steering the error in trajectory tracking onto the manifold origin. This is complemented by an action to compensate for robustness. The algorithmic efficiency of the anti-windup techniques is characterized by applying them to two chemical processes; a stirred tank reactor and a mixing tank, which respectively feature the systems with variable longtime delay and with variable height. Different reference signals are applied considering various disturbances and relevant error benchmarks like integral time square error (ITSE) and integral square error (ISE) are observed. Experimental results evidence that the proposed algorithms circumvent actuators' saturation and offer adequate tracking accuracy despite application of disturbances as demonstrated by the reduced overshoot and settling time achieved in the response. It is shown that robustness is achieved without deteriorating stability of the system. Also, the original schemes of controllers do not need to be changed for realizing the proposed algorithms.

R. Hu et al. in [8] investigate coupling effects and nonlinearity in manipulators by realizing multi-joints coordination in the robot. The main objective here is to exploit the benefits offered by dynamically combining two control techniques namely; super-twisting SMC and fractional-order non-singular terminal SMC. The former control offers fast system response with suppressed chattering while the latter exhibits superior steady-state tracking accuracy. The proposed controller named 'Dynamic Fractional-Order Non-singular Terminal Super-Twisting SMC' gets benefit from these remarkable features and allows the position of the sliding surface to dynamically change. The controller design is based on the derived model of a four degree of freedom low-cost humanoid manipulator. Kinematics is derived using Denavit–Hartenberg (DH) parameters while the derivation of manipulator dynamics considers friction effects. Lyapunov theory is used to prove stability of the manipulator. The higher order term hiding the sign function suppresses chattering phenomenon in SMC. Simulation and experimental results demonstrate that the proposed control law permits accurate tracking accuracy, chattering mitigation and improved error convergence.

In [9], Jnayah, S. et al. have designed a robust direct torque control (DTC) strategy of an induction machine (IM) with a three-level inverter. The speed, flux and torque controllers employ sliding mode control theory, which proves to be more robust as compared to conventional hysteresis and PI based control algorithms. The simulation results show that the proposed control strategy is robust against sudden changes in load torque. Furthermore, the proposed SMC based DTC approach is implemented on a field programmable gate array (FPGA) using the Xilinx system generator. The parallel processing of the FPGA has been demonstrated through hardware co-simulation results.

Liang, Y. et al. in [10] propose a novel sliding mode controller for trajectory tracking in systems with unknown uncertainties. The proposed controller combines a PID type sliding surface with a variable gain hyperbolic reaching law, resulting in improved control performance. This choice of reaching law combines the benefits of integral sliding mode (ISM) and terminal sliding mode (TSM) control algorithms, and provides global finite-time convergence. Utilizing a variable gain hyperbolic term instead of conventional switching term eliminates chattering and ensures variable approaching velocities to the sliding surface for different initial positions. Simulation studies using a two-link robot validate the effectiveness of the proposed controller, demonstrating fast response, and high tracking accuracy even in the presence of time-varying uncertain external disturbances and load variations.

The authors in [11] have designed a dynamic integral sliding mode control (DISMC) strategy to boost and regulate the output voltage of proton exchange membrane fuel cell (PEMFC) system operating under varying loads. The voltage control problem is solved using a 2-loop control strategy, with a PI controller in the outer loop for voltage regulation and the inner current control loop employs the DISMC algorithm. The DISMC offers increased robustness due to the elimination of the reaching phase and it also yields a continuous control effort. The efficacy of the designed technique is validated on a hardware setup in which a portable PEMFC is connected to a boost power converter. The results of voltage regulation are compared for the DISMC and PI controllers in simulation and hardware settings, which show that the proposed scheme exhibits a better transient response and demonstrates superior robustness against sudden changes in load.

Li, Z. et al. in [12] present a sensorless control algorithm for permanent magnet linear synchronous motors (PMLSMs) that enhances tracking capability and estimation accuracy. The system combines a continuous terminal sliding mode controller (CTSMC) with a fuzzy super twisting sliding mode observer (FSTSMO). The CTSMC enables fast and continuous control, achieving rapid tracking of desired speed, while the FSTSMO algorithm improves sensorless estimation accuracy by using adaptive gains instead of fixed gains (as used in conventional SMO). Simulation as well as experimental results demonstrate the effectiveness of the proposed system, with reduced position tracking errors and improved dynamic performance. This research offers a promising solution for high-precision operation in sensorless PMLSM systems, with potential applications in various industries.

**Conflicts of Interest:** The authors declare no conflict of interest.

## References

1. Asif; Ahmad, W.; Qureshi, M.B.; Khan, M.M.; Fayyaz, M.A.B.; Nawaz, R. Optimizing Large-Scale PV Systems with Machine Learning: A Neuro-Fuzzy MPPT Control for PSCs with Uncertainties. *Electronics* **2023**, *12*, 1720. [CrossRef]
2. Yao, G.; Yang, Y.; Wang, Z.; Xiao, Y. Permanent Magnet Synchronous Motor Control Based on Phase Current Reconstruction. *Electronics* **2023**, *12*, 1624. [CrossRef]
3. Vennemann, J.; Brasse, R.; König, N.; Nienhaus, M.; Grasso, E. Super Twisting Sliding Mode Control with Compensated Current Controller Dynamics on Active Magnetic Bearings with Large Air Gap. *Electronics* **2023**, *12*, 950. [CrossRef]
4. Ahmad, S.; Uppal, A.A.; Azam, M.R.; Iqbal, J. Chattering Free Sliding Mode Control and State Dependent Kalman Filter Design for Underground Gasification Energy Conversion Process. *Electronics* **2023**, *12*, 876. [CrossRef]
5. Riaz, S.; Yin, C.-W.; Qi, R.; Li, B.; Ali, S.; Shehzad, K. Design of Predefined Time Convergent Sliding Mode Control for a Nonlinear PMLM Position System. *Electronics* **2023**, *12*, 813. [CrossRef]
6. Ali, S.; Prado, A.; Pervaiz, M. Hybrid Backstepping-Super Twisting Algorithm for Robust Speed Control of a Three-Phase Induction Motor. *Electronics* **2023**, *12*, 681. [CrossRef]
7. Prado, A.J.; Herrera, M.; Dominguez, X.; Torres, J.; Camacho, O. Integral Windup Resetting Enhancement for Sliding Mode Control of Chemical Processes with Longtime Delay. *Electronics* **2022**, *11*, 4220. [CrossRef]
8. Hu, R.; Xu, X.; Zhang, Y.; Deng, H. Dynamic Fractional-Order Nonsingular Terminal Super-Twisting Sliding Mode Control for a Low-Cost Humanoid Manipulator. *Electronics* **2022**, *11*, 3693. [CrossRef]
9. Jnayah, S.; Moussa, I.; Khedher, A. IM Fed by Three-Level Inverter under DTC Strategy Combined with Sliding Mode Theory. *Electronics* **2022**, *11*, 3656. [CrossRef]
10. Liang, Y.; Zhang, D.; Li, G.; Wu, T. Adaptive Chattering-Free PID Sliding Mode Control for Tracking Problem of Uncertain Dynamical Systems. *Electronics* **2022**, *11*, 3499. [CrossRef]
11. Yasin, A.; Yasin, A.R.; Saqib, M.B.; Zia, S.; Riaz, M.; Nazir, R.; Abdalla, R.A.E.; Bajwa, S. Fuel Cell Voltage Regulation Using Dynamic Integral Sliding Mode Control. *Electronics* **2022**, *11*, 2922. [CrossRef]
12. Li, Z.; Wang, J.; Wang, S.; Feng, S.; Zhu, Y.; Sun, H. Design of Sensorless Speed Control System for Permanent Magnet Linear Synchronous Motor Based on Fuzzy Super-Twisted Sliding Mode Observer. *Electronics* **2022**, *11*, 1394. [CrossRef]

 *electronics*

*Article*

# Optimizing Large-Scale PV Systems with Machine Learning: A Neuro-Fuzzy MPPT Control for PSCs with Uncertainties

Asif [1], Waleed Ahmad [1], Muhammad Bilal Qureshi [1], Muhammad Mohsin Khan [2], Muhammad A. B. Fayyaz [3,*] and Raheel Nawaz [4]

[1] Department of Electrical and Computer Engineering, COMSATS University Islamabad, Abbottabad Campus, Abbottabad 22060, Pakistan; bilalqureshi@cuiatd.edu.pk (M.B.Q.)
[2] Sino-Pak Center for Artificial Intelligence (SPCAI), Pak-Austria Fachhochschule-Institute of Applied Sciences and Technology (PAF-IAST), Haripur 22620, Pakistan
[3] Department of Operations Technology, Events and Technology Management, Manchester Metropolitan University, Manchester M15 6BH, UK
[4] Pro Vice Chancellor (Digital Transformation), Staffordshire University, Stoke-on-Trent ST4 2DE, UK
* Correspondence: m.fayyaz@mmu.ac.uk

**Abstract:** The article proposes a new approach to maximum power point tracking (MPPT) for photovoltaic (PV) systems operating under partial shading conditions (PSCs) that improves upon the limitations of traditional methods in identifying the global maximum power (GMP), resulting in reduced system efficiency. The proposed approach uses a two-stage MPPT method that employs machine learning (ML) and terminal sliding mode control (TSMC). In the first stage, a neuro fuzzy network (NFN) is used to improve the accuracy of the reference voltage generation for MPPT, while in the second stage, a TSMC is used to track the MPP voltage using a non-inverting DC—DC buck-boost converter. The proposed method has been validated through numerical simulations and experiments, demonstrating significant enhancements in MPPT performance even under challenging scenarios. A comprehensive comparison study was conducted with two traditional MPPT algorithms, PID and P&O, which demonstrated the superiority of the proposed method in generating higher power and less control time. The proposed method generates the least power loss in both steady and dynamic states and exhibits an 8.2% higher average power and 60% less control time compared to traditional methods, indicating its superior performance. The proposed method was also found to perform well under real-world conditions and load variations, resulting in 56.1% less variability and only 2–3 W standard deviation at the GMPP.

**Keywords:** maximum power point tracking; machine learning; partial shading; terminal sliding mode control

**Citation:** A.; Ahmad, W.; Qureshi, M.B.; Khan, M.M.; Fayyaz, M.A.B.; Nawaz, R. Optimizing Large-Scale PV Systems with Machine Learning: A Neuro-Fuzzy MPPT Control for PSCs with Uncertainties. *Electronics* **2023**, *12*, 1720. https://doi.org/10.3390/electronics12071720

Academic Editor: Enrique Romero-Cadaval

Received: 1 February 2023
Revised: 10 March 2023
Accepted: 28 March 2023
Published: 4 April 2023

## 1. Introduction

The continuous increase in global warming and the decrease in fossil energy sources has led to a sharp inclination towards renewable sources as a substitute source of energy. Among these renewable sources, the solar system has been extensively used for power generation in a variety of applications due to its multiple benefits, such as uninterrupted power, no noise, no pollution, and easy maintenance. This increasing demand for power generation using photovoltaic (PV) systems for both residential and industrial areas requires an appropriate and efficient optimization of energy production systems. To draw maximum PV energy, many methods have been developed in the past, such as incremental conductance, perturb and observe, constant voltage, parasitic capacitance, and constant current. This allows the controller to harvest the maximum available power from the PV system under varying solar irradiance and temperature scenarios; these controllers are usually known as maximum power point tracking (MPPT) controllers. However, most MPPT methods suffer from a lack of strict convergence analysis and are not capable of handling partial shading conditions (PSCs) [1,2].

### 1.1. Literature Review

This research aims to find a more effective solution for PV systems that are partially shaded by analyzing and modeling their performance. By simulating the relationship between environmental conditions and the PV array's output characteristics, we can gain a deeper understanding of the special effects of PS. This knowledge is crucial for developing an MPPT algorithm. Power generation in solar systems is influenced by the intermittence of both temperature and irradiance. The output power can also fluctuate because of the non-linear relationship between current and voltage in conventional PV cells. Therefore, incorporating an MPPT algorithm can help to maximize power output under varying meteorological conditions. However, when PV systems are partially shaded, they exhibit distinct characteristics, and multiple peaks may appear on the P-V characteristic curve. The impact of shading depends on the pattern and placement of the PV arrays, which can decrease the efficiency of the tracking algorithm as the PV array tends to operate at a local MPP [3–5].

MPPT calculations are an essential part of the operation of PV systems. These calculations are designed to optimize the output power of PV panels by determining the point on the P-V curve where maximum power can be obtained. Under ideal conditions where all PV cells receive the same amount of sunlight, the MPPT algorithm can quickly and easily locate a single peak on the P-V curve and adjust the system accordingly. However, there are scenarios where the P-V curve of the panel has multiple local peaks. One of the prevalent reasons is PS, which happens when the cells in a panel are exposed to varying levels of sunlight. This can occur when the panel is partially blocked by an object such as a tree or building, or when the panel is installed in a location with uneven sunlight. In such scenarios, the standard MPPT calculations may only identify the local maximum power point (LMPP) in the previous working point region, as the global maximum power point (GMPP) could be located much farther away on the P-V curve [6]. Therefore, it is crucial to utilize sophisticated MPPT methods and techniques to trace the global peak power point during PS scenarios and augment the general effectiveness and operation of the PV system. This will immensely reduce the competitiveness of the PV panel, in particular during PS. To solve PS, numerous approaches have been proposed to alleviate the effect of PSCs in such PV panels and a lot of research has been carried out, focusing on finding the GMPP by reducing the search area.

Hu et.al presented an idea that states the current of faulted cells or modules increases under the PSC, causing the temperature rise of a few faulted modules or cells [7]. Therefore, there are a lot of LMPPs. The usage of a thermal camera to identify PSCs is also proposed. The proposed approach can detect the cell or module faults and can make use of the thermographical data gathered from panels to split the PV array into healthy and unhealthy segments and also efficiently determine PS. However, this method consumes a lot of computational time for global maximum power point tracking (GMPPT) [8].

Tamir Shaqarin suggested an approach that works well for tracking precision and steady-state error to track the GMPP under any climate condition. The proposed approach is using "particle swarm optimization (PSO) through targeted position-mutated elitism" (PSO-TPME) with a reinitialization mechanism on a PV system under partial shading conditions. The fast-converging and global exploration capabilities of PSO-TPME make it appealing for online optimization. But a significant implementation complexity is associated with PSO-TPME based MPPT [9].

A MPPT approach with minimal complexity is suggested by A. Safari which is founded on an adjustable step size incremental conductance method and a straightforward linear equation [10,11]. The aim of this approach is to relocate the operating point near the GMPP. This technique is based on a variable step size incremental conductance, which automatically adjusts the step size to track the GMPP and minimize energy loss. However, this method has a significant drawback, as the linear function may not be effective when the PV array has multiple LMPPs. Furthermore, this approach may struggle to adapt to system parameters that change over time.

Different "perturb & observe" (P&O) and incremental conductance (InC) MPP techniques have been used in the PV system by the researcher for detecting the GM using P-V & I-V curves. The techniques used in [12] are evaluated based on speed, accuracy, and complexity. Fractional short circuit current (FSCC), P&O, fractional open circuit voltage (FOCV), and InC techniques are conventional and the most commonly employed techniques in PV energy systems. However, the major drawback of these techniques is lack of robustness and frequent oscillations around local optima [13].

An adaptive neuro-fuzzy (NF) interference system strategy is proposed in [14,15] which is applied to extract MMPT for the PV system under intermittent environmental circumstances; however, this technique based on machine learning requires a huge database. Additionally, this technique suffers from the chattering effect, steady-state error, and oscillations in the desired output.

Other researchers have also employed population-based global optimization methods which have been combined with deep neural networks to enhance their global exploration capabilities and reduce their computational complexity [16–18] such as the firefly algorithm (FA) [19], the artificial bee colony algorithm (ABC) [20], and the genetic algorithm (GA) [21] to achieve maximum power, but these algorithms suffer from convergence speed, require lot of tuning parameters (population size, crossover probability, and mutation rate), and are sentinel to noise. Moreover, all these PSO, GA, FA, and ABC population-based algorithms are not efficient when it comes to the control problem, because of their inability to handle uncertainty and nonlinear systems efficiently and are less flexible to changing environment conditions.

In conclusion, it is obvious that there are multiple MPPT techniques available with distinct features, including the type of sensing material, rate of convergence, level of intricacy, efficiency, expense, and suitability, all targeting to locate the GMPP under PSCs. Each method is associated with its advantages and disadvantages but a general drawback that is being observed in all traditional techniques is the challenge of chattering and slow convergence. Focusing on this research problem, a hybrid MPPT technique is proposed that can perform under varying climate conditions and shading patterns with slight chattering, and can assure fast and finite convergence to GMMP.

*1.2. Original Contribution*

The following are in-depth descriptions of the key innovations and significant contributions made through this research.

1. This research presents a strong and sophisticated controller known as the nonlinear terminal sliding mode controller (TSMC) that is specifically designed to track the MPPT of PV arrays' PSCs by utilizing a non-inverting buck-boost converter.
2. To achieve this, the proposed controller utilizes a neuro-fuzzy network (NFN) for reference generation, which is trained using over 22,000 distinct PS scenarios.
3. The proposed controller is designed to ensure fast and finite-time convergence, providing a reliable and efficient solution for MPPT under PSCs.
4. The robustness and chattering minimization around the GMPP are tested by introducing uncertainties in the system, demonstrating the success of the proposed controller in challenging operating conditions.
5. To demonstrate the efficacy of the proposed controller, an experimental setup is established, which allows for a comprehensive evaluation of the controller's performance.
6. To evaluate its performance, a comparison of the proposed controller with other algorithms is already available. The results are presented in Table 1, which illustrates the superiority of the proposed controller in terms of its performance characteristics. The comparison highlights the effectiveness of the proposed controller and its ability to perform better than other existing algorithms.

**Table 1.** Comparison of proposed work with existing techniques.

| References | Methodology | Merits | Demerits |
| --- | --- | --- | --- |
| Punitha et. al. [22] | Incremental conductance | Fast response and good tracking | Offline and large computational time |
| Patel et. al. [23] | Neural networks | High tracking speed | Oscillations at MPPT |
| Koad et. al. [24] | Particle swarm optimization | Reduce steady state error | Large computational time |
| El-Helw et. al. [25] | Perturb and observe | Easy implementation | Slow tracking and oscillation around MPTT |
| Proposed work | Neuro-fuzzy and TSMC | Robust, fast convergence and minimum chattering | Offline technique, requires large dataset |

The present study is organized in a systematic manner to comprehensively address the proposed PV system with a MPPT controller. The following sections outline the methodology and results of the study.

Section 2 delves into the statistical and mathematical modeling of the given PV system, providing a thorough understanding of the system's behavior. The shading effect, a crucial aspect of PV systems, is described in Section 3. In Section 4, the use of a machine learning-based neuro-fuzzy network (MLNFN) is presented for the generation and training of reference voltage. This methodology is applied to improve the performance of the MPPT controller. The average state-space model of the DC—DC buck-boost convertor is explained in Section 5, while Section 6 presents the robust nonlinear TSMC. To evaluate the proposed controller's performance under varying environmental conditions, simulation results are analyzed using MATLAB/Simulink in Sections 7 and 8. The hardware validation of the proposed system and a comprehensive performance analysis are presented in Sections 9 and 10, respectively. Finally, the study concludes in Section 10, summarizing the key findings and highlighting the contributions made to the field of PV systems and MPPT controllers.

## 2. PV System Mathematical Modeling

The PV cell has a p-n junction and produces electric power due to the photons. It consists of a current source $I_{ph}$, a series resistance $R_s$, a shunt resistance $R_p$, and a diode, as shown in Figure 1.

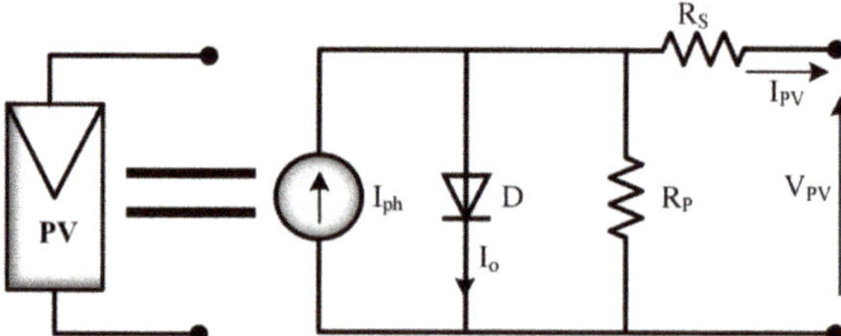


**Figure 1.** PV array equivalent model.

To evaluate the parameters of the photovoltaic (PV) system it is necessary to have knowledge of the PV power-voltage or current-voltage curve under standard conditions of measurement (SCM). This information can be obtained through the testing and characterization of the PV module under SCM, which typically includes a specific irradiance level and cell temperature. This information can then be used to model the PV system and determine the performance and efficiency of the system under different operating conditions.

As $R_p$ and $R_s$ have low values, they can be neglected in some cases. This means that the total current is mainly determined by the photocurrent and the resistance losses are considered to be negligible. However, this will depend on the specific PV system and the operating conditions, and it is always important to consider the actual values of $R_p$ and $R_s$ to make sure they can be neglected. The current of a PV array arranged in a combination of series and/or parallel can be described by [26,27].

$$I = I_{PV}N_P - I_O N_P \left[ \exp\left( \frac{V + R_S\left(\frac{N_S}{N_P}\right)I}{V_T a N_s} \right) - 1 \right] - \frac{V + R_S\left(\frac{N_S}{N_P}\right)I}{R_P\left(\frac{N_S}{N_P}\right)}, \tag{1}$$

where

$I_{PV}$: PV array current
$I_O$: Diode saturation current
$a$: Ideal factor
$R_S$: Resistance in series
$R_P$: Resistance in parallel
$N_S$: Number of series cell
$N_P$: Number of parallel cell
$V_T$: Thermal voltage

PV array thermal voltage is given by,

$$V_T = \frac{N_S K T}{q} \tag{2}$$

where

$q$: Electron charge
$K$: Boltzmann's constant
$T$: Temperature of p-n junction

The PV current $I_{PV}$ is given by,

$$I_{PV} = (I_{PVN} + K_i \Delta T)\frac{G}{G_N} \tag{3}$$

where

$I_{PVN}$: Nominal condition PV current
$G$: Irradiance at the surface of panel
$G_N$: Nominal condition irradiance
$K_i$: Temperature coefficient of short circuit current
$\Delta T$: Difference of nominal and actual temperature

Saturation current of the diode is represented by,

$$I_O = \frac{I_{SCN} + K_i \Delta T}{\exp\left(\frac{V_{ocn} + K_i \Delta T}{a V_t}\right) - 1} \tag{4}$$

where

$I_{SCN}$: Nominal condition short circuit current
$V_{OCN}$: Nominal condition open circuit voltage

MATLAB/Simulink is used to model and simulate the above-mentioned equations and corresponding results are presented in Figures 2, 3, 9 and 10.

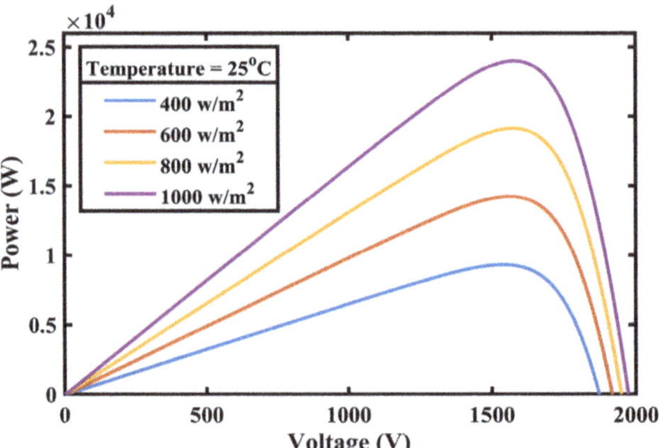

**Figure 2.** PV curve at different irradiances.

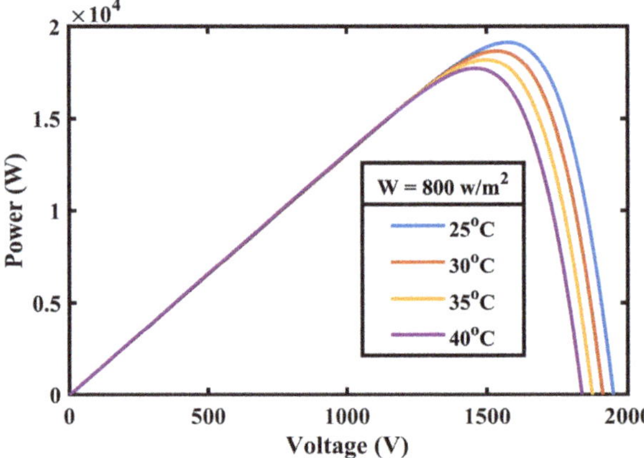

**Figure 3.** PV curve at different temperatures.

These figures illustrate the I-V and P-V attributes of the PV system, which are used in the current work and subject to different levels of irradiances and temperatures. Four PV arrays are linked in a series such that each array comprises of fifteen series modules and a total output of 24 KW power of the other two parallel strings.

The P-V curve of the system is presented in Figure 2 with a constant temperature that is directly proportional to power. While Figure 3 shows the P-V curve of the system at different temperatures by keeping irradiance constant, which is inversely proportional to power.

## 3. Influence of Shading Effect on PV Array

It is a common practice to connect multiple solar panels in a series or parallel configuration to meet power requirements. However, when certain panels are blocked from sunlight by passing clouds, nearby trees or buildings, the shaded cells absorb some of the power generated by unshaded cells, resulting in the generation of heat which can cause damage to the shaded cells [28].

To prevent this, bypass diodes can be employed as depicted in Figure 4, which help to prevent a negative voltage across the shaded cells. From Figure 4, the bypass diode begins to conduct when the condition $V_i - \sum_{i=1}^{n} V_i \geq V_D$ is satisfied where $i \neq 2$ and $V_D$ is the forward voltage drop of the diode [29,30].

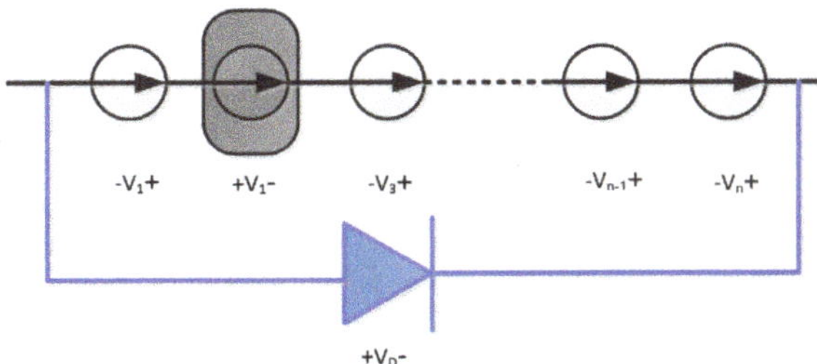


**Figure 4.** Bypass diode parallel with PV cells.

However, activating diodes at different voltage levels alters the characteristics of the PV system and results in multiple peaks, as shown in Figure 5. This trait can cause the system to operate at a local peak instead of the optimal global peak, resulting in a decrease in PV efficiency.

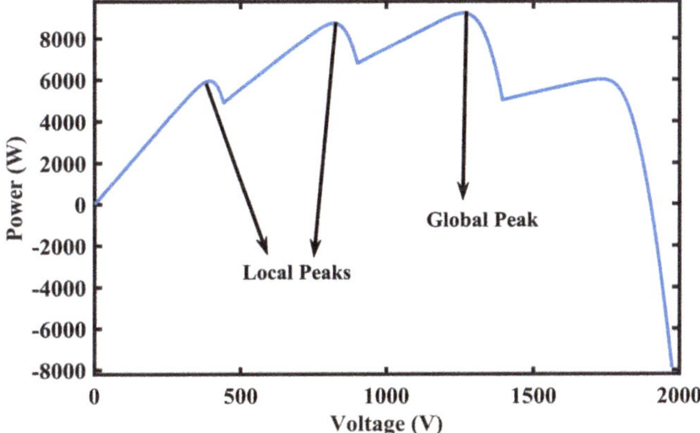


**Figure 5.** Global and local peaks due to the shading effect.

## 4. Machine Learning Based Proposed System Framework

Figure 6 illustrates the MLNF technique used in this research, which is based on the Takagi-Sugeno (TS) fuzzy inference system. TS and Mamdani are two popular types of fuzzy inference systems used in the field of fuzzy logic. The main difference between the two is the way they handle fuzzy rules and fuzzy outputs. In Mamdani-type systems, the output of each rule is a fuzzy set, which is then combined to form the final output using fuzzy logic operations such as union or centroid defuzzification. In contrast, TS-type systems use a linear combination of the inputs to generate a crisp output. There are several advantages of using TS-type systems as they are computationally efficient, accurate, provide better interpolation, and are easy to tune compared to Mamdani-type systems [31].

The proposed framework has five inputs, including four different irradiance values (G1, G2, G3, and G4) and one temperature (T) parameter. The fuzzification layer, which is the input layer, consists of three triangular membership functions for each parameter. The output layer comprises a linear equation for each rule. The MLNFN generates a reference voltage for the peak power from the PV arrays under PS and varying environmental conditions.

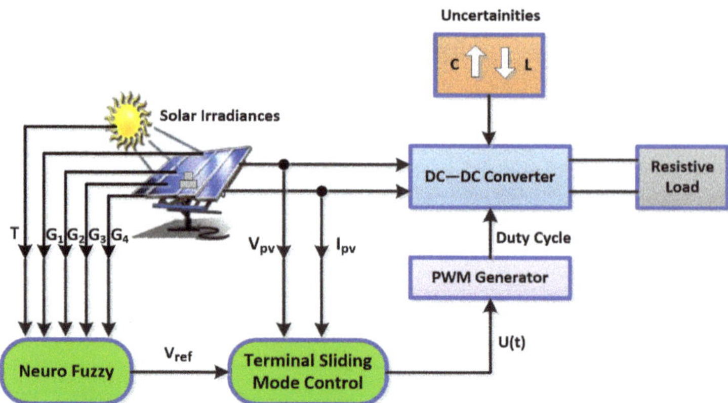



**Figure 6.** Proposed MLNF control scheme.

Large datasets are essential for training the MLNFN. These datasets are generated by modeling the PV array system in MATLAB/Simulink under different (PSCs). To train the NFN, around 22,000 scenarios of PSCs are randomly generated for MPP voltages to be used as reference voltages for the controller.

Figure 7 illustrates the reference voltages generated using the MLNFN against three different membership functions at shading pattern SP1, which lasts from 0 to 0.5 s, and SP2, which lasts from 0.5 to 1 s. The voltage changes abruptly from 1645 V to 1265 V when irradiances change from SP1 to SP2 in the case of triangular membership functions. Table 2 provides a comparative analysis of reference voltages generated using different NF techniques. The table helps in understanding the effectiveness of the proposed technique over the existing ones.

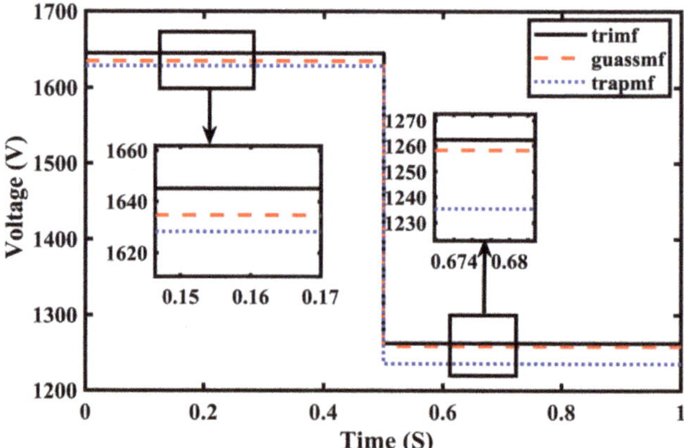



**Figure 7.** Reference voltage generation using various NF techniques.

**Table 2.** Comparison of reference voltage prediction for various NF techniques.

| FIS Membership Functions | Shading Pattern (SP) | Reference Voltage (V) | Actual Voltage Generated (V) | Error (V) |
|---|---|---|---|---|
| TRIMF | SP1 | 1262 | 1262.1 | 0.1 |
| TRAPMF | SP1 | 1262 | 1235 | 27 |
| GUASSMF | SP1 | 1262 | 1258 | 4 |

## 5. Non-Inverting Average State-Space Model of Buck-Boost Converter

The DC—DC buck-boost converter in non-inverting configuration moves up or down the voltages to force the PV array to operate at the MPP, from input (PV array) to output (load). With the help of the switching period $T$, the converter is periodically controlled where: $T = T_{on} + T_{off}$, $T_{on}$ is the ON time and $T_{off}$ is the OFF time. The converter's duty ratio is defined by $u = T_{on}/T$.

To reduce the waves in the converter, input voltage capacitor $C_1$ is used; while for limiting the output voltage capacitor, $C_2$ is used. In this work, it is assumed that the converter is operating in continuous conduction mode (CCM).

Figure 8 shows the approximate circuit of the non-inverting buck-boost converter with two switching intervals. The first switching interval has both switches, $S_1$ and $S_2$ active while the diodes $D_1$ and $D_2$ are inactive. In the second switching interval, both the diodes $D_1$ and $D_2$ are active while the switches; $S_1$ and $S_2$ are inactive [32,33]. In the first interval, according to Kirchhoff's current and voltage law, we have:

$$I_{PV} = I_{PV} - I_L \tag{5}$$

$$V_L = V_{C1} \tag{6}$$

$$I_{C2} = -\frac{V_{C2}}{R} \tag{7}$$

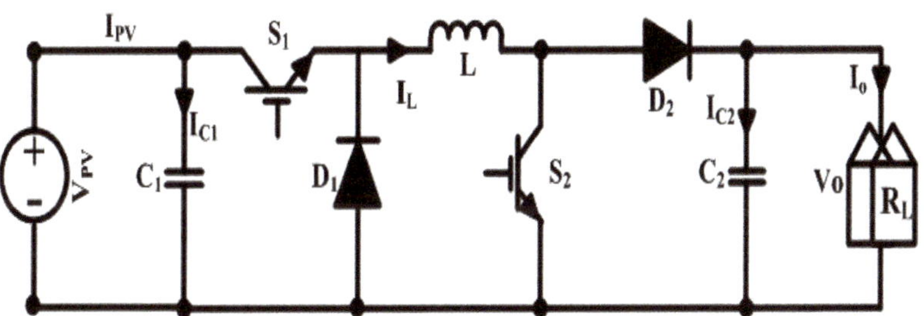


**Figure 8.** Non−inverting configuration of DC-DC buck-boost converter.

Whereas in the second interval $S_1$ and $S_2$ are inactive and $D_1$ and $D_2$ are forward-biased. Using Kirchhoff's current and voltage law, we have:

$$I_{C1} = I_{PV} \tag{8}$$

$$V_L = -V_{C2} \tag{9}$$

$$I_{C2} = I_L - \frac{V_{C2}}{R} \tag{10}$$

In light of the volt second balance of the inductor and charge balance of the capacitor, we can express:

$$\frac{dv_{c1}}{dt} = \frac{I_{PV}}{C_1} - \frac{I_L}{C_1}u \tag{11}$$

$$\frac{di_L}{dt} = \frac{V_{C1}}{L}u - \frac{V_{C2}}{L}(1-u) \tag{12}$$

$$\frac{dv_{c2}}{dt} = \frac{I_L}{C_2}(1-u) - \frac{V_{C2}}{R_{C2}} \tag{13}$$

Averaging the model for one switching duration and considering $x_1$, $x_2$ and $x_3$ to be the average value of $V_{C1}$, $I_L$ and $V_{C2}$, we can write, $x_1 = <v_{c1}>$, $x_2 = <i_L>$, $x_3 = <v_{C2}>$ and $u = <u>$.

Hence Equations (11)–(13) can be rearranged as,

$$\dot{x}_1 = \frac{i_{pv}}{c_1} - \frac{x_2}{c_1}u \tag{14}$$

$$\dot{x}_2 = \frac{x_1}{L}u - \frac{x_3}{L_1}(1-u) \tag{15}$$

$$\dot{x}_3 = \frac{x_2}{c_2}(1-u) - \frac{x_3}{R_{C2}} \tag{16}$$

The final Equations (14)–(16) are utilized in the designing of PV system control law.

## 6. Design of Terminal Sliding Model Control

A controller design based on the nonlinear robust terminal sliding mode (TSM) is proposed for tracking the MPPT of PV arrays under PSCs using a non-inverting buck-boost converter. In this controller, the error, $e1$, is defined as the discrepancy between the desired output voltage of the PV array and the actual one and is given in Equation (17). The controller uses this error to adjust the reference voltage generated by the MLNFN algorithm and to extract the maximum power from the PV array,

$$e_1 = x_1 - x_{1d} \tag{17}$$

where $x_1$ is $V_{pv}$ and $x_{1d}$ refers to $V_{ref}$. The derivative of Equation (17) with its dynamics reported in (15) becomes,

$$\dot{e}_1 = \frac{i_{pv}}{c_1} - \frac{x_2}{c_1}u - x_{1d} \tag{18}$$

The first stage is to design a sliding surface and the next stage is the selection of a control law for holding the system trajectory on the sliding surface making the tracking error zero. The equation of TSMC is given by,

$$s = \left[\frac{d}{dt} + \propto\right]^{n-1} e_1 \tag{19}$$

$$s = e_2^r + \propto e_1 \tag{20}$$

where $e_2 = \dot{e}_1$, $n$ represents the relative system degree, $e_1$ is the error among the $\propto$, desired reference voltage, a positive parameter chosen randomly or by some approach whichever is more suitable choice and the output voltage, $r$ is rational power equal to $p/q$.

By taking the derivative of sliding surface we have,

$$\dot{s} = re_2^{r-1}\dot{e}_2 + \propto \dot{e}_1 \tag{21}$$

Control law is given by Equation (22),

$$u(t) = u_{eq} + u_{dis} \tag{22}$$

$$u_{dis} = -k_1(s) - k_2 sign(s) \tag{23}$$

where $u_{eq}$ is the equivalent control vector while $u_{dis}$ is the discontinuous control (the correction factor) vector which is given by $u_{dis} = ksign(s)$, where k is a controlled gain. To obtain $u_{eq}$, Equation (22) will be simplified, the parameters used in the equation are given as,

$$e_1 = x_1 - x_{1d} \tag{24}$$

$$\dot{e_1} = \frac{i_{pv}}{c_1} - \frac{x_2}{c_1}u - x_{1d} \tag{25}$$

$$e_2 = \propto^{\frac{1}{r}} e_1^{\frac{1}{r}} \tag{26}$$

Taking derivative of $e_2$ yields,

$$\dot{e_2} = \frac{1}{r} \propto^{\frac{1}{r}} e_1^{\frac{1}{r}-1} \dot{e_1} \tag{27}$$

By substituting the parameters in Equation (22) we obtain,

$$\dot{s} = r \left[ \propto^{\frac{1}{r}} e_1^{\frac{1}{r}} \right]^{r-1} \left[ \propto^{\frac{1}{r}} e_1^{\frac{1}{r}} \left( \frac{i_{pv}}{c_1} - \frac{ux_2}{c_1} - \dot{x}_{1d} \right) \right] + \\ \propto \left( \frac{i_{pv}}{c_1} - \frac{ux_2}{c_1} - \dot{x}_{1d} \right) \tag{28}$$

By taking $\dot{s} = 0$ we obtain, $u_{eq}$ as,

$$u_{eq} = \frac{i_{pv}(r-1) + \dot{x}_{1d}c_1(1-r)}{x_2(r-1)} \tag{29}$$

Finally, by adding Equations (23) and (29) we obtain the control law $u(t)$,

$$u(t) = \frac{i_{pv}(r-1) + \dot{x}_{1d}c_1(1-r)}{x_2(r-1)} - k_1 k_2 sign(s) \tag{30}$$

## 7. Proposed Control System Performance Validation

This section explains the validation of the proposed system for MPPT under various PSCs using MATLAB/Simulink.

### 7.1. Simulation Setttings

MATLAB 2017Ra is used to perform the simulations where powergui and a constant time solvent are opted for to run the simulation. The information about the PV array being used in this research is given in Table 3. The parameter values used in converter and controller designing are given in Table 4. Three distinct shading patterns (SPs) namely SP1, SP2, and SP3 are subjected to varying irradiance, while SP4 exhibits uniform irradiance, as mentioned in Table 5. For the SP4 case with regular irradiance, only a single PV curve is produced, which in turn keeps a simpler detection for the MPP through any traditional methods. However, in the remaining three cases of PS, it is difficult to identify the global peak, as the PV characteristic curve now exhibits multiple local maxima points instead of just one GMPP. These local maxima points appear due to the shading of certain portions of the PV array, which can cause the power output of the shaded portions to decrease, creating multiple peaks in the PV curve, making it challenging to pinpoint the GMPP accurately.

**Table 3.** Parameters of PV system.

| Parameters | Symbols | Values |
|---|---|---|
| Maximum power | $P_{MAX}$ | 200 W |
| Open circuit voltage | $V_{OC}$ | 32.90 V |
| Optimum voltage | $V_{MP}$ | 26.30 V |
| Short circuit current | $I_{SC}$ | 8.210 A |
| Optimum current | $I_{MP}$ | 7.610 A |
| Temperature coefficient of $I_{SC}$ | $T_{SC}$ | 0.00318 A/°C |
| Temperature coefficient of $V_{OC}$ | $T_{OC}$ | −0.123 V/°C |
| Parallel resistance | $RP$ | 601.33 Ohms |
| Series resistance | $RS$ | 0.23 Ohms |

**Table 4.** Parameters of converter and controller.

| Parameters | Symbols | Values |
| --- | --- | --- |
| Capacitor at input | $C1$ | 1 mF |
| Capacitor at output | $C2$ | 48 µF |
| Inductor | $L$ | 20 mH |
| Switching frequency | $FS$ | 5000 Hz |
| Constant | $K1$ | 10 |
| Constant | $K2$ | 100 |
| Rational power | $R$ | 0.501 |

**Table 5.** SP for P-V and I-V curves.

| SP | $G_1$ (w/m²) | $G_2$ (w/m²) | $G_3$ (w/m²) | $G_4$ (w/m²) |
| --- | --- | --- | --- | --- |
| SP1 | 200 | 400 | 600 | 1000 |
| SP2 | 400 | 500 | 800 | 800 |
| SP3 | 600 | 600 | 1000 | 1000 |
| SP4 | 1000 | 1000 | 1000 | 1000 |

The proposed MLNFN TSMC strategy is evaluated from three distinct characteristics i.e., (1) robustness to environmental changes (2) PSCs, and (3) controller uncertainties.

Initially, the simulations were carried out to achieve the P-V and I-V attribute curves of the four SPs mentioned above in Table 5. Figure 9 shows the P-V curve attributes under SP1 to SP4 while Figure 10 shows I-V curve under SP1 to SP4.

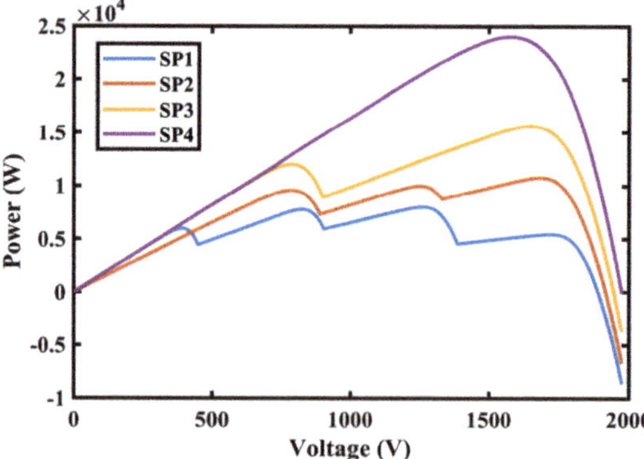

**Figure 9.** P-V curves at different SPs.

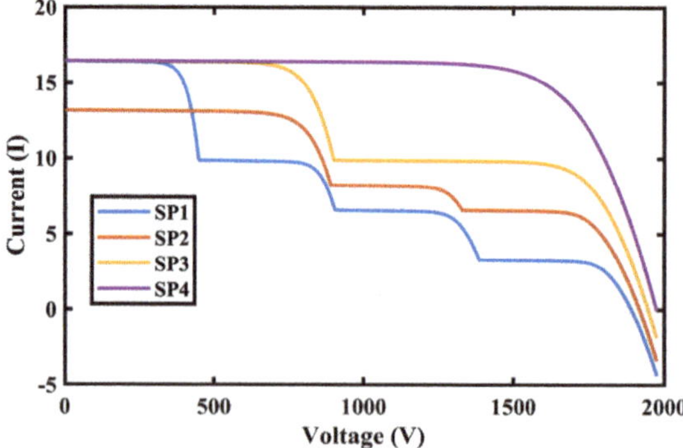

**Figure 10.** I-V curves at different SPs.

It can be noted from the results that irradiances are directly proportional to the current/voltage; however, when SPs are different then various local peaks have happened.

*7.2. Test Case Senarios*

Two different sets of SPs with different temperatures taken for the comparative performance analysis of the proposed technique are shown in Table 6.

**Table 6.** SP for test case scenarios.

| SP | $G_1$ (w/m$^2$) | $G_2$ (w/m$^2$) | $G_3$ (w/m$^2$) | $G_4$ (w/m$^2$) | T ($^\circ$C) |
|---|---|---|---|---|---|
| SP1 | 200 | 400 | 600 | 1000 | 30 |
| SP2 | 400 | 500 | 800 | 800 | 25 |

It was observed in the case of the SP1 scenario that the GMPP was situated at 1645 V and the associated output power was 15,000 W, as shown in Figure 11. While for the case of SP2, Figure 12 shows that MPP was at 1250 V and the associated output power was 8000 W. These results clearly show that the proposed MLNFN TSMC efficiently tracks the GMPP in the existence of local peaks.

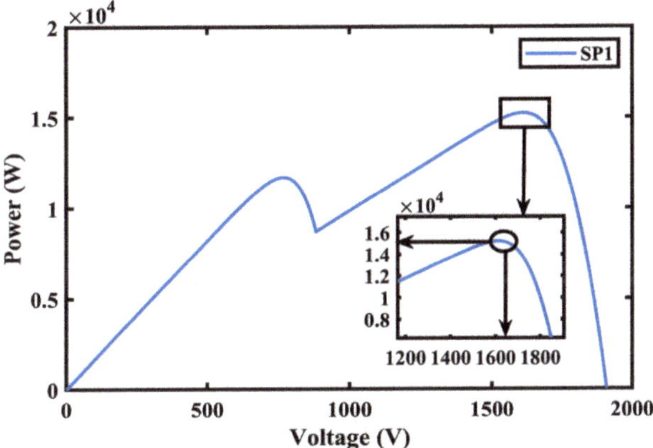

**Figure 11.** Voltage and power at SP1.

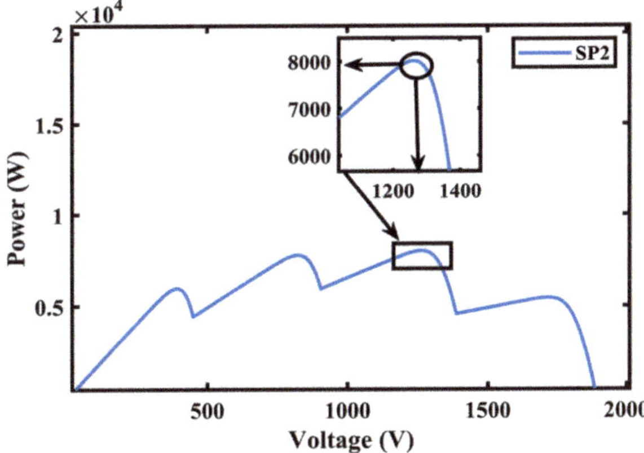

**Figure 12.** Voltage and power at SP2.

## 8. Comparative Analysis of Proposed Technique

For detailed analysis regarding the performance improvement achieved with the proposed strategy, a comparative study with existing conventional techniques was carried out and can be found in the subsequent sections.

### 8.1. Performance Analysis without Uncertainities

Initially, the proposed MPPT technique's efficiency was evaluated under both steady-state and dynamic conditions. The dynamic response of the technique is depicted in Figure 13, which shows the tracking of voltage when there was a sudden change in the set point from SP1 to SP2. The PV terminal voltage at 1645 V was regulated by the proposed MPPT technique, and at SP1, the output power reached 15,000 W, which was the MPP. In the proposed MPPT technique, at $t = 0.5$ s, the SP suddenly changed from SP1 to SP2. This resulted in locating the new MPP at 1265 V, where the array power output changed to 8000 W, as shown in Figure 13 successfully. Furthermore, the planned MPPT technique instantly controlled the duty ratio of the buck-boost converter.

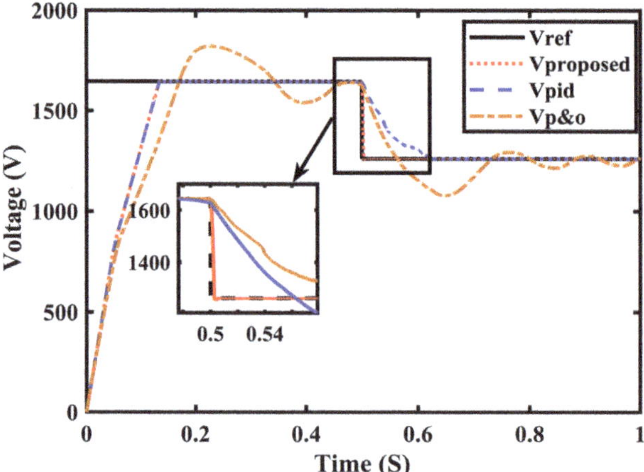


**Figure 13.** Voltage tracking without uncertainties.

Moreover, at the same time for the same system, the simulations run for conventional techniques (P&O and PID). As shown in Figure 13, the conventional techniques failed to track instantly the voltage at the time of dynamic change in the shading pattern; that was at 0.5 s when SP changed from SP1 to SP2. It can be noted that in this scenario the simulations were carried out without adding uncertainties in the system and it is clear from the results that the proposed method obtained success in tracking the MPP with a rapid response under different shading patterns.

### 8.2. Performance Analysis with Uncertainities

The Simulink model which is described already is used by adding some uncertainties into the system to check the robustness and to make a comparison of it with other existing techniques shown in Figure 14. Uncertainties in the shape of capacitance and inductance were added to the parameters of the buck-boost converter. A capacitance of 58 $u$F was added in parallel with C2 and an inductance of 30 mH was added in series with an inductor which was conducted for 0.7 to 0.8 s by mean of signal builder and switches.

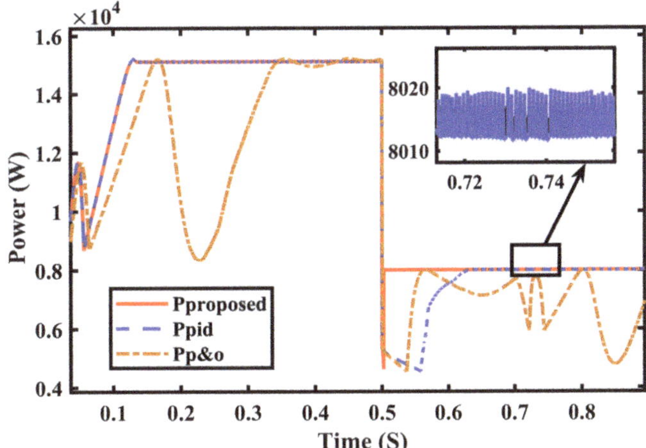

**Figure 14.** Power analysis without uncertainties.

The voltage tracking and the changing shading pattern from SP1 to SP2 for the proposed technique and other existing methods (P&O and PID) are displayed in Figures 15 and 16. Initially, SP1 was applied where the MPP was located at 1645 V and its subsequent output power was 15,000 W. This proposed methodology successfully operated at this point. The P&O worked at 1630 V and the calculated value of the output was 14,500 W. The proposed technique was able to accurately locate the new MPP at 1265 V and 8000 W despite the uncertainties caused by changes in the shading patterns from SP1 to SP2. However, it is important to note that traditional methods failed to perform well under these conditions, often oscillating around local peaks and experiencing significant chattering, as seen in Figure 16. The proposed technique, however, was able to precisely identify the optimal operating voltage, resulting in a significant enhancement in system efficiency. Both the proposed and traditional methods are illustrated in Figures 15 and 16, respectively, showing the variations in PV terminal voltage, output power, and the duty cycle during the transition from SP1 to SP2.

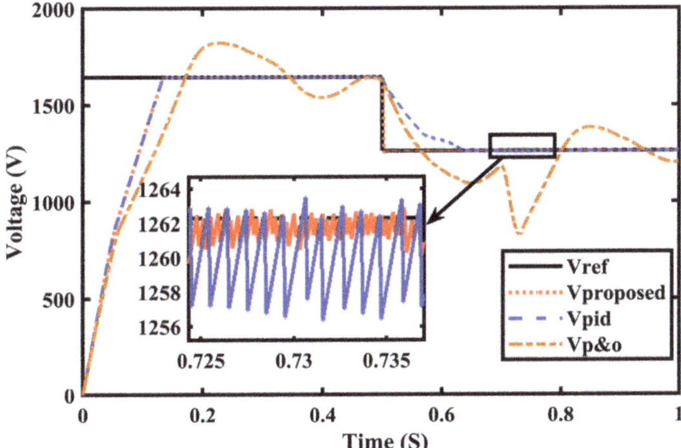

**Figure 15.** Voltage tracking with uncertainties.

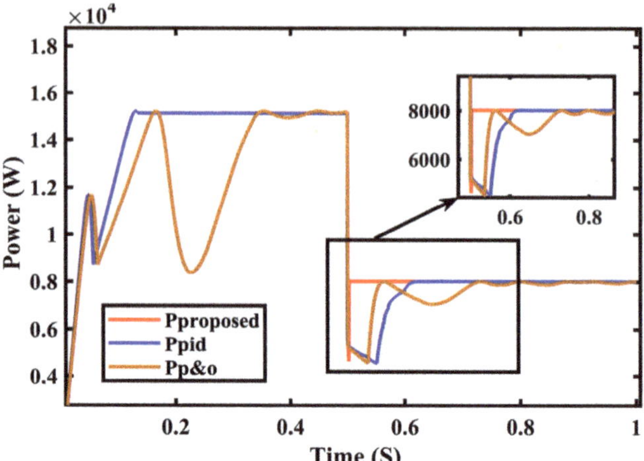

**Figure 16.** Power analysis with uncertainties.

The results indicate clearly that the proposed technique showed minimum chattering as compared to P&O and PID for 0.7 to 0.8 s when uncertainties were added; it could bear abrupt changes which demonstrated control robustness.

### 8.3. Duty Cycle and Fault Analysis

Figure 17 demonstrates that the proposed technique was successful in the instantaneous control of the duty cycle. When the shading pattern changes from SP1 to SP2 at 0.5 s then the duty cycle changes from 0.21 to 0.35. On the other hand, the error convergence was tested when the shading patterns changed from SP1 to SP2, which is shown in Figure 18.

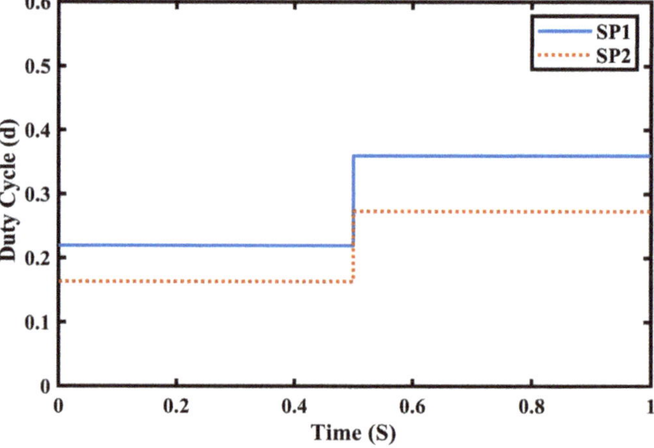

**Figure 17.** Duty cycle at SP1 and SP2.

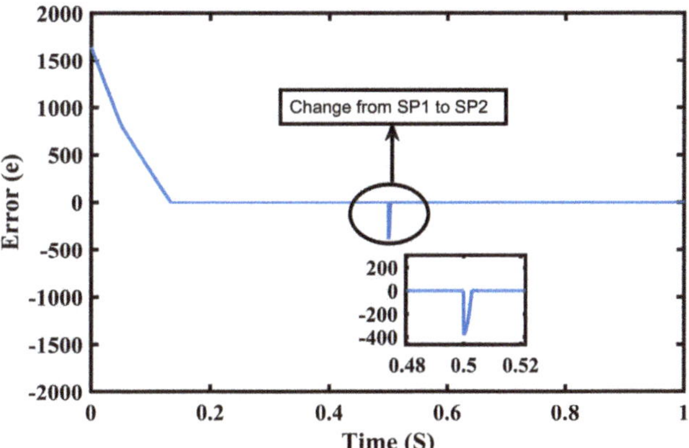



**Figure 18.** Error in SP1 and SP2.

## 9. Comparative Analysis of Proposed Technique

The proposed system has been validated through a practical experimental setup. The PV emulator used in this study is based on the single diode model, where the PV system acts as a current source with an antiparallel diode and intrinsic resistances. Literature shows that a PV cell is essentially a voltage source dependent on current, and its output current and open-circuit voltage ($V_{OC}$) vary with changes in irradiance and temperature. Therefore, a DC power source with high current and low sensitivity can emulate the electrical characteristics of a PV system. The superposition of constant current, constant voltage, and diode activation effectively mimics the electrical behavior of a PV cell. The load was changed using a 500 W variable resistor, which in practical terms alters the maximum power delivery to the load and forces the algorithm to maximize the power at a changing load following the maximum power delivery theorem. The impact of irradiance was evaluated by suddenly changing the voltage of the DC current source. Figure 19 illustrates the physical connections among the PV system components such as the DC—DC buck-boost converter, sensors, microcontrollers for MPPT control, data acquisition, and load. The performance of the proposed MLNFN TSMC was compared to that of a ZN-tuned PID controller. The emulation function was limited by the variable current generation for the PV module. Table 7 lists the values of the components used for the practical application of control. Figure 20 shows the layout of the experimental setup.

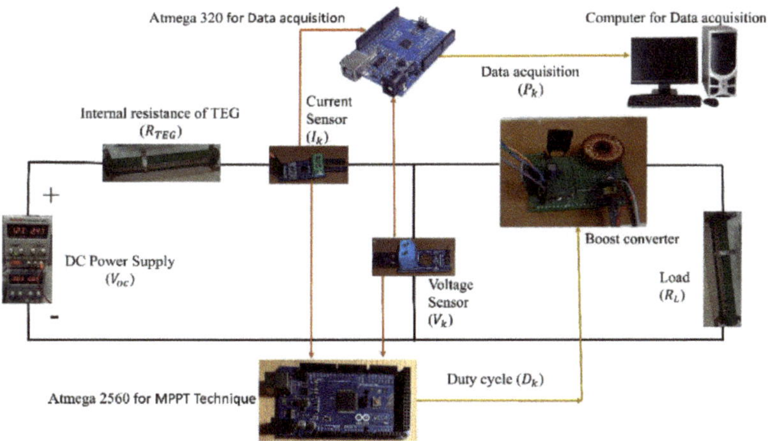



**Figure 19.** An experimental setup with a low-cost PV emulator.

**Table 7.** Specifications of hardware components.

| Parameters | Symbols |
| --- | --- |
| Load, ($R_L$) | 5,10 $\Omega$, 300 W |
| DC source 1 | PS305 |
| DC source 2 | MS305-D dual channel |
| Switching frequency, ($f$) | 61 kHz |
| Inductor ($L$) | 1 mH |
| Output capacitor ($C_{in}$) | 1000 $\mu$F |
| Input capacitor ($C_{out}$) | 100 $\mu$F |
| Oscilloscope | Tektronix TDS-3052B |
| Power diode | PHY 10SQ04 |
| Voltage sensor | B25 voltage sensor |
| Current sensor module | ACS172 |
| Micro-controller | ATmega 2560/328 |
| MOSFET | IRF730 |

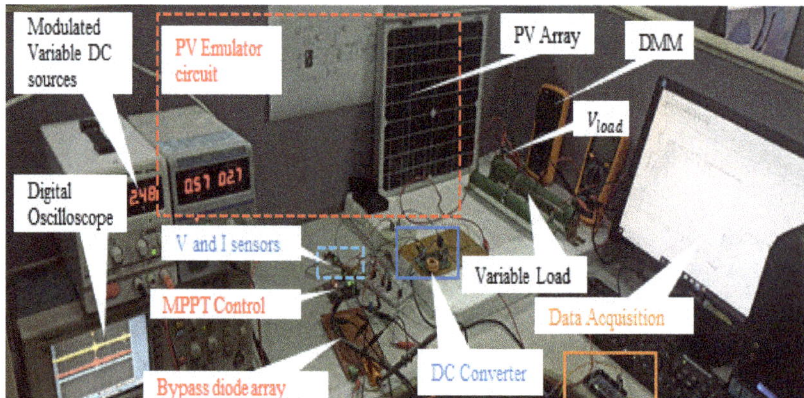


**Figure 20.** Experimental setup for MPPT.

Figure 21 shows tracked power with dynamic load conditions by MLNFN TSMC and PID controllers. PID takes up to 250 ms for final GMPP settling as shown in Figure 21B. Oscillations which were unavoidable were noticed after 250 ms represented in magnification window, in addition to this less maximum power was achieved with PID. It added to power loss and decreased inefficiency. MLNFN TSMC tracks GM faster as compared to PID and settles at GMPP in 100–120 ms as shown in Figure 21A. MLNFN TSMC showed minimum oscillations after GMPP detection and consequently generated the least power loss in steady and dynamic states showing an 8.2% higher average power and 60% less control time. The performance in transition and final steady states was observed in the experiment, similar to the anticipated behavior in the mathematical model. Negligible fluctuations and the least settling time in experimental results reiterated better performance of MLNFN TSMC. Figure 22A,B, shows the reference voltages computed by the MLNFN TSMC and PID controller, respectively.

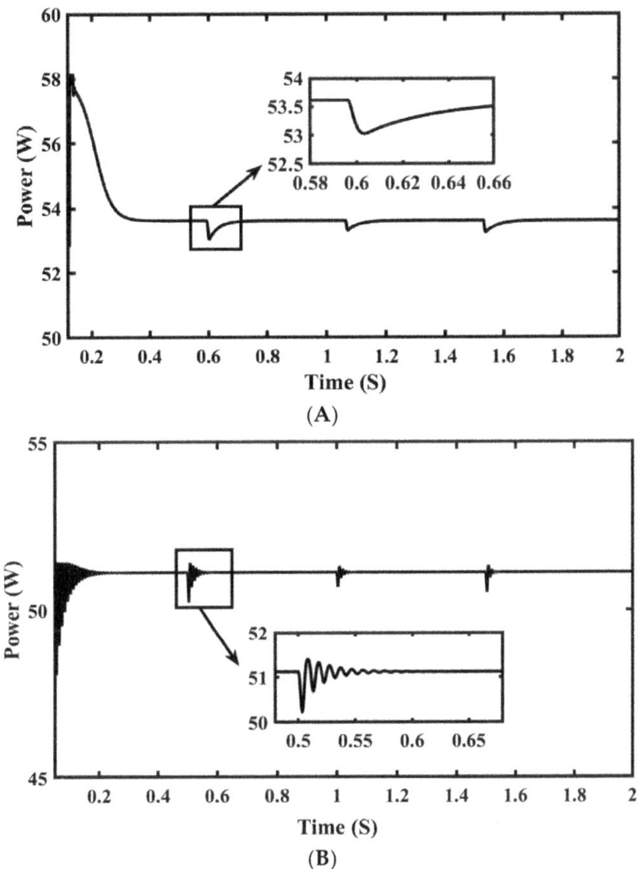



**Figure 21.** Experimental power transient of PV System (A) MLNFN TSMC (B) PID.

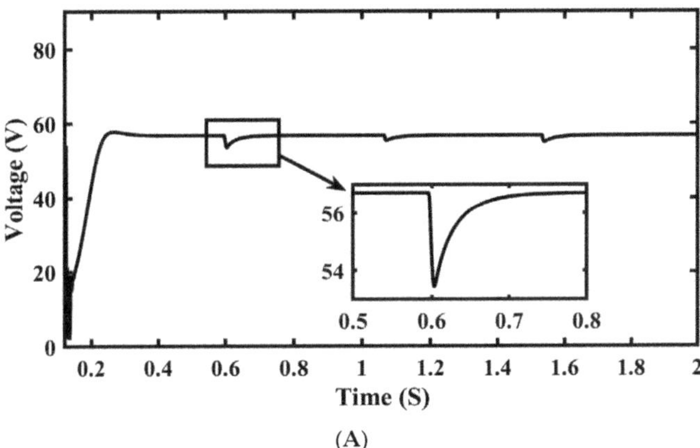



**Figure 22.** *Cont.*

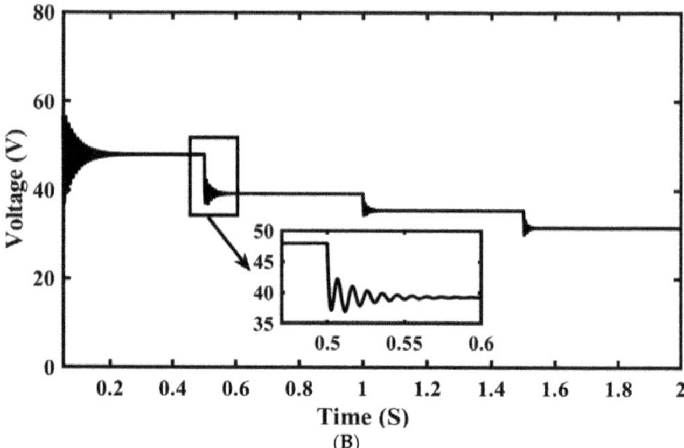



**Figure 22.** Experiment voltage transient of PV System (**A**) MLNFN TSMC (**B**) PID.

## 10. Common Performance Analysis and Discussions

This study delves into the shortcomings of conventional PID control-based methods for tracking the MPP in PV systems during PSCs. These techniques are unable to circumvent local maxima (LM) traps and often lead to voltage fluctuations that impede the integration of large-scale PV systems into power grids. To address these challenges, in recent years the researchers have proposed a new technique called the intelligent modified shuffled frog leaping algorithm (IMSFLA) which has been found to be more effective in tracking the GMPP with an efficiency of 99%. IMSFLA also demonstrates minimal oscillations at the GMPP resulting in increased power output to the load. The IMSLFA algorithm operates by dividing the problem into several sub-problems, which are then optimized in parallel. The algorithm uses a combination of local and global search strategies to explore the search space efficiently. The local search strategy helps the algorithm to converge quickly, while the global search strategy helps it to avoid local optima and find the global MPP. Compared with conventional techniques, the reported IMSFLA provides improved accuracy, faster tracking speed, and robustness to noise.

Our proposed method of MLNFN TSMC has been verified as robust in dynamic as well as static operating conditions through hardware experiments. These experiments show that the proposed technique performs well under real-world conditions and load variations, resulting in 56.1% less variability and only a 2–3 W standard deviation at the MPP and found to mimic the IMSFLA performance.

The purposed technique combines the advantages of fuzzy logic control, neural network-based control, and terminal sliding mode control to achieve optimal performance. In real-world partial shading conditions and load variation, the performance of the neuro-fuzzy terminal sliding mode MPPT technique depends on several factors, such as the complexity of the system, the quality of the sensors, and the accuracy of the modeling of the PV system. However, the technique has been shown to be effective in improving the performance of PV systems under partial shading conditions and load variation. The purposed technique improves the efficiency of the PV system under partial shading conditions and load variation, compared to other MPPT techniques. The technique was able to adjust the duty cycle of the DC—DC converter to ensure that the PV system operated at the MPP, even under partial shading conditions. The results also showed that the proposed technique was able to maintain a stable output voltage and current under partial shading conditions and load variation. The neuro-fuzzy logic was able to adjust the controller parameters in real-time based on the input signals, which ensured the stability of the system. Furthermore, the terminal sliding mode control was able to provide fast and accurate tracking of the MPP, even under dynamic partial shading conditions.

## 11. Conclusions

This study proposes a two-stage MPPT approach for PV systems operating under PSCs. A NFN is employed to enhance the reference voltage generation for MPPT. The proposed strategy guarantees finite-time convergence of MPP voltage tracking and resolves the issue of multiple peaks caused by shading conditions, in contrast to traditional TSMC methods. The method is tested through real-time experiments and numerical simulations and was found to significantly improve MPPT performance, even under rapidly changing irradiance and temperature conditions. The proposed robust TSMC approach is also more resilient, as demonstrated by simulations under varying weather conditions and uncertainties. The results show that the proposed method can quickly and accurately track the MPP, even in the presence of uncertainties, when compared to P&O, PID and incremental conductance conventional controllers.

**Author Contributions:** Data curation, A.; formal analysis, M.B.Q. and W.A.; funding acquisition, M.A.B.F.; investigation, A., W.A. and M.M.K.; methodology, A. and M.B.Q.; Project administration, M.B.Q. and M.A.B.F.; Software, A. and W.A.; supervision, M.B.Q. and R.N.; validation, M.M.K.; visualization, R.N.; writing—original draft, A. and M.B.Q.; writing—review and editing, M.M.K. and M.A.B.F. All authors have read and agreed to the published version of the manuscript.

**Funding:** This research received no external funding.

**Data Availability Statement:** Not applicable.



## References

1. Sampaio, P.G.V.; González, M.O.A. Photovoltaic solar energy: Conceptual framework. *Renew. Sustain. Energy Rev.* **2017**, *74*, 590–601. [CrossRef]
2. Kittner, N.; Lill, F.; Kammen, D.M. Energy storage deployment and innovation for the clean energy transition. *Nat. Energy* **2017**, *2*, 17125. [CrossRef]
3. Verma, D.; Nema, S.; Shandilya, A.M.; Dash, S.K. Maximum power point tracking (MPPT) techniques: Recapitulation in solar photovoltaic systems. *Renew. Sustain. Energy Rev.* **2016**, *54*, 1018–1034. [CrossRef]
4. Rezk, H.; Fathy, A.; Abdelaziz, A.Y. A comparison of different global MPPT techniques based on meta-heuristic algorithms for photovoltaic system subjected to partial shading conditions. *Renew. Sustain. Energy Rev.* **2017**, *74*, 377–386. [CrossRef]
5. Liu, J.; Li, J.; Wu, J.; Zhou, W. Global MPPT algorithm with coordinated control of PSO and INC for rooftop PV array. *J. Eng.* **2017**, *13*, 778–782. [CrossRef]
6. Yeung, R.S.C.; Chung, H.S.H.; Tse, N.C.F.; Chuang, S.T.H. A global MPPT algorithm for existing PV system mitigating suboptimal operating conditions. *Sol. Energy* **2017**, *141*, 145–158. [CrossRef]
7. Hu, Y.; Cao, W.; Wu, J.; Ji, B.; Holliday, D. Thermography-based virtual MPPT scheme for improving PV energy efficiency under partial shading conditions. *IEEE Trans. Power Electron.* **2014**, *11*, 5667–5672. [CrossRef]
8. Mohanty, S.; Subudhi, B.; Ray, P.K. A new MPPT design using grey wolf optimization technique for photovoltaic system under partial shading conditions. *IEEE Trans. Sustain. Energy* **2015**, *7*, 181–188. [CrossRef]
9. Tamir, S. Particle Swarm Optimization with Targeted Position-Mutated Elitism (PSO-TPME) for Partially Shaded PV Systems. *Sustainability* **2023**, *15*, 3993.
10. Loukriz, A.; Haddadi, M.; Messalti, S. Simulation and experimental design of a new advanced variable step size Incremental Conductance MPPT algorithm for PV systems. *ISA Trans.* **2016**, *62*, 30–38. [CrossRef]
11. Safari, A.; Mekhilef, S. Incremental Conductance MPPT Method for PV Systems. In Proceedings of the 2011 24th Canadian Conference on Electrical and Computer Engineering (CCECE), Niagara Falls, ON, Canada, 8–11 May 2011.
12. Saravanan, S.; Babu, N.R. Maximum power point tracking algorithms for photovoltaic system system—A review. *Renew Sustain. Energy Rev.* **2016**, *57*, 192–204. [CrossRef]
13. Mao, M.; Cui, L.; Zhang, Q.; Guo, K.; Zhou, L.; Huang, H. Classification and summarization of solar photovoltaic MPPT techniques: A review based on traditional and intelligent control strategies. *Energy Rep.* **2020**, *6*, 1312–1327. [CrossRef]
14. Abadi, I.; Imron, C.; Noriyati, R.D. Noriyati. Implementation of Maximum Power Point Tracking (MPPT) Technique on Solar Tracking System Based on Adaptive Neuro-Fuzzy Inference System (ANFIS). *EDP Sci.* **2018**, *43*, 01014.
15. Armghan, H.; Ahmad, I.; Armghan, A.; Khan, S.; Arsalan, M. Backstepping based non-linear control for maximum power point tracking in photovoltaic system. *Sol. Energy* **2018**, *159*, 134–141.
16. Katche, M.L.; Makokha, A.B.; Zachary, S.O.; Adaramola, M.S. A Comprehensive Review of Maximum Power Point Tracking (MPPT) Techniques Used in Solar PV Systems. *Energies* **2023**, *16*, 2206. [CrossRef]

17. Craciunescu, D.; Fara, L. Investigation of the Partial Shading Effect of Photovoltaic Panels and Optimization of Their Performance Based on High-Efficiency FLC Algorithm. *Energies* **2023**, *16*, 1169. [CrossRef]
18. Abo-Khalil, A.G.; El-Sharkawy, I.I.; Radwan, A.; Memon, S. Influence of a Hybrid MPPT Technique, SA-P&O, on PV System Performance under Partial Shading Conditions. *Energies* **2023**, *16*, 577.
19. Farayola, A.M.; Sun, Y.; Ali, A. Global maximum power point tracking and cell parameter extraction in Photovoltaic systems using improved firefly algorithm. *Energy Rep.* **2022**, *8*, 162–186. [CrossRef]
20. Gong, L.; Hou, G.; Huang, C. A two-stage MPPT controller for PV system based on the improved artificial bee colony and simultaneous heat transfer search algorithm. *ISA Trans.* **2023**, *132*, 428–443. [CrossRef]
21. Hassan, A.; Bass, O.; Masoum, M.A. An improved genetic algorithm based fractional open circuit voltage MPPT for solar PV systems. *Energy Rep.* **2023**, *9*, 1535–1548. [CrossRef]
22. Punitha, K.; Devaraj, D.; Sakthivel, S. Artificial neural network based modified incremental conductance algorithm for maximum power point tracking in photovoltaic system under partial shading conditions. *Energy* **2013**, *62*, 330–340. [CrossRef]
23. Hiren, P.; Agarwal, V. Maximum power point tracking scheme for PV systems operating under partially shaded conditions. *IEEE Trans. Ind. Electron.* **2008**, *55*, 1689–1698.
24. Koad, R.B.; Zobaa, A.F.; El-Shahat, A. A novel MPPT algorithm based on particle swarm optimization for photovoltaic systems. *IEEE Trans. Sustain. Energy* **2016**, *8*, 468–476. [CrossRef]
25. El-Helw, H.M.; Magdy, A.; Marei, M.I. A hybrid maximum power point tracking technique for partially shaded photovoltaic arrays. *IEEE Access* **2017**, *5*, 11900–11908. [CrossRef]
26. Sun, Y.; Peng, Y.; Deng, F. Improved SPSO-based Parameter Identification of Solar PV Cells IV Model. In Proceedings of the 2017 International Conference on Computer Systems, Electronics and Control (ICCSEC), Dalian, China, 25–27 December 2017; IEEE: Piscataway, NJ, USA, 2017; pp. 1738–1742.
27. Nazri, N.S.; Fudholi, A.; Ruslan, M.H.; Sopian, K. Mathematical modeling of photovoltaic thermal-thermoelectric (PVT-TE) air collector. *Int. J. Power Electron. Drive Syst.* **2018**, *9*, 795.
28. Bressan, M.; Gutierrez, A.; Gutierrez, L.G.; Alonso, C. Development of a real-time hot-spot prevention using an emulator of partially shaded PV systems. *Renew. Energy* **2018**, *127*, 334–343. [CrossRef]
29. Bingöl, O.; Özkaya, B. Analysis and comparison of different PV array configurations under partial shading conditions. *Sol. Energy* **2018**, *160*, 336–343. [CrossRef]
30. Ahsan, S.; Niazi, K.A.K.; Khan, H.A.; Yang, Y. Hotspots and performance evaluation of crystalline-silicon and thin-film photovoltaic modules. *Microelectron. Reliab.* **2018**, *88*, 1014–1018. [CrossRef]
31. Rajurkar, S.; Verma, N.K. Developing deep fuzzy network with Takagi Sugeno fuzzy inference system. In Proceedings of the 2017 IEEE International Conference on Fuzzy Systems (FUZZ-IEEE), Naples, Italy, 9–12 July 2017; IEEE: Piscataway, NJ, USA, 2017; pp. 1–6.
32. Arsalan, M.; Iftikhar, R.; Ahmad, I.; Hasan, A.; Sabahat, K.; Javeria, A. MPPT for photovoltaic system using nonlinear backstepping controller with integral action. *Sol. Energy* **2018**, *170*, 192–200. [CrossRef]
33. Kaouane, M.; Boukhelifa, A.; Cheriti, A. Regulated output voltage double switch Buck-Boost converter for photovoltaic energy application. *Int. J. Hydrog. Energy* **2016**, *41*, 20847–20857. [CrossRef]

 *electronics*

*Article*

# Permanent Magnet Synchronous Motor Control Based on Phase Current Reconstruction

**Guozhong Yao, Yun Yang, Zhengjiang Wang * and Yuhan Xiao**

Faculty of Transportation Engineering, Kunming University of Science and Technology, Kunming 650504, China
* Correspondence: wangzhengjiang@kust.edu.cn; Tel.: +86-158-7792-4502

**Abstract:** The traditional single current sensor control strategy of a permanent magnet synchronous motor (PMSM) often adopts the DC bus method, which makes it difficult to eliminate the blind area of current reconstruction. Therefore, a current reconstruction method based on a sliding mode observer is proposed. Based on the current equation of the motor, the method takes the $\alpha$-axis and $\beta$-axis currents as the observation objects and shares the same synovial surface, so that the $\alpha$-axis current observation value and the $\beta$-axis current observation value converge to the actual current value at the same time and the unknown $\beta$-axis current information is obtained. The control system first tests the performance of the motor under different working conditions when the parameters are matched, and then tests the current reconstruction ability of the parameter mismatch. The results show that the current observer with a matched parameter can accurately and quickly reconstruct the $\beta$-axis current under various operating conditions, and the maximum current error does not exceed 4 mA. When the parameters are mismatched, high-performance control of the motor can still be achieved. The proposed method has excellent robustness.

**Keywords:** PMSM; single current sensor; current reconstruction; sliding mode control; robustness

**Citation:** Yao, G.; Yang, Y.; Wang, Z.; Xiao, Y. Permanent Magnet Synchronous Motor Control Based on Phase Current Reconstruction. *Electronics* **2023**, *12*, 1624. https://doi.org/10.3390/electronics12071624

Academic Editors: Jamshed Iqbal, Ali Arshad Uppal and Muhammad Rizwan Azam

Received: 5 March 2023
Revised: 23 March 2023
Accepted: 28 March 2023
Published: 30 March 2023



## 1. Introduction

A permanent magnet synchronous motor (PMSM) has the advantages of high power density and fast dynamic response. It is widely used in industrial production, new energy vehicles, ship propulsion, and other fields [1–4].

The control of the motor usually uses constant volts per hertz (V/f), direct torque control (DTC), field-oriented control (FOC), and other methods. V/f control has matured for a long time [5]. It can generate the required torque by coordinating the amplitude and frequency of the stator voltage, which is the simplest motor control strategy. However, the performance of V/f control under low frequency and high load conditions is far less than that of other control methods [6], which leads to the limitation of V/f control applications in situations where transient performance is high [7], and high-performance motor control often uses other methods. Compared with FOC, DTC is less dependent on rotor position [8], but has larger torque and flux fluctuations [9–11]. The research on sensorless FOC has a history spanning more than 30 years [12–14]. The dependence on the rotor position cannot limit the development of FOC. FOC control has become the mainstream control method for the motor. Current feedback is an indispensable part of FOC. The traditional FOC control strategy often uses two or three current sensors to obtain current information. The single current sensor motor has received extensive attention due to its small system size and low production cost [15–18].

The traditional PMSM single current control strategy generally adopts two methods: the DC bus method and the current observer method. As early as 1989, the DC bus method was proposed [19]; its essence is to use the instantaneous current of the inverter fixed point, which contains multi-phase current information, to reconstruct the current. However, it takes time to collect current information, and the effective voltage vector has a short action time, which leads to the existence of the current reconstruction blind area. In recent decades, many improvement strategies have been proposed to overcome the current blind spot. Reference [20] proposed a three-state pulse width modulation technology that divides the space vector into two regions. In the low modulation ratio region, two basic voltages with a difference of 120° are used to synthesize the reference voltage. This technology reduces the blind area of current reconstruction and suppresses the common mode voltage, but its performance is not as good as the traditional seven-segment space vector pulse width modulation control method. In reference [21], it is proposed to use an isolated current sensor instead of a DC bus current sensor to sample twice in each PWM cycle. However, this method requires additional leads, which will introduce parasitic parameters and lead to system performance degradation. At the same time, the current reconstruction blind area in the overmodulation region is too large, which is also an important factor limiting the method. The PWM phase shift method proposed in reference [22] can increase the observable area of the current by moving the PWM waveform when the non-zero voltage vector action time is too short, but this method still has a voltage vector action time after phase shift that is less than the minimum sampling time. Reference [23] proposed a new phase current reconstruction technique in which a single current sensor is installed on a branch rather than a DC bus. The blind area of the current reconstruction in the sector boundary region and the low modulation region can be reduced without introducing any additional algorithm compensation strategy, but there is a blind area in the high modulation ratio region. Reference [24] proposed a new phase current reconstruction scheme without using a zero-switching state. The current reconstruction dead zone is divided into six sectors. Each sector is divided into three parts by the corresponding vector synthesis method to obtain enough effective switching states, thereby reducing the current reconstruction dead zone, but it will increase the current ripple and switching loss and reduce the service life of the system. Reference [25] proposed the idea of "substitution." When the reference voltage vector is located in the blind area of overmodulation reconstruction, it is replaced by the adjacent reconfigurable maximum voltage vector to broaden the operating range of the phase current reconstruction technology, but this does not solve the blind area problem in other areas.

The core idea of the above improvement measures is to be able to collect the required current information in time by changing the space voltage vector or switching states within a limited time. However, due to factors such as the high dynamic response of motor control, the requirements of space voltage vectors and switching are very high. Once these measures are adopted, it is not only difficult to completely eliminate the blind area of the current reconstruction, but it also has a negative impact on the control performance of the motor. Therefore, the current observer becomes another direction in the current reconstruction. The current observer method without a low modulation ratio region and a sector transition region has certain advantages.

In [26], an adaptive observer is proposed to realize the single current control of the motor. The observer equation and adaptive law are not complicated, but their robustness depends on parameter identification. Incorrect parameter estimation will bring errors to current reconstruction. The ESO observer designed in reference [27] can realize the accurate control of a single current sensor, but the parameter design of ESO is very difficult. It is usually obtained by the trial-and-error method, which not only increases the workload of the algorithm but also makes it difficult to obtain the optimal parameters. Verma et al. [28] reconstructed the $\beta$-axis current based on the mathematical model of the motor and the PI loop, but the anti-interference ability of the PI loop needs to be improved. Reference [29] designed a single current sensor control algorithm based on Kalman filter, but the calculation of the correction current is not accurate enough, the control performance is poor, the motor torque accuracy is low, and there is a non-periodic pulse spike. Reference [30] can realize single current sensor control without a position sensor, but the robustness of the system to the stator resistance is not very good. Reference [31] combines the DC bus method and the observer method and uses the Luenberger observer to compensate for the limitation of the sector boundary region, but there is still a blind area in the low modulation region.

The above current observers all introduce complex structures, which increase the computational burden and have a certain impact on the real-time performance of motor control. In addition, the error of current reconstruction and external interference cannot achieve the ideal current reconstruction effect. Table 1 briefly analyzes the current reconstruction strategy.

**Table 1.** Brief comparisons of single current control strategies.

| Method | References | Advantages | Disadvantages |
|---|---|---|---|
| DC bus method | Reference [19] | The DC bus method is proposed. | The current blind area. |
| | Reference [20] | Reduces the blind area and the common-mode voltage. | Not as good as the traditional 7-segment SVPWM. |
| | Reference [21] | Needs an isolated current sensor. | Additional lead wire. |
| | Reference [22] | PWM Phase Shift. | Cannot eliminate all blind area. |
| | Reference [23] | One Branch. | A new blind area. |
| | Reference [24] | Switching state phase shift method. | Current ripple and switch damage. |
| | Reference [25] | Substitution method. | Ignore the sector boundary. |
| Current observer method | Reference [26] | Observer equations and adaptive laws are not complicated. | Robustness depends on parameter identification. |
| | Reference [27] | Accurate. | ESO parameters. |
| | Reference [28] | Based on the PI loop. | PI ring is not robust. |
| | Reference [29] | EKF. | Low accuracy. |
| | Reference [30] | The estimation technique is independent of machine parameters. | The robustness of stator resistance is not very good. |
| | Reference [31] | The DC bus method and observer method are combined. | A blind area in the low modulation region. |

At present, regarding the PMSM single current sensor control system, there is no system that can completely eliminate the blind area of current reconstruction and has the characteristics of a simple structure and strong robustness. The elimination of the blind area in the current reconstruction will inevitably make the motor run more smoothly. Without introducing complex structure, the anti-interference ability is greatly enhanced, the real-time control effect is better, and the motor can adapt to a more complex working environment. Therefore, the purpose of this paper is to first realize the current reconstruction with no blind area, high precision, and good robustness by a very simple method, and then realize the high performance control of PMSM by a single current sensor.

## 2. Mathematical Model of PMSM

The control strategy is based on the α-β two-phase stationary coordinate system. The voltage equation of PMSM can be expressed [32] as follows:

$$\begin{bmatrix} u_\alpha \\ u_\beta \end{bmatrix} = \begin{bmatrix} R_S + pL_d & \omega_e(L_d - L_q) \\ -\omega_e(L_d - L_q) & R_S + pL_q \end{bmatrix} \begin{bmatrix} i_\alpha \\ i_\beta \end{bmatrix} + \begin{bmatrix} E_\alpha \\ E_\beta \end{bmatrix} \tag{1}$$

Among them, $u_\alpha$ and $u_\beta$ represent the voltage component of the α-β two-phase stationary coordinate system, where $R_S$ is the stator resistance and $\omega_e$ is the angular velocity of the motor. $L_d$ and $L_q$ are the inductance components on the two-phase rotating coordinate axis. $i_\alpha$ and $i_\beta$ are the current components on the α axis and the β axis, respectively, and $E_\alpha$ and $E_\beta$ represent the extended back electromotive force.

The voltage equation of Equation (1) can be written as the current equation:

$$\frac{d}{dt}\begin{bmatrix} i_\alpha \\ i_\beta \end{bmatrix} = \frac{1}{L_d}\begin{bmatrix} -R_S & -\omega_e(L_d - L_q) \\ \omega_e(L_d - L_q) & -R_S \end{bmatrix} \begin{bmatrix} i_\alpha \\ i_\beta \end{bmatrix} + \begin{bmatrix} u_\alpha \\ u_\beta \end{bmatrix} - \frac{1}{L_d}\begin{bmatrix} E_\alpha \\ E_\beta \end{bmatrix} \tag{2}$$

$\psi_f$ is the stator flux, and the extended back electromotive force calculation equation can be expressed as follows:

$$\begin{bmatrix} E_\alpha \\ E_\beta \end{bmatrix} = \begin{bmatrix} -\psi_f \omega_e sin\theta \\ \psi_f \omega_e cos\theta \end{bmatrix} \tag{3}$$

### 3. Current Reconstruction

#### 3.1. Current Sliding Mode Observer Design

Chakraborty [33] proposed that in the dual current sensor control system, if the a-phase current is measured in error, the $\alpha$-axis and $\beta$-axis currents will be wrong; if only the b-phase current is measured incorrectly, the $\beta$-axis current will be wrong, while the $\alpha$-axis current is still correct.

Equation (4) is obtained by the Clark transformation, and Kirchhoff's current law fully shows that the $\alpha$-axis current is only related to the a-phase current, while the $\beta$-axis current is related to the a and b two-phase currents. Obviously, if the single current sensor measures the b-phase current, the current of the $\alpha$-axis and $\beta$-axis cannot be obtained; if the single current sensor measures the a-phase current, the $\alpha$-axis current will be measurable, and the current that needs to be reconstructed is only the $\beta$-axis current. The three-phase reconstruction current can be obtained by a further inverse Clark transformation.

$$\begin{bmatrix} i_\alpha \\ i_\beta \end{bmatrix} = \begin{bmatrix} 1 & 0 \\ \frac{\sqrt{3}}{3} & \frac{2\sqrt{3}}{3} \end{bmatrix} \begin{bmatrix} i_A \\ i_B \end{bmatrix} \tag{4}$$

The traditional sliding mode observer knows the current and voltage information in the stationary two-phase coordinate system. The purpose is to obtain the extended back electromotive force information, that is, so that Equation (3) can further analyze the rotor position and speed information. The traditional sliding mode observer can be expressed by (2) as follows:

$$\frac{d}{dt}\begin{bmatrix} \hat{i}_\alpha \\ \hat{i}_\beta \end{bmatrix} = \frac{1}{L_d}\begin{bmatrix} -R_S & -\omega_e(L_d - L_q) \\ \omega_e(L_d - L_q) & -R_S \end{bmatrix}\begin{bmatrix} \hat{i}_\alpha \\ \hat{i}_\beta \end{bmatrix} + \frac{1}{L_d}\begin{bmatrix} u_\alpha - ksgn(\hat{i}_\alpha - i_\alpha) \\ u_\beta - ksgn(\hat{i}_\beta - i_\beta) \end{bmatrix} \tag{5}$$

In Equation (5), considering that the $\alpha$-axis current is obtained indirectly by the a-phase current sensor, while the $\beta$-axis actual current is unknown, and the $\beta$-axis actual current is only applied in the sliding mode surface $(\hat{i}_\beta - i_\beta)$, if it is replaced by other known information and the sliding mode observer can still operate normally, then the $\beta$-axis current will be reconstructed, that is as follows:

$$\begin{bmatrix} -R_S & -\omega_e(L_d - L_q) \\ \omega_e(L_d - L_q) & -R_S \end{bmatrix}\begin{bmatrix} \hat{i}_\alpha \\ \hat{i}_\beta \end{bmatrix} + \frac{1}{L_d}\begin{bmatrix} u_\alpha - ksgn(\hat{i}_\alpha - i_\alpha) \\ u_\beta - ksgn(\hat{i}_\alpha - i_\alpha) \end{bmatrix} \tag{6}$$

However, if according to this design, the $\alpha$-axis observation current of the sliding mode observer still converges normally to the current value measured by the sensor, but the $\beta$-axis current is completely out of control. Considering that the $\beta$-axis current does not converge to the actual measurement value due to the lack of $\beta$-axis current error information $(\hat{i}_\beta - i_\beta)$, in the traditional sliding mode observer, in order to obtain the back electromotive force $E_\beta$, $ksgn(\hat{i}_\alpha - i_\alpha)$ is used instead of $E_\beta$ in the sliding mode observer. When the sliding mode observer converges, the two are equal. In fact, $E_\beta$ also contains the $\beta$-axis current error information $(\hat{i}_\beta - i_\beta)$. Therefore, on the basis of Equation (5), the following observer can be designed by introducing the back electromotive force information into the current equation of the motor as follows:

$$L_d \dot{\hat{i}}_s = \begin{bmatrix} -R_S & -\omega_e(L_d - L_q) \\ \omega_e(L_d - L_q) & -R_S \end{bmatrix}\hat{i}_s + \begin{bmatrix} u_\alpha - E_\alpha - q \cdot sigmoid(\tilde{i}) \\ u_\beta - E_\beta - t \cdot sigmoid(\tilde{i}) \end{bmatrix} \tag{7}$$

where q and t are constants, $\tilde{i} = \hat{i}_\alpha - i_\alpha$, $\hat{i}_s = \begin{bmatrix} \hat{i}_\alpha \\ \hat{i}_\beta \end{bmatrix}$

The current observer mainly has the following characteristics:

(1)  $\alpha$-axis and $\beta$-axis share the same sliding surface, namely $sigmoid(\tilde{i})$. This is due to the lack of important information on the traditional $\beta$-axis sliding mode surface, namely

the actual value of the β-axis current. After introducing the back-EMF containing the β-axis information, the α-axis current observation value and the β-axis current observation value will converge to the actual value at the same time, so the same sliding mode surface can be used for observation.

(2)　This design reconstructs the β-axis current instead of obtaining the back electromotive force. The reconstruction of the current requires the back electromotive force. If the acquisition method of the electromotive force is non-inductive, then it will be a single current sensing PMSM control strategy without a position sensor.

(3)　The quasi-sliding function $sigmoid(\tilde{i})$ is used to replace the traditional symbol function $sgn(\tilde{i})$. The smooth, continuous characteristic is more stable than the step characteristic, which can improve the performance of the sliding mode observer.

### 3.2. Sliding Mode Convergence Verification

Whether the sliding mode observer meets the expectation depends on whether it can converge to the sliding mode surface. The sliding mode convergence is proved below, the sliding mode surface $s = \hat{i}_\alpha - i_\alpha$ is defined, and Equation (8) is selected as the candidate function according to Lyapunov theory [34].

$$F(s) = \frac{s^2}{2} \tag{8}$$

The sliding mode observer is proved to be stable by satisfying the following conditions.

$$F(s) = \frac{s^2}{2} > 0 \tag{9}$$

$$\dot{F}(s) = s \cdot \dot{s} < 0, (s \neq 0) \tag{10}$$

The α-axis current observer value is infinitely close to the actual value, fluctuating around the actual value, but it will not be equal to the actual value, so s will not be equal to 0. The condition (9) is obviously established, and then whether the synovial membrane converges or not can be judged by whether the condition (10) is true.

By taking (4) minus (1), we can obtain the following:

$$\dot{s} = \frac{1}{L_d}\begin{bmatrix} R_S + pL_d & \omega_e(L_d - L_q) \\ -\omega_e(L_d - L_q) & R_S + pL_q \end{bmatrix}\tilde{i} - \frac{1}{L_d}sigmoid(\tilde{i}) \tag{11}$$

If $s > 0$, then $\dot{s} < 0$ satisfies the condition that follows:

$$-R_S|s| - \omega_e(L_d - L_q)\tilde{i}_\beta - qsigmoid(s) < 0 \tag{12}$$

In the surface-mounted permanent magnet synchronous motor, if the d-axis and q-axis inductance components are equal, and the constant q is designed to be positive, then the Equation (12) holds.

If $s < 0$, then $\dot{s} > 0$ satisfies the condition that:

$$R_S|s| - \omega_e(L_d - L_q)\tilde{i}_\beta - qsigmoid(s) > 0 \tag{13}$$

Similarly, if the constant q is designed to be positive, then Equation (13) holds.

In summary, when $q$ and $t$ are positive, the sliding mode will converge, but the specific parameters need to be adjusted according to the different motors to reduce the error. Two different parameters are set to adjust the current error, and the two parameters can be equal.

### 4. Experimental Verification

Through the above analysis, the control system shown in Figure 1 is built to verify the effect of the β-axis current reconstruction. The motor parameters are shown in Table 2.

Based on the control system of the current observer, the a-phase current is collected to reconstruct the three-phase current for current feedback, and the high-performance motor control of a single current sensor is realized.

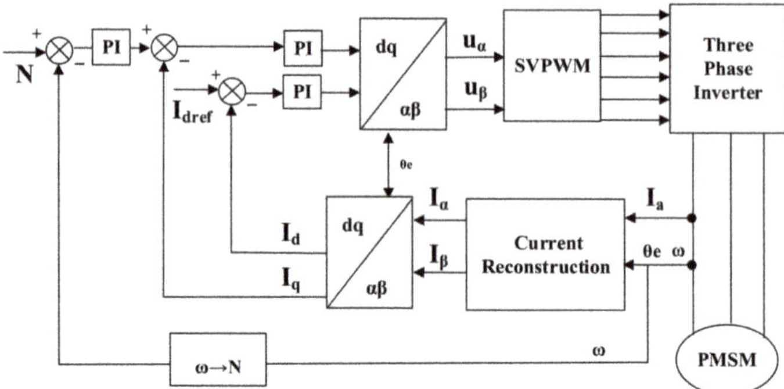

**Figure 1.** Control block diagram.

**Table 2.** System Parameters.

| Parameters | Value | Unit |
| --- | --- | --- |
| Polar logarithm | 4 | – |
| Stator resistance | 2.875 | $\Omega$ |
| Stator inductance | 8.5 | mH |
| Flux linkage | 0.175 | Wb |
| Moment of inertia | 0.001 | $Kg \cdot m^2$ |
| PWM frequency | 10 | kHz |

*4.1. Analysis of Steady-State and Dynamic Reconstructed Current under Matched Parameter*

The working condition W is no-load at a speed of 1000 rpm. According to the a-phase current and rotor position information, the sliding mode current observer is used to reconstruct the $\beta$-axis current, as shown in Figures 2 and 3, which show the error between the reconstructed $\beta$-axis current and the actual $\beta$-axis current. The reconstructed $\beta$-axis current is almost consistent with the measured current of the sensor. After the start-up phase, the $\beta$-axis current error is periodic. Most of the errors are concentrated within $\pm 2$ mA, and a few errors exceed the fluctuation range of $\pm 2$ mA. The maximum current error is only 4 mA in the start-up phase. After entering the stable phase, the maximum current error is reduced to 3 mA, and the maximum current error is only 0.85% of the steady-state current amplitude.

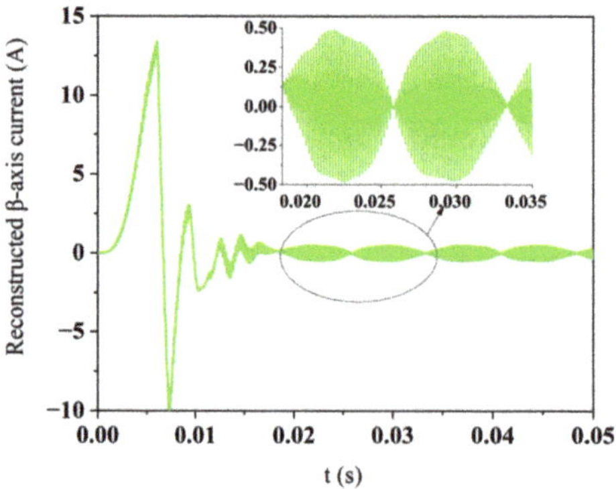

**Figure 2.** Reconstructed $\beta$-axis current.

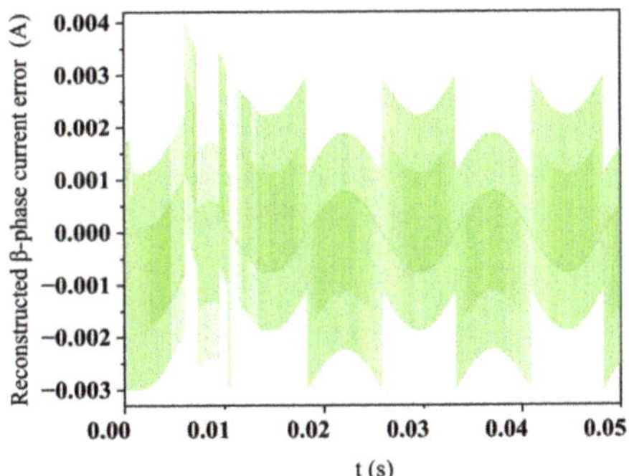

**Figure 3.** Reconstructed β-axis current error.

The motor control system shown in Figure 1 aims to replace two or three current sensors with a single current sensor, so it is necessary to analyze the error of reconstructing three-phase current. The a-phase current is collected by the sensor, and the reconstruction of the b- and c-phase currents requires the participation of the β-axis current. The β-axis current must have a reconstruction error, so that the reconstructed b- and c-phase currents must also have errors, as shown in Figures 4 and 5. Because the error of b and c two-phase currents comes from the β-axis current, it can be found that the distribution of b and c two-phase current error is almost the same as that of β-axis current error, but the two-phase current error is smaller than the β-axis current error as a whole. The current error is 3.6 mA before entering the stable stage, and the current error is only 2.6 mA after stabilization. This is because the β-axis current coefficient in the Clark transform is less than 1.

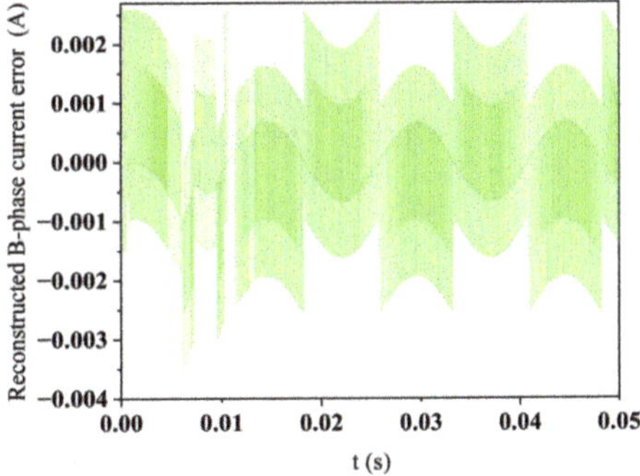

**Figure 4.** Reconstructed B-phase current error.

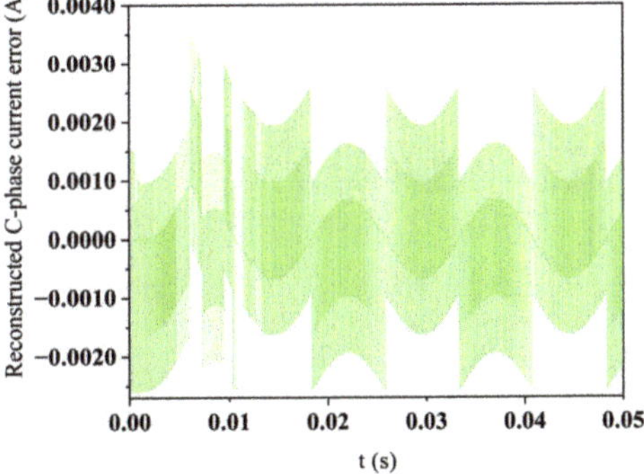

**Figure 5.** Reconstructed C-phase current error.

After applying the reconstructed current to the motor operation, the motor speed and torque shown in Figure 6 can be obtained. The motor speed overshoot of the control system in Figure 1 is only 50 rpm, and the speed tends to be stable at 16 ms. When it is stable, the difference from the set speed is within 1 rpm, and the maximum starting torque is 22 N·m. After the speed is stabilized at 16 ms, the torque is also stabilized and finally fluctuates around 0 N·m positive and negative 0.5 N·m.

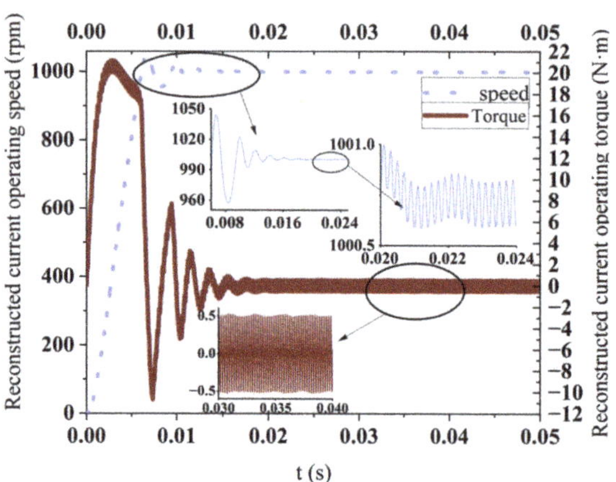

**Figure 6.** The steady-state operating conditions of the reconstructed current.

Based on the above simulation results, it can be confirmed that the current reconstruction system in Figure 1 can achieve fast, accurate, and stable control of the motor under 1000 rpm in a no-load environment. In order to further measure the operation status of the system under the dynamic condition of the motor, the dynamic condition analysis is carried out below. Now assume a complex working condition M: The load at 0 s is 2 N·m, and the speed is set to 600 rpm. The load increases to 5 N·m in 0.02 s, and the speed increases to 1000 rpm. In 0.07 s, the load is reduced to 2 N·m and the speed is reduced to 800 rpm. The speed curve under dynamic working condition m is shown in Figure 7, in the three different set speed changes, the system overshoot is below 50 rpm, and the response time is not prolonged compared with the steady-state condition, but the steady speed becomes larger due to the existence of the load and the deviation of the set speed. In the case of 1000 rpm and 5 N·m, the speed is always 5 rpm different from the set speed. Although the load will increase the deviation between the motor speed and the set speed, it can achieve high performance control of the motor.

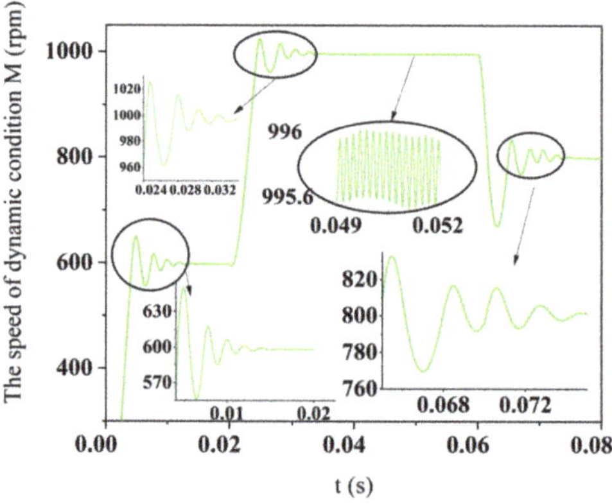

**Figure 7.** The speed of dynamic condition M.

Figure 8 is the torque diagram under operating condition M. In order to verify the current reconstruction ability of the reconstructed current under extremely complex conditions, the speed is changed while the load is changed, this is already a complicated working condition, and the designed system can control the torque from a large fluctuation state to a stable working state in an instant, which fully shows the real-time performance of the system.

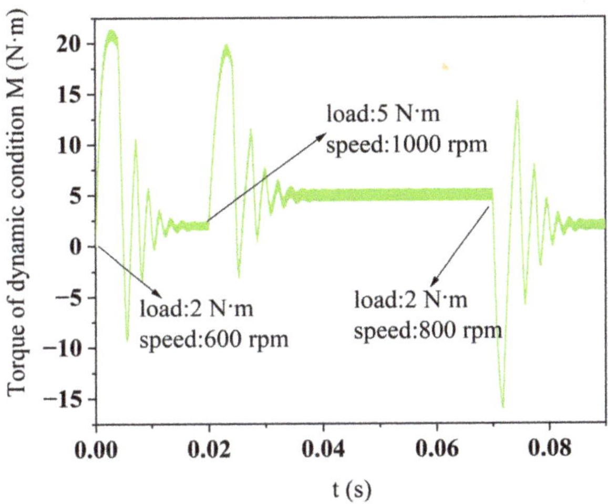


**Figure 8.** Torque of dynamic condition M.

The reconstructed β-axis current under the M condition is shown in Figure 9. Due to the simultaneous jump of speed and load, the current changes at the three time nodes of 0, 0.02, and 0.07 are more severe, but the reconstructed β-axis current can still quickly track the actual β-axis current. The error between the reconstructed β-axis current and the actual β-axis current under dynamic conditions is shown in Figure 10. The maximum error of the reconstructed current is still less than 4 mA, and the position where the maximum current error occurs is the position where the set speed changes. When the motor runs at 1000 rpm and 5 N·m, the current error of 4 mA only accounts for 0.08% of the steady-state current amplitude. Compared with the steady-state, no-load environment, the current error is further reduced.

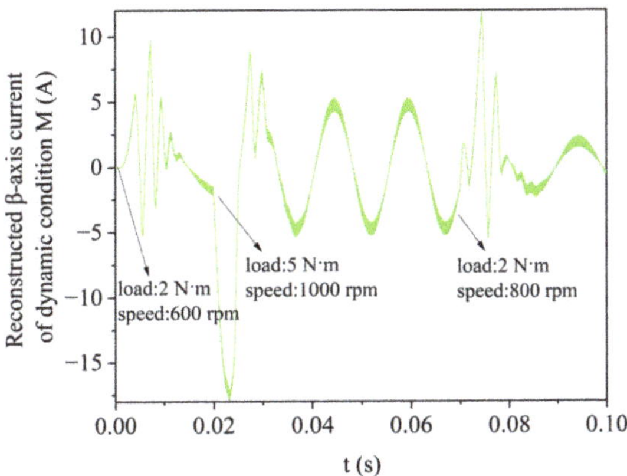


**Figure 9.** Reconstructed β-axis current of dynamic condition M.

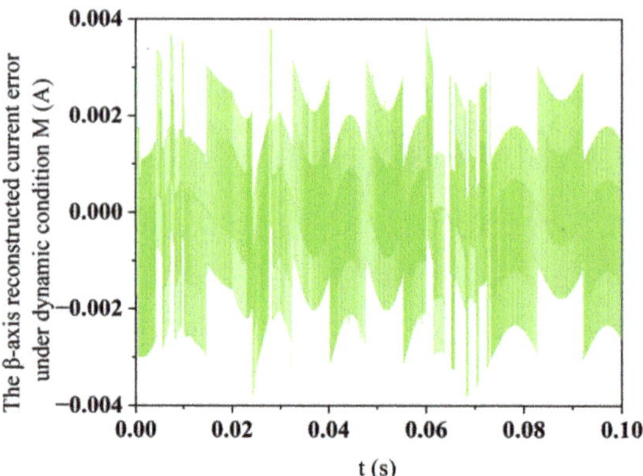

**Figure 10.** The β-axis reconstructed the current error under dynamic condition M.

According to the above analysis, the position of the maximum error of 4 mA of the β-axis reconstruction current appears in two cases: one is the stage where the motor load remains unchanged and the speed rises, as shown in Figure 2; the second is the stage of motor load change and speed change, as shown in Figure 10. According to the above phenomena, it can be judged that the change in motor speed will lead to a sudden change in the reconstructed current error, and then the maximum current error will occur, but whether the load change will lead to the maximum error has not been determined. To this end, increase the working condition N: set the speed at 1000 rpm unchanged, 5 N·m load start, 0.05 s into 15 N·m load.

Figure 11 is the β current reconstruction error under the operating condition N. Under this operating condition, the load of 15 N·m causes the speed to fluctuate at 15 rad at the set speed. The maximum error of the β-axis current is still 4 mA and only occurs once, which proves that the maximum error in the above analysis only appears in the speed adjustment stage. The increase in load will affect the error in the stable stage, such as the spike error in Figure 10. As the load increases, the spike error accumulates into a sharp angle error, but it still does not exceed 4 mA. After further verification, in the normal operating range of the motor, the adjustment of the speed leads to an increase in the local current error, and the introduction of the load leads to an increase in the current error in the stable phase, but the reconstructed current error under the parameter adaptation will not exceed 4 mA.

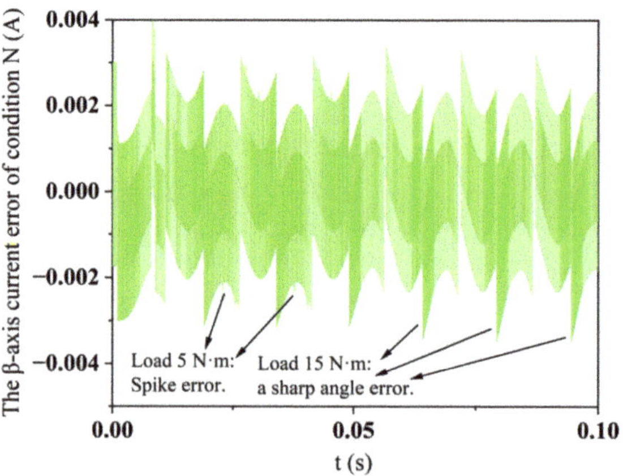


**Figure 11.** The β-axis current error of condition N.

In order to better demonstrate the ability of current reconstruction, the maximum current error in each stage of each working condition accounts for the amplitude of the stable current in this stage, as shown in Table 3. Under various working conditions, the maximum proportion of current error is less than 1%. When the load is added, the proportion of current error is even smaller due to the increase in stable current amplitude, and the lowest proportion is only 0.073%. The proportion of maximum current error reflects local error, which shows that the average error of current will be smaller, and the smaller the current error, the more feasible it is to replace the sensor current value. Table 3 shows the strong current reconstruction ability under steady and dynamic conditions, and the single current sensor FOC control is completely feasible.

**Table 3.** Brief Comparison of Reconstructed Current Error.

| Working Condition | Process | Percentage of Maximum Current Error | Purpose of Working Condition |
|---|---|---|---|
| W | 1000 rpm without load | 0.85% | Steady-state condition verification |
| M | 0 s: 600 rpm with 2 N·m<br>0.02 s: 1000 rpm with 5 N·m<br>0.07 s: 800 rpm with 2 N·m | 0.32%<br>0.073%<br>0.17% | Dynamic condition verification |
| N | 0 s: 1000 rpm with 5 N·m<br>0.05 s: 1000 rpm with 2 N·m | 0.076%<br>0.13% | Locate the position where the maximum error of the current occurs. |

### 4.2. Robustness Verification of Reconstructed Current

In the actual operation of PMSM, due to factors such as temperature and magnetic saturation, parameters such as resistance and inductance will be perturbed, and the performance of the motor control system with poor robustness will be affected. Under the complex working condition M, in order to verify the robustness of the current observer to the resistance, inductance, and flux linkage, the parameter perturbation is introduced for analysis.

Figure 12 is the β-axis current reconstruction error of $\hat{R}_S = 1.3R_S$, M and M without load. The speed changes at three moments of 0, 0.02, and 0.07, resulting in a large current error, but it quickly converges to ±1 A, and the no-load current error only fluctuates within 5 mA. Although the current error will increase to 1 A when the load is 5 N·m in 0.02 s, the average current error is 0, and the motor can still operate at high performance, which indicates that the current observer has good robustness to the stator resistance.

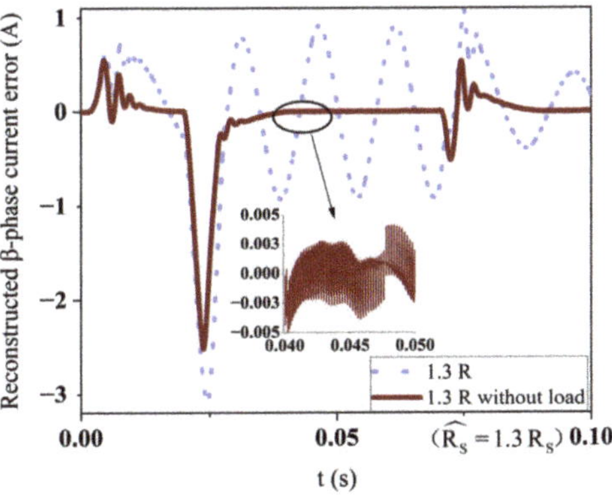


**Figure 12.** Mismatched β-axis current error.

The mismatch of $L_d$ will also affect the operating performance of the motor. The β-axis current reconstruction error under M condition and M condition without load is shown in Figure 13. Compared with the current reconstruction performance when the stator resistance $R_S$ is mismatched, the mismatch of $L_d$ causes the current error to fluctuate violently when the speed is adjusted, but the current error can still be stabilized quickly. Even at 1000 rpm and a 5 N·m load, the stable current error peak does not exceed 1 A. Obviously, the current observer has good robustness for the stator inductance.

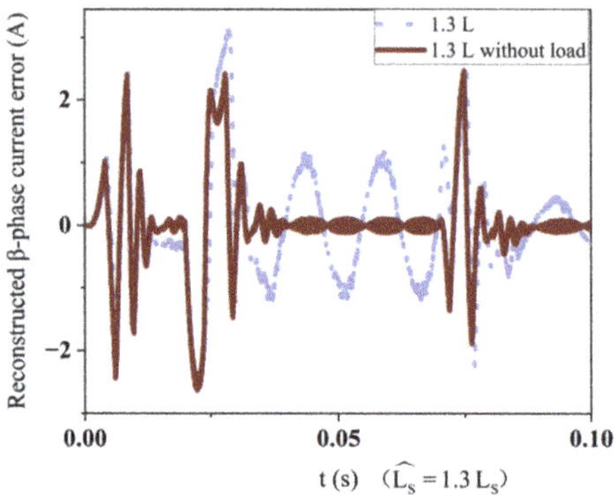


**Figure 13.** Mismatched β-axis current error.

Figure 14 still shows the β-axis current reconstruction error of M condition and M condition without load at this time $\hat{\psi}_f = 1.05\,\psi_f$. Different from the inductance and resistance mismatch, the applied load has little effect on the reconstructed current error, and the peak current error is below 0.9 A. However, the degree of flux mismatch is much smaller than the resistance and inductance in Figures 12 and 13. Obviously, the flux mismatch has a greater impact on the current error. When $\hat{\psi}_f = 1.5\psi_f$, the no-load double current sensor speed of M condition, the double current sensor speed of M condition, and the single current sensor speed of M condition are shown in Figure 15. When controlled by double current sensors, the fluctuation of no-load speed is equivalent to that of loaded speed controlled by a single current sensor. However, when controlled by double current sensors with load, speed fluctuation is much higher than when controlled by a single current sensor with load. At 1000 rpm, the speed fluctuation of a dual-current sensor with load control is up to 40 rpm, and the fluctuation of a dual-sensor no-load and a single-sensor load control is only 10 rpm. Even in the case of parameter mismatch, the reconstructed current can achieve the actual current control effect. So far, the current observer has shown excellent robustness to motor parameters.

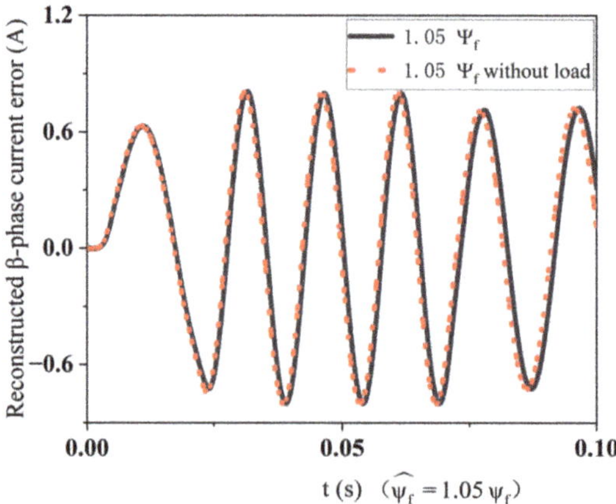


**Figure 14.** Mismatched β-axis current error.

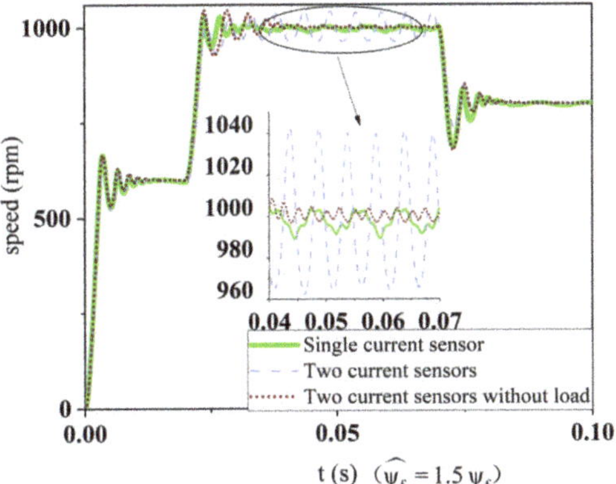



**Figure 15.** Mismatched speed.

*4.3. Comparison of Control Effect between Reconstructed Current and Actual Current*

This design is designed to replace PMSM multi-current sensor control. In order to verify that the reconstructed current can replace the actual current, the effects of FOC dual-current sensor control, V/f control, and improved direct torque control (IDTC) are compared. The control effects are shown in Figure 16, in which Figure 16a shows the speed curve and Figure 16b shows the torque curve. The dynamic performance of V/f control alone is far worse than that of other control methods, and the response time is relatively long, so it can be considered to be combined with other algorithms in special circumstances. IDTC has a fast dynamic response and a good control effect, but the speed and torque overshoot are large. The control effect of the FOC single current sensor reaches the control effect of the actual current even better because the error of the current observer is small, which is not obvious compared with the control effect of the FOC with position sensor. The FOC double current sensor used here has no position sensor. Generally speaking, the more mature control method of reducing sensors is not to use the position sensor, and this control effect comparison diagram shows that in the FOC control method, the single current sensor is better than the double current sensor without the position sensor to some extent. When the effect of position-free control using a single current sensor is not ideal, a single current sensor with a position sensor can be selected for control, which can reduce the number of sensors while still achieving high performance control.

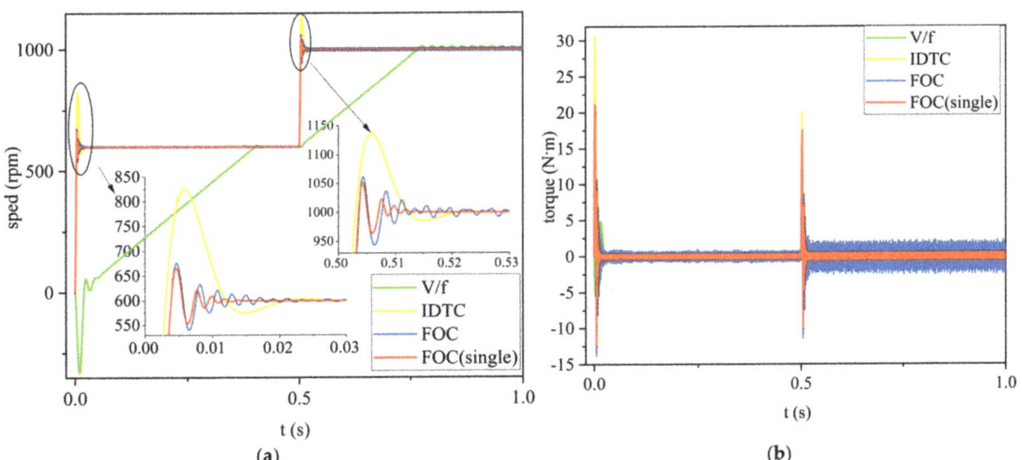



**Figure 16.** The performance of PMSM under four control methods. (**a**) Speed; (**b**) Torque.

## 5. Conclusions

After analyzing the current acquisition scheme and the current equation of the motor, a sliding mode current observer is proposed to realize the single current sensor control of the PMSM. Under steady-state conditions, the maximum error of the reconstructed current is less than 4 mA, accounting for only 0.85% of the steady-state current amplitude. Under the dynamic condition, the speed and load change suddenly at the same time, and the current tracking ability is still very strong. Due to the addition of the load, the stable current rises, accounting for only 0.17% of the steady-state current amplitude. In the robustness test, the mismatch of inductance, resistance, and flux linkage leads to an increase in the local error of the reconstructed current, which can reach 3 A, but the average error fluctuates around 0 A, and the motor can still maintain high performance control. Compared with V/f, DTC, and FOC, the single current FOC algorithm has low overshoot, a fast dynamic response, and stable operation. Through a series of tests and analyses, the following conclusions can be drawn:

(1) There is no connection between the designed current observer and the DC bus method. There is no blind area in the current reconstruction. There is no need to obtain other current information except the a-phase current, and no additional hardware measures are required.

(2) The current observer does not introduce complex structure. It has simple calculation, low dependence on parameters, high robustness, and high precision in the current reconstruction. It can achieve high-performance control of the motor.

(3) The disadvantage of this design is that a position sensor is needed. Further research considers the current reconstruction without a position sensor, but the position acquisition does not consider the use of observers. Multiple observers will increase the complexity of the system. This design can replace position sensorless FOC control. FOC without a position sensor generally adopts double current sensors and reduces one position sensor at the same time, but the current observer designed in this paper can reduce one current sensor. The current sensor can achieve better control in the case of reducing one current sensor, as shown in Figure 16.

In addition, in the error analysis, we found that the maximum local error occurs in the speed adjustment stage, which is independent of the load; however, what is related to the load is the stab error in the stable stage. As the load increases, the spike error will accumulate into a sharp angle error. Due to the small current error, the local maximum error is not much different from the spike error, but this finding may be of great significance in trying to further apply the current observer.

**Author Contributions:** Conceptualization, G.Y. and Z.W.; methodology, Y.X.; software, Y.Y.; validation, G.Y., Z.W. and Y.X.; formal analysis, Y.Y.; investigation, Y.Y.; resources, Y.Y.; data curation, Y.Y.; writing—original draft preparation, Y.Y.; writing—review and editing, G.Y.; visualization, Y.Y.; supervision, Y.X.; project administration, G.Y.; funding acquisition, G.Y. All authors have read and agreed to the published version of the manuscript.

**Funding:** This work is supported in part by the National Natural Science Foundation of China under Grant 52066008, the Development of domestic Electronic Control System (ECU) for the China VI diesel engine under Grant 202104BN050007, the Key Technology Research and Development of a methanol/diesel dual fuel engine under Grant 202103AA080002, and the Research and Application of Key Technologies for extended-range commercial electric vehicles under Grant 202102AC080004.

**Data Availability Statement:** Data are available upon request from the authors.

## References


1. Fernández, D.; Kang, Y.; Laborda, D.; Gómez, M.; Briz, F. Permanent Magnet Synchronous Machine Torque Estimation Using Low Cost Hall-Effect Sensors. *IEEE Trans. Ind. Appl.* **2021**, *57*, 3735–3743. [CrossRef]
2. Elyazid, A.; Yanis, H.; Koussaila, I.; Kaci, G.; Djamal, A.; Azeddine, H. New Fuzzy Speed Controller for Dual Star Permanent Magnet Synchronous Motor. In Proceedings of the 2021 IEEE 1st International Maghreb Meeting of the Conference on Sciences and Techniques of Automatic Control and Computer Engineering MI-STA, Tripoli, Libya, 25–27 May 2021; pp. 69–73.
3. Zhang, M.; Xiao, F.; Shao, R.; Deng, Z. Robust Fault Detection for Permanent-Magnet Synchronous Motor via Adaptive Sliding-Mode Observer. *Math. Probl. Eng.* **2020**, *11*, 1–6. [CrossRef]
4. Kang, L.; Jiang, D.; Xia, C.; Xu, Y.; Sun, K. Research and Analysis of Permanent Magnet Transmission System Controls on Diesel Railway Vehicles. *Electronics* **2021**, *10*, 173. [CrossRef]
5. Munoz-Garcia, A.; Lipo, T. A new induction motor V/f control method capable of high-performance regulation at low speeds. *IEEE Trans. Ind. Appl.* **1998**, *34*, 813–821. [CrossRef]
6. Zhe, Z.; Liu, Y.; Bazzi, A. An improved high-performance open-loop V/f control method for induction machines. In Proceedings of the 2017 IEEE Applied Power Electronics Conference and Exposition (APEC) IEEE, Tampa, FL, USA, 26–30 March 2017.
7. Zhang, Z.; Bazzi, A. Robust Sensorless Scalar Control of Induction Motor Drives with Torque Capability Enhancement at Low Speeds. In Proceedings of the 2019 IEEE International Electric Machines & Drives Conference (IEMDC), San Diego, CA, USA, 5 August 2019.
8. Zhang, X.; Foo, G. Overmodulation of Constant-Switching-Frequency-Based DTC for Reluctance Synchronous Motors Incorporating Field-Weakening Operation. *IEEE Trans. Ind. Electron.* **2019**, *66*, 37–47. [CrossRef]
9. Foo, G.; Zhang, X. Robust Direct Torque Control of Synchronous Reluctance Motor Drives in the Field-Weakening Region. *IEEE Trans. Power. Electron.* **2017**, *32*, 1289–1298. [CrossRef]
10. Cho, Y.; Lee, K.-B.; Song, J.-H.; Lee, Y. Torque-ripple minimization and fast dynamic scheme for torque predictive control of permanent-magnet synchronous motors. *IEEE Trans. Power. Electron.* **2015**, *30*, 2182–2190. [CrossRef]
11. Fan, Y.; Zhang, L.; Cheng, M.; Chau, K. Sensorless SVPWM-FADTC of a new flux modulated permanent-magnet wheel motor based on a wide-speed sliding mode observer. *IEEE Trans. Ind. Electron.* **2015**, *62*, 3143–3151. [CrossRef]
12. Tian, B.; Molinas, M.; An, Q.; Zhou, B.; Wei, J. Freewheeling Current-Based Sensorless Field-Oriented Control of Five-Phase Permanent Magnet Synchronous Motors Under Insulated Gate Bipolar Transistor Failures of a Single Phase. *IEEE Trans. Ind. Electron.* **2022**, *69*, 213–224. [CrossRef]
13. Zhang, G.; Zhou, H.; Wang, G.; Li, C.; Xu, D. Current sensor fault tolerant control for encoderless IPMSM drives based on current space vector error reconstruction. *IEEE J. Emerg. Sel. Top Power. Electron.* **2020**, *8*, 3658–3668. [CrossRef]
14. Zhang, G.; Wang, G.; Wang, H.; Xiao, D.; Li, L.; Xu, D. Pseudorandom-frequency sinusoidal injection based sensorless IPMSM drives with tolerance for system delays. *IEEE Trans. Power Electron.* **2019**, *34*, 3623–3632. [CrossRef]
15. Bierhoff, M.; Göllner, M. A current sensor less speed control algorithm for induction motors. In Proceedings of the ECON 2016—42nd Annual Conference of the IEEE Industrial Electronics Society, Florence, Italy, 23–26 October 2016; pp. 2606–2611.
16. Jankowska, K.; Dybkowski, M. A Current Sensor Fault Tolerant Control Strategy for PMSM Drive Systems Based on Cri Markers. *Energies* **2021**, *14*, 3443. [CrossRef]
17. Rahim, N.; Rahman, M.; Elias, M.; Ping, H. Implementation of Speed Sensorless Drives for IPMSM Based on Simplified Stator Flux Observer. *Int. Rev. Electr. Eng.-IREE* **2012**, *7*, 4879.
18. Kommuri, S.; Lee, S.; Veluvolu, K. Robust Sensors-Fault-Tolerance with Sliding Mode Estimation and Control for PMSM Drives. *IEEE ASME Trans. Mechatron.* **2017**, *23*, 17–28. [CrossRef]
19. Green, T.; Williams, B. Derivation of motor line-current waveforms from the DC-link current of an inverter. *IEE Proc. B Electric. Power Appl.* **1989**, *136*, 196–204. [CrossRef]
20. Lu, H.; Cheng, X.; Qu, W.; Sheng, S.; Li, Y.; Wang, Z. A three-phase current reconstruction technique using single DC current sensor based on TSPWM. *IEEE Trans. Power Electron.* **2014**, *29*, 1542–1550.
21. Xu, Y.; Yan, H.; Zou, J.; Wang, B.; Li, Y. Zero Voltage Vector Sampling Method for PMSM Three-Phase Current Reconstruction Using Single Current Sensor. *IEEE Trans. Power Electron.* **2017**, *32*, 3797–3807. [CrossRef]
22. Gu, Y.; Ni, F.; Yang, D.; Liu, H. Switching- state phase shift method for three-phase-current reconstruction with a single DC-link current sensor. *IEEE Trans. Ind. Electron.* **2011**, *58*, 5186–5194.
23. Wang, W.; Yan, H.; Zou, J.; Zhang, X.; Zhao, W.; Buticchi, G.; Gerada, C. New Three-Phase Current Reconstruction for PMSM Drive with Hybrid Space Vector Pulse width Modulation Technique. *IEEE Trans. Power Electron.* **2021**, *36*, 662–673. [CrossRef]
24. Lu, J.; Zhang, X.; Hu, Y.; Liu, J.; Gan, C.; Wang, Z. Independent phase current reconstruction strategy for IPMSM sensorless control without using null switching states. *IEEE Trans. Ind. Electron.* **2018**, *65*, 4492–4502. [CrossRef]
25. Sun, K.; Wei, Q.; Huang, L.; Matsuse, K. An overmodulation method for PWM-inverter-fed IPMSM drive with single current sensor. *IEEE Trans. Ind. Electron.* **2010**, *57*, 3395–3404. [CrossRef]
26. Salmasi, F.; Najafabadi, T. An Adaptive Observer with Online Rotor and Stator Resistance Estimation for Induction Motors With One Phase Current Sensor. *IEEE Trans. Energy Convers.* **2011**, *26*, 959–966. [CrossRef]
27. Teng, Q.; Cui, H.; Duan, J.; Zhu, J.; Guo, Y.; Lei, G. Extended state observer-based vector control for PMSM drive system with single phase current sensor. In Proceedings of the 2017 20th International Conference on Electrical Machines and Systems (ICEMS), Sydney, NSW, Australia, 11–14 August 2017; pp. 1–6.

28. Verma, V.; Chakraborty, C.; Maiti, S. Speed sensorless vector controlled induction motor drive using single current sensor. *IEEE Trans. Energy Convers.* **2013**, *28*, 938–950. [CrossRef]
29. Beng, G.; Zhang, X.; Vilathgamuwa, D. Sensor Fault-Resilient Control of Interior Permanent-Magnet Synchronous Motor Drives. *IEEE ASME Trans. Mechatron.* **2015**, *20*, 855–864. [CrossRef]
30. Badini, S.; Kumar, C.; Verma, V. Speed Sensorless Vector Controlled PMSM Drive with A Single Current Sensor. In Proceedings of the 2020 IEEE International Conference on Power Electronics, Drives and Energy Systems (PEDES), Jaipur, India, 16–19 December 2020; pp. 1–6.
31. Hafez, B.; Abdel-Khalik, A.; Massoud, A.; Ahmed, S.; Lorenz, R. Single-Sensor-Based Three-Phase Permanent-Magnet Synchronous Motor Drive System with Luenberger Observers for Motor Line Current Reconstruction. *IEEE Trans. Ind. Appl.* **2014**, *50*, 2602–2613. [CrossRef]
32. Yang, W.; Guo, H.; Sun, X.; Wang, Y.; Riaz, S.; Zaman, H. Wide-Speed-Range Sensorless Control of IPMSM. *Electronics* **2022**, *11*, 3747. [CrossRef]
33. Chakraborty, C.; Verma, V. Speed and current sensor fault detection and isolation technique for induction motor drive using axes transformation. *IEEE Trans. Ind. Electron.* **2015**, *62*, 1943–1954. [CrossRef]
34. Wang, G.; Zhang, H. A new speed adaptive estimation method based on an improved flux sliding-mode observer for the sensorless control of PMSM drives. *ISA Trans.* **2021**, *126*, 675–686. [CrossRef]

*Article*

# Super Twisting Sliding Mode Control with Compensated Current Controller Dynamics on Active Magnetic Bearings with Large Air Gap

Jonah Vennemann *, Romain Brasse, Niklas König, Matthias Nienhaus and Emanuele Grasso

Laboratory of Actuation Technology, Saarland University, 66123 Saarbrücken, Germany
* Correspondence: vennemann@lat.uni-saarland.de

**Abstract:** Due to their unique properties, like no mechanical contact and therefore no wear and no lubrication needed, Active Magnetic Bearings (AMBs) have been a dynamic field of research in the past decades. The high non-linearities of AMBs generate many challenges for the control of the otherwise unstable system, thus they need to be addressed to deliver the performance that modern applications require. Integrating the current controller dynamics into the model of a position controller in a cascading control loop helps to improve the performance of the control loop compared to a plain current controlled schema. Further, this nested control loop guarantees the predefined current dynamics of the current controller, tuned according to an industrial criterion. The systems dynamics are modelled and the proposed controller is validated experimentally on a physical test bench. The experimental results show a performant position control with a nested and explicit current controller on an AMB, even with a large air gap and star-connected coils. The trajectory range of the rotor was reduced by 87% to $\pm20$ μm, compared to a plain current-controlled model. The proposed control strategy lays the foundation for further research, especially concerning sensorless position estimation techniques since these usually have limited bandwidth and benefit from a predefined current dynamic.

**Keywords:** active magnetic bearing; AMB; second order sliding mode control; SOSMC; current control; super twisting; large air gap

**Citation:** Vennemann, J.; Brasse, R.; König, N.; Nienhaus, M.; Grasso, E. Super Twisting Sliding Mode Control with Compensated Current Controller Dynamics on Active Magnetic Bearings with Large Air Gap. *Electronics* **2023**, *12*, 950. https://doi.org/10.3390/electronics12040950

Academic Editors: Jamshed Iqbal, Ali Arshad Uppal and Muhammad Rizwan Azam

Received: 27 January 2023
Revised: 8 February 2023
Accepted: 9 February 2023
Published: 14 February 2023

## 1. Introduction

Compared to classical mechanical bearings, active magnetic bearings (AMBs) offer unique properties due to lack of mechanical contact and have, therefore, attracted immense interest in research and industry. They offer advantages like drastically reduced friction on the rotor, no mechanical wear, reduced noise, low maintenance cost, and no lubrication required as well as increased efficiency and higher rotational speed when applied to electrical machines [1]. Thus, the economic and ecological impact of those machines can be improved significantly. The here-considered AMBs are based on magnetic reluctance forces, and according to Earnshwas theorem, a purely passive suspension is therefore not feasible [2]. Hence, reluctant-force-based AMBs require control algorithms with position feedback in order to stabilise the rotor at the centre position [1]. The position feedback is realised conventionally by additional sensors such as laser- or hall-effect sensors. However, sensors often are costly and require installation space and maintenance effort of the overall system, therefore weakening the economic merits of AMBs. In the past decades, sensorless control has been widely applied to AMBs to retrieve position information from electrical quantities already being measured in the system. Such a self-sensing approach can replace position sensors in cost-critical applications or provide further redundancy with existing position sensors in cases where high functional safety is required.

In the field of electrical machines, sensorless algorithms developed over the past decades are either based on the induced back-EMF or machine anisotropies. The review works [3–5] provide a good overview of existing techniques. In particular, there is a technique, which successfully exploits the star-point voltage and modified pulse width modulation (PWM). In some literature this technique is called Direct Flux Control [6–11].

These works demonstrate the robustness, accuracy and increased signal-to-noise ratio (SNR) of this approach. More in detail, there is another circuitry based on a re-settable integrator circuit that allows to amplify the star-point voltage with increased SNR. This approach seems interesting for future research works. Nevertheless, the technique requires not only an accessible star-point but also the modification of the PWM pattern, used to introduce zero voltage vectors and explicit, active voltage vectors for measurement. This puts several requirements on the switching power amplifier used. In a previous publication, such a power amplifier was introduced and validated [12]. The experimental validation of this driving approach showed satisfactory results. A detailed report can be found in [12].

The aim of this work is to provide an elaborate position control strategy for the star-connected AMB. Since the test bench utilised in this application will also be used for teaching and demonstrating porpuses, a relatively large air gap has been chosen. This allows to observe the motion of the rotor with bare eyes and makes the technology easier accessible for students. In addition, the production of mechanical parts with higher tolerances is cost-effective and even enables the use of rapid prototyping. Thus, the control algorithm must be robust enough to compensate for the strong non-linearities of the AMB system. The most employed position controller in industrial applications is a Proportional Integral Derivative (PID) controller [1]. This type of controller is generally widely spread, well-documented and well-known by many engineers, making it easy to apply and tune. The PID controller delivers suitable and sufficient control performance for many fundamental applications. However, a review of active magnetic bearing control strategies finds that the PID controller cannot keep up with the demands of the most recent applications [13]. That is especially true for high-speed applications. This issue calls for novel control strategies for active magnetic bearings, which can stabilise the rotor at high speeds and reduce vibrations in the system. One way to achieve a better performance of the PID controller is the latter's extension. The PID controller can be extended using fuzzy logic [14]. A comparison between PID and fuzzy-PID finds a better performance and lower energy consumption in the case of the fuzzy-PID controller [15]. Another way of improving the PID controller's performance is using a neural network to update the controller's gain during operation. A study finds good control performance and robustness to uncertainties compared to a plain PID controller [16]. A third method to extend the PID controller is fractional order control [17]. A study investigated the performance of a fractional order PID controller and found a distinct reduction of vibration compared to the plain PID controller [18]. A different control strategy is Flatness-Based control [19]. Experiments show a good performance considering trajectory tracking problems where the rotor is not held at a constant reference position but follows a given reference trajectory [20,21]. A further control strategy that gained attention is sliding mode control [22]. Experiments show a good disturbance rejection ability and good robustness to model uncertainties [23,24].

Comparing the different possibilities, the sliding mode control strategy seems most interesting for further research, because of its good robustness to model and parameter uncertainties, which is especially important with regard to the large air gap, since it induces strong non-linearities. Further, its lower computational and tuning effort, compared to strategies like fractional order PID control, makes it more easily applicable. Sliding mode control in all its varieties has been widely deployed to active magnetic bearings. Recent developments in sliding mode control focus on the theoretical and practical development of higher order sliding mode control algorithms, the discrete time implementation of sliding mode control and the application of sliding mode control to multi agent systems [25,26]. Similar to the PID controller, there are techniques that extend the sliding mode controller by adjusting its gains in order to ensure robustness and a finite time convergence on AMBs [26–28].

Many state of the art control strategies for AMBs rely on voltage control, where the cascading control loop, consisting of a position and a current controller, is integrated into one controller. This strategy is known to be beneficial for the overall performance of the AMB system [1]. On the other hand, integrating the current controller into the position controller makes it impossible to influence the current reference directly. However, the exertion of influence on the current reference can be beneficial for safety reasons, as the maximum current can easily be restricted. Further, the current controller can be tuned using a well-known industrial design criterion, which guarantees a specific and predefined current behaviour. An additional current limiting controller could also provide additional safety in the case of voltage control, but the current dynamics will still not feature a predefined behaviour.

To not leave behind the potential performance benefits, the aim is to integrate the dynamics of the current controller into the position controller model. This forms a cascading control loop consisting of a current controller with predefined behaviour, according to an industrial criterion and a position controller, that is aware of the current controller's dynamics. A further potential benefit of the explicit current controller is the smoothing of the current reference from the sliding mode controller. Since sliding mode is known for its high-frequency switching nature, it seems beneficial to have a predefined bandwidth for the current dynamics. Especially since this control pattern will be used in further research regarding bandwidth-limited sensorless techniques. Tables A1 and A2 provide a list of the used symbols and indices.

## 2. Mathematical Modeling

### 2.1. Modelling of AMBs

The here-considered active magnetic bearings utilise reluctance forces. Reluctance forces are generated at the boundaries between materials with differing permeabilities $\mu$. Figure 1 shows schematically the structure of a simple single-phase AMB and the magnetic flux $\Phi$ along the iron with the length $l_{fe}$.

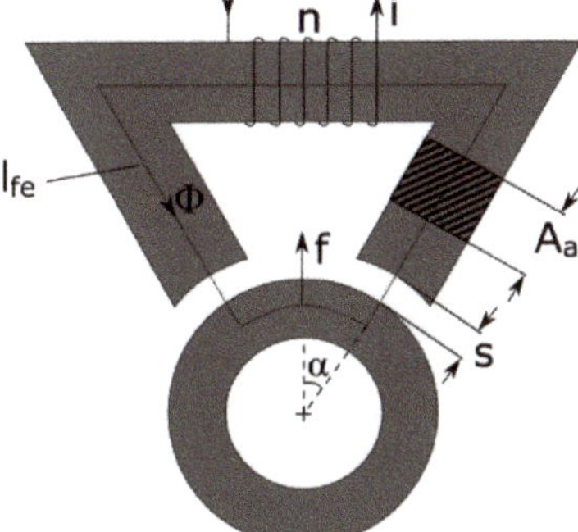


**Figure 1.** Force and geometry of a radial magnet, adopted from [12].

The calculation of reluctance forces is based on the field energy $W_a$. In a homogeneous field, the energy stored in an air gap with size $s$, a cross-sectional area $A$ and a volume $V_a = 2sA_a$ can be written as [1,29]:

$$W_a = \frac{1}{2}B_a H_a V_a = \frac{1}{2}B_a V_a A_a 2s = B_a V_a A_a s. \tag{1}$$

Here, $B_a$ denotes the magnetic flux density in the air gap and $H_a$ denotes the magnetic field in the air gap, respectively. The force $f$ equals the partial derivative of $W_a$ with respect to the air gap (displacement) $s$ [1,29,30]:

$$f = -\frac{\partial W_a}{\partial s} = B_a H_a A_a.$$ (2)

Inserting the constitutive law $B = \mu_0 \mu_r H$ yields:

$$f = \frac{B_a{}^2 A_a}{\mu_0}.$$ (3)

According to [1,29,30], the magnetic circuit in Figure 1 can be described using the number of the coil windings $n$ and the current $i$ flowing through the coils:

$$B = \mu_0 \frac{ni}{2s}.$$ (4)

Assuming the flux density remains constant along the iron with length $l_{fe}$ and the air gap $s$, the force of an AMB can be obtained by inserting Equation (4) into Equation (3):

$$f = \mu_0 A_a \left(\frac{ni}{2s}\right)^2 = \frac{1}{4}\mu_0 n^2 A_a \frac{i^2}{s^2}.$$ (5)

A new variable $k$ is introduced to simplify the equation:

$$f = k\frac{i^2}{s^2},$$ (6)

with

$$k = \frac{1}{4}\mu_0 n^2 A_a.$$ (7)

As the poles of the electromagnet affect the rotor with an angle $\alpha$, the force $f$ of the AMB in Figure 1 can be obtained by [1,29]:

$$f = k\frac{i^2}{s^2}\cos\alpha.$$ (8)

Equation (8) indicates, that the force depends quadratically on the current and inverse quadratically on the air gap. Since the force depends quadratically on the current, the force is independent of the sign of the current and always acts in the same direction, namely towards the electro magnet. However, in control theory, linear relations are preferred. Often, non-linear relationships are approximated in the operating point. For an active magnetic bearing, the operating point is defined by a bias current $i_0$ and the nominal air gap $s_0$.

Further, there are often two electromagnets opposite to each other, as shown in Figure 2. In order to generate counteracting forces, the upper magnet is driven with the sum of the bias current $i_0$ and the control current $i_x$. Instead, the lower magnet is driven by the difference between the bias current $i_0$ and the control current $i_x$. This configuration is called differential driving mode [1,29,30].

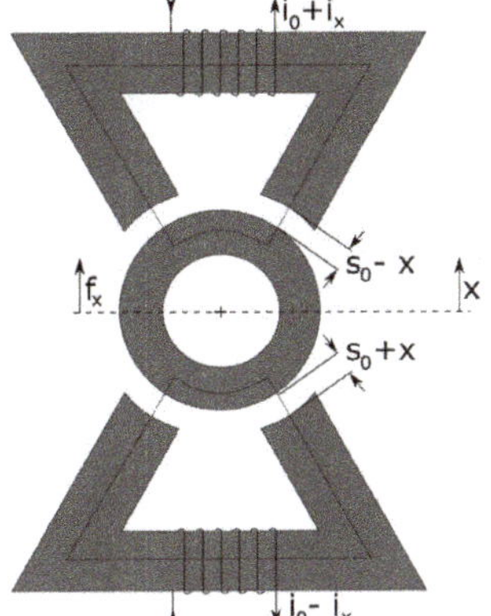

**Figure 2.** Differential driving mode of an axial bearing.

The force $f_x$ can now be represented as the difference between the forces of the two magnets $f_+$ and $f_-$, respectively [1,29]:

$$f_x = f_+ - f_- = k\left(\frac{(i_0 + i_x)^2}{(s_0 - x)^2} - \frac{(i_0 - i_x)^2}{(s_0 + x)^2}\right)\cos\alpha. \tag{9}$$

Equation (9) can be simplified and linearised with respect to $x \ll s_0$ [1]:

$$f_x = \frac{4ki_0}{s_0^2}(\cos\alpha)i_x + \frac{4ki_0^2}{s_0^3}(\cos\alpha)x, \tag{10}$$

$$f_x = k_i i_x - k_s x, \tag{11}$$

with

$$k_i = \frac{4ki_0}{s_0^2}\cos\alpha, \tag{12}$$

and

$$k_s = -\frac{4ki_0^2}{s_0^3}\cos\alpha. \tag{13}$$

Equations (12) and (13) show that the force/current and the force/displacement factors depend quadratically and cubic, respectively, on the air gap. As mentioned, a linear relationship is preferred for the control. Hence, both factors are linearised in the operating point as illustrated in Figure 3. Due to this linearisation, large parameter deviations can be expected when leaving the operating point.

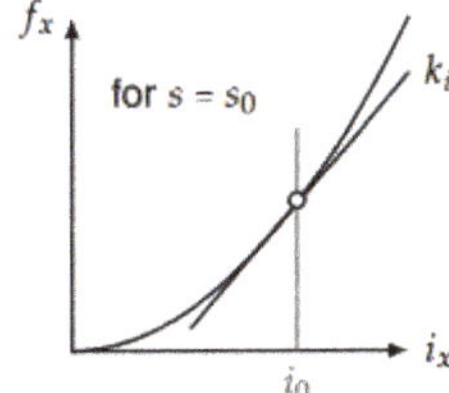

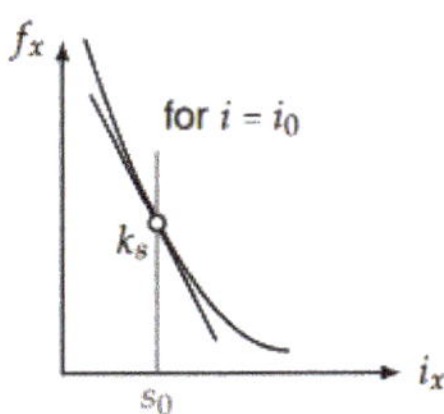


**Figure 3.** Linearised force-current factor $k_i$ and force-displacement factor $k_s$; adopted from [1].

While deriving this model, approximations have been made to simplify the physical correlations. The surface of the ferromagnetic material $A_{fe}$ and the projected surface of the air gap $A_a$ will not be equal in an actual application. Further, flux leakage along the air gap must be expected. Moreover, material properties, such as hysteresis, saturation and losses in copper and iron, such as eddy current and hysteresis losses, have been neglected. Mechanical losses due to friction between the rotor and the medium surrounding it have not been modelled. The considered rotor is rigid, symmetric and gyroscopic effects are neglected. With these approximations, the error calculating the force lies within the range from 5 to 10% [1].

However, in order to design a sliding mode controller, the description must be available in the state space and ideally, in the controllability form [31]. The controllability form allows to easily read the relationship between the eigenvalues of the system and the model parameters. Further, the existence of the controllability form shows that the system is indeed controllable. The previously derived differential equations can be transformed into a state space description by introducing the displacement $x$ of the rotor and its velocity $v = \dot{x}$ as the two states of the system:

$$x = \begin{bmatrix} x \\ v \end{bmatrix}. \tag{14}$$

For a system in the state space:

$$\dot{x}(t) = Ax(t) + bu(t), \tag{15}$$
$$y(t) = cx(t) + du(t), \tag{16}$$

this yields the following system in matrix notation:

$$A = \begin{bmatrix} 0 & 1 \\ -\frac{ks}{m} & 0 \end{bmatrix}, \qquad b = \begin{bmatrix} 0 \\ \frac{ki}{m} \end{bmatrix}, \tag{17}$$
$$c^\mathsf{T} = \begin{bmatrix} 1 & 0 \end{bmatrix}, \tag{18}$$
$$d = 0, \tag{19}$$
$$x(0) = x_0. \tag{20}$$

This system is not yet in the desired controllability form, but can be transferred into the following system [31,32]:

$$A_R = \begin{bmatrix} 0 & 1 \\ -\frac{ks}{m} & 0 \end{bmatrix}, \qquad b_R = \begin{bmatrix} 0 \\ 1 \end{bmatrix}, \tag{21}$$
$$c^\mathsf{T}_R = \begin{bmatrix} \frac{ki}{m} & 0 \end{bmatrix}, \tag{22}$$
$$d = 0, \tag{23}$$
$$x_R(0) = x_{R0}, \tag{24}$$

which represents the controllability form.

## 2.2. Integrating the Current Controller Dynamics

Figure 4 shows the current control loop and its components. The PI controller, voltage amplifier and coils are in the forward path and the current sensor is in the backwards path, all with their corresponding transfer functions:

- PI Controller: $K_{pi}\frac{1+\tau_{ii}s}{\tau_{ii}s}$ with the controller gain $K_{pi}$ and the time constant $\tau_{ii}$ tuned according to the modulus optimum criterion.
- H-bridge Voltage Amplifier: $\frac{1}{1+\tau_A s}$ with the time constant $\tau_A = \frac{1}{2f_A}$ where $f_A$ is the PWM frequency of the H-bride.

- Current Sensor: $\frac{1}{1+\tau_s s}$ with its time constant $\tau_s = \frac{1}{f_s}$, where $f_s$ is the sensors sampling frequency.
- Coil: $\frac{1}{R(1+\tau_a s)}$ with is resistance $R$, inductance $L$, gain $\frac{1}{R}$ and its electric time constant $\tau_a = \frac{L}{R}$.

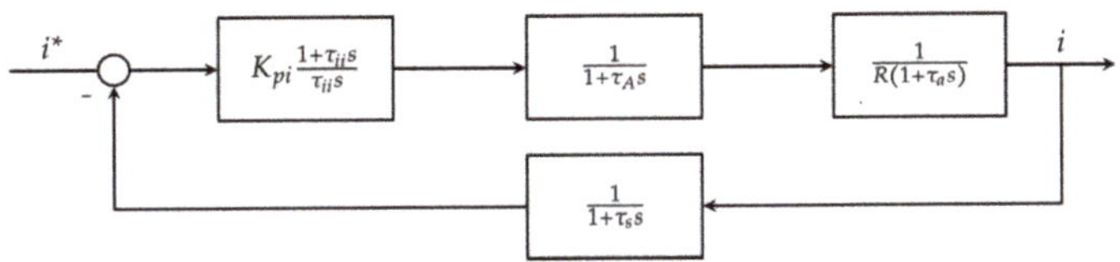


**Figure 4.** Current controller for AMB with feedback loop.

As mentioned, the PI current controller is tuned according to the industrial modulus optimum criterion, in order to achieve a predefined behaviour [12,33]. The PI controller's time constant $\tau_{ii}$ can be set as $\tau_{ii} = \tau_a$, which then cancels the slowest time constant. The PI controllers gain $K_{pi}$ is tuned according to the modulus optimum criterion:

$$K_{pi} = \frac{R_a \tau_a}{2\tau_\Sigma}. \tag{25}$$

The current control loop can be simplified with respect to the modulus optimum criterion, so the feedback loop can be expressed as [33]

$$G_{PI}(s) = \frac{1}{2\tau_\Sigma^2 s^2 + 2\tau_\Sigma s + 1}. \tag{26}$$

Here, $\tau_\Sigma$ is the sum of the time constants of the current sensor $\tau_s$ and the time constant of the voltage amplifier $\tau_A$. This transfer function now features a maximum overshoot of 4%, a rise time of $4.5\tau_\Sigma$, a settling time of $8.4\tau_\Sigma$ and a control error of $\pm 2\%$ [33].

The transformation of Equation (26) yields the following state space description [31,32]:

$$A_{pi} = \begin{bmatrix} 0 & 1 \\ -\frac{1}{2\tau_\Sigma^2} & -\frac{1}{\tau_\Sigma} \end{bmatrix}, \qquad b_{pi} = \begin{bmatrix} 0 \\ 1 \end{bmatrix}, \tag{27}$$

$$c_{pi}^\mathsf{T} = \begin{bmatrix} \frac{1}{2\tau_\Sigma^2} & 0 \end{bmatrix},$$

$$d = 0.$$

The states are choosen as:

$$x = \begin{bmatrix} i \\ \frac{di}{dt} \end{bmatrix}. \tag{28}$$

Now, both cascading subsystems have been derived and need to be integrated. This yields the following controllable system in the state space:

$$A = \begin{bmatrix} 0 & 1 & 0 & 0 \\ 0 & 0 & 1 & 0 \\ 0 & 0 & 0 & 1 \\ -\frac{k_s}{2m\tau_\Sigma^2} & -\frac{k_s}{m\tau_\Sigma} & \frac{2k_s\tau_\Sigma^2 + m}{2m\tau_\Sigma^2} & -\frac{1}{\tau_\Sigma} \end{bmatrix}, \qquad b = \begin{bmatrix} 0 \\ 0 \\ 0 \\ 1 \end{bmatrix}, \tag{29}$$

$$c^\mathsf{T} = \begin{bmatrix} \frac{k_i}{2m\tau_\Sigma} & 0 & 0 & 0 \end{bmatrix},$$

$$d = 0.$$

With the state vector:

$$x = T_R^{-1} \begin{bmatrix} i \\ \frac{di}{dt} \\ x \\ v \end{bmatrix}, \qquad x_{R0} = T_R^{-1} x_0 = 0, \tag{30}$$

and

$$T_R = \begin{bmatrix} \frac{k_s}{m} & 0 & 1 & 0 \\ 0 & \frac{k_s}{m} & 0 & 1 \\ \frac{k_i}{2m\tau_\Sigma^2} & 0 & 0 & 0 \\ 0 & \frac{k_i}{2m\tau_\Sigma^2} & 0 & 0 \end{bmatrix}. \tag{31}$$

Here, $T_R$ denotes the transformation matrix into the controllability from.

## 3. Controller Design

The goal is now, to design a controller that drives the rotor to the referenced position and stabilises it. As mentioned in the previous section, the AMB model is fraught with uncertainties. Further, the AMB has a highly nonlinear behaviour that has been linearised in the operating point. However, since the operating point is left in phases such as the startup, large parameter deviations must be taken into account. For these reasons, sliding mode control has been chosen for its robustness against model and parameter uncertainties.

After the desired dynamics were derived previously, the sliding variable can be introduced:

$$\sigma = \sigma(x). \tag{32}$$

Sliding mode control aims to drive $\sigma$ to zero in finite time by employing a control law $u = u(x)$. In the case of the AMB, the controller output is equivalent to the current reference $u = u(x) = i^*$. Thus, sliding mode control can be considered a two-part controller design, consisting of a switching function $\sigma(x)$ and a control law $u(x)$.

For a system described in the state space

$$\dot{x}(t) = Ax(t) + bu(t), \tag{33}$$

a switching function can be defined as

$$\sigma(x) = Sx(t), \tag{34}$$

with

$$S \in \mathbb{R}^n. \tag{35}$$

$S$ is a free parameter. Literature states that it is not entirely transparent, how to select a value for $S$ in order to ensure a specific design criterion [34]. However, there are well-studied systematic and traceable methods that provide a range of values for the design of $S$, one of which is the pole-placement method [34,35]. If a system is given in the previously mentioned controllability form, a suitable switching function is given by [34]:

$$\sigma(x) = s_1 x_1 + s_2 x_2 + \dots + s_{n-1} x_{n-1} + x_n. \tag{36}$$

However, to counteract the widely differing magnitudes of the states, the switching function for the AMB is now chosen in a more general form as

$$\sigma(\mathbf{x}) = s_1 i + s_2 \frac{di}{dt} + s_3 x + s_4 v. \tag{37}$$

This switching function results in a high-frequency switching motion in the control action, causing the so-called chattering phenomenon. On the other hand, the switching function is also the root of the robustness against parameter and model uncertainties. Having the current and its derivative integrated into the switching function, allows the sliding mode controller to counteract the otherwise unknown disturbances induced by the current control loop.

For the part of the control law, the conventional sliding mode is especially known for the chattering phenomenon. Chattering can be a severe issue in cascade control systems, such as a magnetic bearing with an explicit current controller [35]. For this reason, the conventional sliding mode is considered unsuitable and second-order sliding mode is employed. Figure 5 shows that, in contrast to the conventional sliding mode, second order sliding mode drives not only $\sigma$ to zero but also its derivative $\dot{\sigma}$. In theory, second-order sliding mode is free of chattering. In reality, however, discretisation and disturbances add residual chattering. Several algorithms induce a second-order sliding motion in the system. Since $\dot{\sigma}$ needs to be driven to zero as well, additional measurements are often necessary. In reality, however, this is often not feasible because additional sensors would be needed. This narrows down the choice for a second-order sliding mode control algorithm. The only algorithm that does not require a measurement of $\dot{\sigma}$ is the so-called super-twisting control algorithm. The algorithm is given by [34,35]:

$$u = c \sqrt{|\sigma|}\, sign(\sigma) + w, \qquad c = 1.5\sqrt{C}; b = 1.1C, \tag{38}$$
$$\dot{w} = b\, sign(\sigma).$$

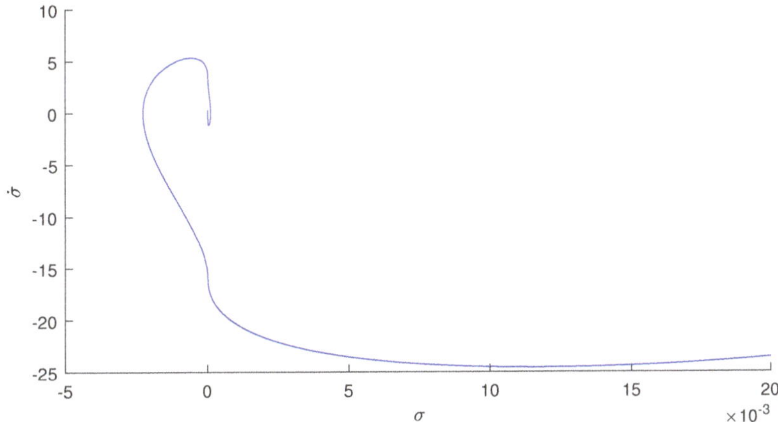

**Figure 5.** Trajectory of $\sigma$ over $\dot{\sigma}$ with super twisting controller from Equation (38) applied.

The parameters $c$ and $b$ are positive scalar controller gains. The scalar $C$ symbolises the Lipschitz constant for unknown but bounded disturbances. Other than the conventional sliding mode, super twisting control is a continuous control algorithm. The high-frequency switching function is now hidden under the integral

$$w = b \int sign(\sigma)dt. \tag{39}$$

The proposed controller design introduces the dynamics of the current control loop into the sliding mode controller without having to abandon the explicit current controller. This way, the proposed design tries to combine the advantages of the a current controlled schema, like easy exertion of influence on the current, with the performance boost of a voltage control schema.

## 4. Results and Discussion

### 4.1. Test Bench Design

In order to validate the derived control approach, a test bench consisting of two individual two-degree of freedom (DOF) bearings is engineered. Both bearings work independently of each other and shift in axial direction is neglected in this prototype application. Thus, no additional thrust bearing is provided and the rotor is only supported in x- and y-direction. However, due to the reluctant forces, the rotor is still supported passively in z-direction. Figure 6 shows the control loop for one single bearing. The control loop consists of two independent position controllers for x- and y-axis. The current references $i_x^*$ and $i_y^*$ respectively, are fed into the current controller. The current controller supplies the demanded current across the star connected coils of the AMB utilizing a PWM pattern. The bearing consists of 4 coil pairs in differential driving mode, two per axis, as displayed in Figure 2. The current control loop, as described earlier, and the corresponding hardware are explained in detail in the previous publication [12].

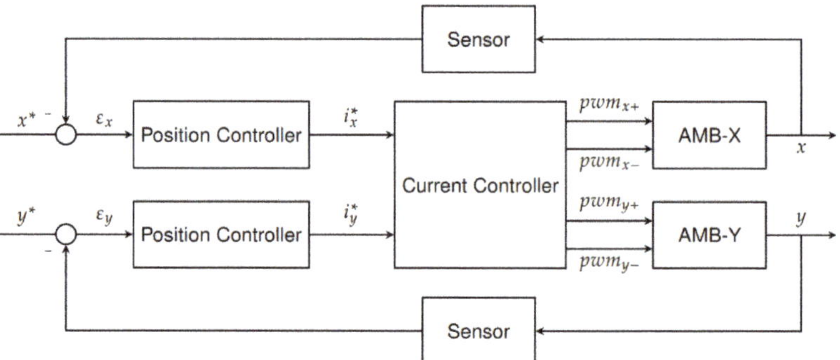


**Figure 6.** Block diagram of the overall control loop for one bearing, adopted from [12].

Figure 7 shows a 3D model of the test bench. Each bearing consists of an 8-pole stator with a heteropolar coil arrangement (blue). The 8 coils (red) have 96 windings each and are interconnected to pairs. The position of the rotor is measured by a total of four position sensors (orange), that operate at a sampling frequency of 2 kHz. The rotor is composed of an aluminium hull (light grey) and 2 ferromagnetic cores (darker grey), which are embedded in the hull at each end. This construction prevents magnetic "sticking" in the event of a touchdown or in the startup phase. Figures 8 and 9 illustrate this design with a 3D Model of the used rotor and a cross-section. This design leads to a nominal mechanical air gap of 1 mm and a nominal magnetic air gap of 2 mm, meaning the outer aluminium hull has a thickness of 1mm. Figure 10 shows the front view on the bearing, with clearly visible coils, stator and rotor.

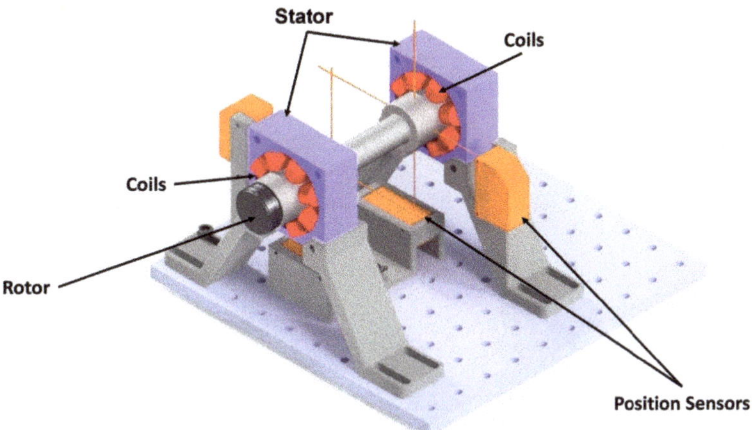


**Figure 7.** Assembly of the test bench with stator (blue), coils (red), position sensors (orange) and rotor (grey) [12].

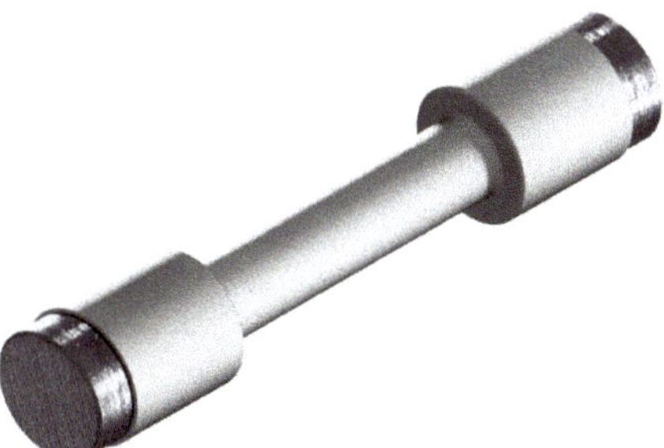

**Figure 8.** 3D model of the rotor used in the test bench.

**Figure 9.** Cross section of the rotor, revealing the outer aluminium hull and the inner ferromagnetic cores.

The current reference from the derived sliding mode controller is fed into the modulus optimum tuned PI current controller. Utilizing PWM, the current controller supplies the requested current across the star-connected coils. The current controller and the star connection are investigated and explained in [12]. Parameters of the test bench and controller gains can be found in Appendix B.

The build test bench and the model from Section 2 were validated against each other with respect to the force current relation ship. The force current relationship from Equation (11) can be easily validated by comparing the actual current necessary to levitate the rotors weight to the theoretical current according to Equation (11). For a rotor weighting 406 g and an average current of 0.77 A this yields an error of 5.97% for the force current factor. This is within the 5–10% error expectation due to the unmodelled losses in the system, explained in Section 2 [1].

**Figure 10.** Front view of the inactive AMB with clearly visible coils, rotor and stator and the nominal air gap of 2 mm.

### 4.2. Experiment Current Controlled

In the first experiment, only one bearing is active and the current controlled system from Equations (22)–(24) is considered. Thus, there is no compensation for the current controller dynamics. The second bearing is turned off and remains idle. The experiment starts from the rest position, where the rotor is in contact with the lower part of the stator, and a constant reference displacement $x_{ref} = 0$. Figure 11 shows the error in the displacement and the current reference during the experiment. During this experiment, the rotor can lift off without any external help. After the rotor position is stable, an increasing manual force in the negative y-direction (downwards) followed by force in the positive y-direction (upwards) is applied from seconds two to six and seven to ten. The current reference returns to the unloaded level and sinks according to the force. During the changes in the load, the controller manages to keep the rotor in a stable position. At second 11.5, the rotor contacts the lower stator, enforced by a large external force. At second 14.5, the external force is applied in the opposite direction, forcing the rotor to contact the upper stator. Just as at the beginning of the experiment, the rotor can recover from rest-position and forced touch-down without external help or a specific startup or touchdown sequence. The controller remains stable, despite the relatively large air gap and the associated leaving of the operating point. This recovery underlines the robust controller design. The rotor moves within ±0.15 mm range, except for the forced touchdowns. Table 1 shows the error metric for this experiment. The defined error metrics are the mean error in % of the nominal mechanical air gap (1 mm) after initial rotor lift-off, the absolute maximum error in % of the nominal mechanical air gap (1 mm) after the initial rotor lift-off, the maximal error in mm, and the approximate settling time of the rotor at the beginning of the experiment in seconds. All metrics are provided for the x- and y-axis and measured without external disturbance forces. The maximum error in x direction is higher compared to the error in y direction because mechanical vibrations are influencing the measurement in this particular case.

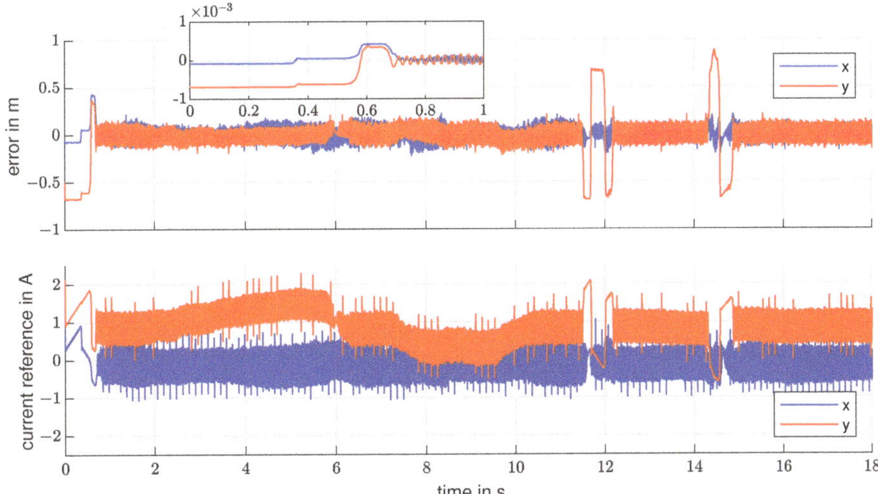

**Figure 11.** Displacement error and current reference $i^*$ without current controller dynamics for one single AMB with constant reference positions.

**Table 1.** Error metrics of the experiment neglecting current controller dynamics, with one bearing, constant reference positions and external disturbance forces.

| Axis | Mean Error in % | Max Error in % | Max Error in mm | Settling Time in s |
|---|---|---|---|---|
| x | 0.9 | 21.18 | 0.21 | 0.7 |
| y | 0.44 | 15.42 | 0.15 | 0.7 |

The next experiment considers the system from Equation (30). The additional states, current and current derivative, were added to the sliding surface to improve the position controllers performance. This experiment is conducted with only one bearing operating. Again, the error in the displacement and the current reference are displayed in Figure 12. The rotor can lift off without external help and starts vibrating in the range of $\pm 20\ \mu m$ after stabilising. This is already a notable improvement compared to the controller without current controller dynamics. Table 2 provides the error metrics for this experiment.

On the other hand, chattering in the current reference increased considerably. The dynamic of the current is much faster than the one of the position. Thus the sliding motion attained a high-frequency component. The chattering frequency rose from 40 Hz to 300 Hz, and the amplitude gained approximately $\pm 0.2$ A.

**Table 2.** Error metrics of the experiment with current controller dynamics and constant reference positions.

| Axis | Mean Error in % | Max Error in % | Max Error in mm | Settling Time in s |
|---|---|---|---|---|
| x | 0.007 | 1.86 | 0.018 | 0.6 |
| y | 0.03 | 1.86 | 0.018 | 0.6 |

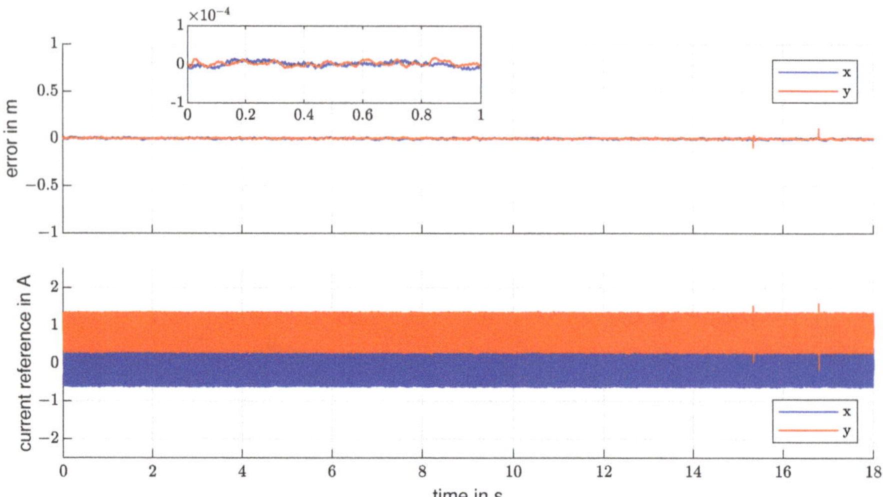

**Figure 12.** Displacement error and current reference $i^*$ with current controller dynamics for one single AMB with constant reference positions.

The experiment is conducted again, now with both bearings operating. The displacement error oscillates around the reference position in the range of $\pm 0.1$ mm as Figure 13 shows. Again the addition of the high-frequency dynamic of the current results in more chattering in the current reference. It is clearly visible, that the unmodelled rotor dynamics add further disturbances. However, the rotor remains stable around the reference position in the presence of the mentioned, unmodelled dynamics, which speaks in favour of a robust controller. The rotor can still lift off from the initial rest position without external help. Table 3 provides the error metrics for this experiment as defined earlier.

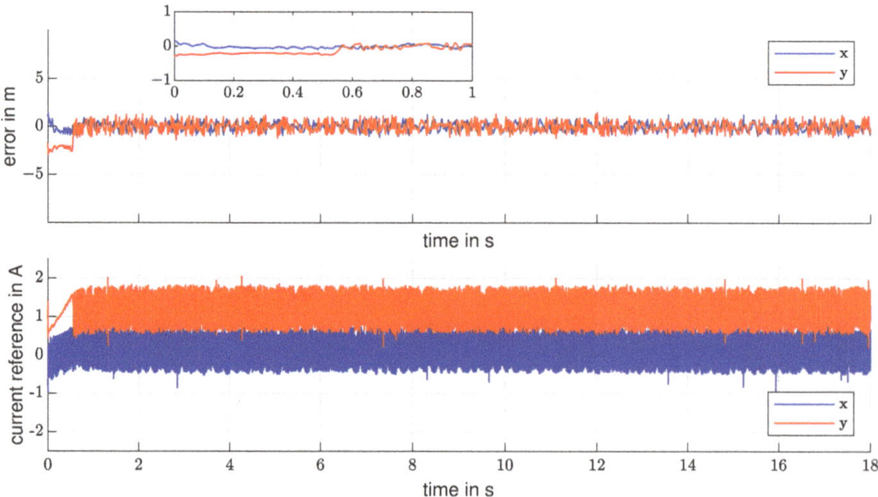

**Figure 13.** Displacement error and current reference $i^*$ with current controller dynamics for both bearings operating at constant reference positions.

**Table 3.** Error metrics of experiment with current controller dynamics and both bearings operating.

| Axis | Mean Error in % | Max Error in % | Max Error in mm | Settling Time in s |
|---|---|---|---|---|
| x | 0.14 | 12.32 | 0.12 | 0.6 |
| y | 0.11 | 14.48 | 0.14 | 0.6 |

## 5. Conclusions

This work proposes a position control strategy for a four-phase star-connected AMB, that integrates the dynamics of the nested, explicit current controller into the position controller. In the experiments, the controller is able to lift from an initial rest position and to levitate the rotor around the reference position which proves to be robust against external disturbance forces. Further, the controller is robust enough to withstand the strong non-linearities and the left of the operating point during the touchdown and startup phases. This is particularly noteworthy because of the large air gap, which introduces large model parameter variations. Integrating the current controller dynamics improves the controller accuracy by about 87% letting the rotor oscillate around the reference position within the range of $\pm 20$ μm or 1% of the magnetic air gap size, considering only one bearing is operating. At the same time, the other is in rest position. On the other hand, the chattering frequency increased distinctly, rising up by 650% to 300 Hz. The amplitude of the chattering motion in the control reference $i^*$ also gained approximately 40%. Considering both bearings operating, the rotor oscillates within $\pm 1$ mm or 5% of the total air gap. The controller is able to recover from touchdown and rest position without external help. A study carried out with a similar model and similar model parameters for sliding mode control for AMBs gave consistent results for the system without current controller dynamics [23]. Compared to this study the integration of the current controller dynamics reduces the trajectory range of the rotor in a stable position by about 83%.

The proposed controller provides increased performance for active magnetic bearing systems with star-connected coils and therefore allows the exploitation of the star point voltage for sensorless position estimation for AMBs. Latter will be the focus of further research. Further, the advancement of the position control strategy, for example by considering the now unmodelled dynamics of the rotor, will be studied.

**Author Contributions:** Conceptualization, E.G., N.K. and J.V.; methodology, E.G., J.V. and N.K.; validation, R.B. and J.V.; formal analysis, N.K. and J.V.; software, R.B. and J.V.; writing—original draft preparation, N.K. and J.V.; writing—review and editing, J.V., E.G., N.K., R.B. and M.N; funding acquisition, M.N. All authors have read and agreed to the published version of the manuscript.

**Funding:** This research received no external funding.

**Data Availability Statement:** Not applicable.

**Acknowledgments:** We acknowledge support by the Deutsche Forschungsgemeinschaft (DFG, German Research Foundation) and Saarland University within the "Open Access Publication Funding" program.

**Conflicts of Interest:** The authors declare no conflict of interest.

## Appendix A

**Table A1.** List of Symbols.

| | |
|---|---|
| $x^*$ | reference displacement in x-direction |
| $x$ | displacement in x-direction |
| $x_0$ | initial rotor displacement in x-direction |
| $y^*$ | reference displacement in y-direction |
| $y$ | displacement in y-direction |
| $y_0$ | initial rotor displacement in y-direction |
| $\epsilon$ | control error |
| $u$ | controller output |
| $i$ | current |
| $i_0$ | bias current |
| $f$ | force |
| $\mu_0$ | permeability of vacuum |
| $\mu_r$ | permeability of iron |
| $n$ | number of windings in a coil |
| $A_a$ | cross-section area |
| $\alpha$ | angle between the legs of the AMB |
| $l_{fe}$ | length of the iron |
| $s$ | magnetic air gap |
| $s_0$ | nominal magnetic air gap |
| $\Phi$ | magnetic flux |
| $W_a$ | field energy |
| $V_a$ | volume in the air gap |
| $B_a$ | magnetic flux density in the air gap |
| $H_a$ | magnetic field in the air gap |
| $k$ | machine constant |
| $k_i$ | force/current factor |
| $k_s$ | force/displacement factor |
| $v$ | velocity |
| $m$ | mass of the rotor |
| $L$ | inductance |
| $R$ | resistance |
| $\tau$ | time constant |
| $x$ | state vector |
| $\sigma$ | sliding variable |
| $S$ | controller gain |
| $c$ | controller gain |
| $b$ | controller gain |

**Table A2.** List of Indices.

| | |
|---|---|
| $*$ | control reference |
| $x$ | in x direction |
| $y$ | in x direction |
| $+$ | in positive direction |
| $-$ | in negative direction |
| $R$ | state space description in controllability form |

## Appendix B

**Table A3.** Test bench and controller parameter.

| Parameter | Symbol | Value | Unit |
| --- | --- | --- | --- |
| number of poles | $N_b$ | 8 | - |
| angle between pole and axis | $\alpha$ | 0.39 | rad |
| coil windings | $n$ | 96 | - |
| maximum voltage | $V_{max}$ | 24 | V |
| maximum current | $I_{max}$ | 5 | A |
| bias current | $i_0$ | 2.5 | A |
| sampling frequency position sensor | $f_{pos}$ | 2 | kHz |
| sampling frequency current sensor | $f_{cur}$ | 10 | kHz |
| PWM frequency | $f_{pwm}$ | 20 | kHz |
| controller gain | $C$ | 1.8 | - |
| sliding surface gains | $S$ | [0.1 0.0001 700 1.5] | - |


## References

1. Maslen, E.H.; Schweitzer, G. (Eds.) *Magnetic Bearings*; Springer: Berlin/Heidelberg, Germany, 2009. [CrossRef]
2. Earnshaw, S. On the nature of the molecular forces which regulate the constitution of the luminferous ether. *Trans. Camb. Philos. Soc.* **1842**, *7*, 97.
3. Benjak, O.; Gerling, D. Review of position estimation methods for IPMSM drives without a position sensor part II: Adaptive methods. In Proceedings of the XIX International Conference on Electrical Machines—ICEM 2010, Rome, Italy, 6–8 September 2010; IEEE: Piscataway, NJ, USA, 2010. [CrossRef]
4. Benjak, O.; Gerling, D. Review of Position Estimation Methods for PMSM Drives Without a Position Sensor, Part III: Methods based on Saliency and Signal Injection. In Proceedings of the 2010 International Conference on Electrical Machines and Systems, Incheon, Republic of Korea, 10–13 October 2010.
5. Benjak, O.; Gerling, D. Review of position estimation methods for IPMSM drives without a position sensor part I: Nonadaptive methods. In Proceedings of the The XIX International Conference on Electrical Machines—ICEM 2010, Rome, Italy, 6–8 September 2010; IEEE: Piscataway, NJ, USA, 2010. [CrossRef]
6. Strothmann, R. Fremderregte Elektrische Maschine. EP Patent EP1005716B1, 14 November 2001.
7. Mantala, C. Sensorless Control of Brushless Permanent Magnet Motors. Ph.D. Thesis, University of Bolton, Bolton, UK, 2013.
8. Schuhmacher, K.; Grasso, E.; Nienhaus, M. Improved rotor position determination for a sensorless star-connected PMSM drive using Direct Flux Control. *J. Eng.* **2019**, *2019*, 3749–3753. [CrossRef]
9. Werner, T. *Geberlose Rotorlagebestimmung in Elektrischen Maschinen*; Springer Fachmedien Wiesbaden: Wiesbaden, Germany, 2018. [CrossRef]
10. Grasso, E.; Mandriota, R.; König, N.; Nienhaus, M. Analysis and Exploitation of the Star-Point Voltage of Synchronous Machines for Sensorless Operation. *Energies* **2019**, *12*, 4729. [CrossRef]
11. Grasso, E. *Direct Flux Control—A Sensorless Technique for Star-Connected Synchronous Machines An Analytic Approach*; Shaker Verlag: Düren, Germany, 2021.
12. Brasse, R.; Vennemann, J.; König, N.; Grasso, E.; Nienhaus, M. Design and Implementation of a Driving Strategy for Star-Connected Active Magnetic Bearings with Application to Sensorless Driving. *Energies* **2023**, *16*, 396. [CrossRef]
13. Huang, T.; Zheng, M.; Zhang, G. A Review of Active Magnetic Bearing Control Technology. In Proceedings of the 2019 Chinese Control And Decision Conference (CCDC), Nanchang, China, 3–5 June 2019; IEEE: Piscataway, NJ, USA, 2019; pp. 2888–2893. [CrossRef]
14. Zadeh, L. Fuzzy logic. *Computer* **1988**, *21*, 83–93. [CrossRef]
15. Psonis, T.; Mitronikas, E.; Nikolakopoulos, P. Comparison of PID and Fuzzy PID Controller for a Linearised Magnetic Bearing. *Tribol. Ind.* **2017**, *39*, 349–356. [CrossRef]
16. Chen, S.Y.; Lin, F.J. Decentralized PID neural network control for five degree-of-freedom active magneticbearing. *Eng. Appl. Artif. Intell.* **2013**, *26*, 962–973. [CrossRef]
17. Shah, P.; Agashe, S. Review of fractional PID controller. *Mechatronics* **2016**, *38*, 29–41. [CrossRef]
18. Shata, A.M.A.H.; Hamdy, R.A.; Abdelkhalik, A.S.; El-Arabawy, I. A fractional order PID control strategy in active magnetic bearing systems. *Alex. Eng. J.* **2018**, *57*, 3985–3993. [CrossRef]
19. Rudolph, J. *Flatness-Based Control An Introduction*; Shaker Verlag: Düren, Germany, 2021.
20. Lévine, J.; Lottin, J.; Ponsart, J.C. Control of Magnetic Bearings: Flatness with Constraints. *IFAC Proc. Vol.* **1996**, *29*, 2798–2803. [CrossRef]

21. von Löwis, J.; Rudolph, J.; Thiele, J.; Urban, F. Flatness-Based Trajectory Tracking Control of a Rotating Shaft. In Proceedings of the 7th International Symposium on Magnetic Bearings, Zürich, Switzerland, 23–25 August 2000; pp. 299–304.

22. Utkin, V. Variable structure systems with sliding modes. *IEEE Trans. Autom. Control.* **1977**, *22*, 212–222. [CrossRef]

23. Wang, X.; Zhang, Y.; Gao, P. Design and Analysis of Second-Order Sliding Mode Controller for Active Magnetic Bearing. *Energies* **2020**, *13*, 5965. [CrossRef]

24. Kang, M.S.; Lyou, J.; Lee, J.K. Sliding mode control for an active magnetic bearing system subject to base motion. *Mechatronics* **2010**, *20*, 171–178. [CrossRef]

25. Mehta, A.; Bandyopadhyay, B. (Eds.) *Emerging Trends in Sliding Mode Control: Theory and Application*; Springer: Singapore, 2021; Volume 318. [CrossRef]

26. Saha, S.; Amrr, S.M.; Saidi, A.S.; Banerjee, A.; Nabi, M. Finite-Time Adaptive Higher-Order SMC for the Nonlinear Five DOF Active Magnetic Bearing System. *Electronics* **2021**, *10*, 1333. [CrossRef]

27. Saha, S.; Amrr, S.M.; Nabi, M. Adaptive Second Order Sliding Mode Control for the Regulation of Active Magnetic Bearing. *IFAC-PapersOnLine* **2020**, *53*, 1–6. [CrossRef]

28. Amrr, S.M.; Alturki, A. Robust Control Design for an Active Magnetic Bearing System Using Advanced Adaptive SMC Technique. *IEEE Access* **2021**, *9*, 155662–155672. [CrossRef]

29. Schweitzer, G.; Traxler, A.; Bleuler, H. *Magnetlager: Grundlagen, Eigenschaften und Anwendungen berüHrungsfreier, Elektromagnetischer Lager*; Springer: Berlin/Heidelberg, Germany, 1993. [CrossRef]

30. Chiba, A.; Fukao, T.; Ichikawa, O.; Oshima, M.; Takemoto, M.; Dorrell, D. (Eds.) *Magnetic Bearings and Bearingless Drives*; Elsevier: Amsterdam, The Netherlands; Newnes: London, UK, 2005.

31. Lunze, J. *Regelungstechnik: Systemtheoretische Grundlagen, Analyse und Entwurf Einschleifiger Regelungen*; Springer: Berlin/Heidelberg, Germany, 2020.

32. Lunze, J. *Regelungstechnik 2: Mehrgrößensysteme, Digitale Regelung*; Springer: Berlin/Heidelberg, Germany, 2020. [CrossRef]

33. Schroeder, D. *Elektrische Antriebe—Regelung von Antriebssystemen*; Springer: Berlin/Heidelberg, Germany, 2015.

34. Shtessel, Y.; Edwards, C.; Fridman, L.; Levant, A. *Sliding Mode Control and Observation*; Control Engineering; Springer: New York, NY, USA, 2014. [CrossRef]

35. Utkin, V.; Poznyak, A.; Orlov, Y.V.; Polyakov, A. *Road Map for Sliding Mode Control Design*; SpringerBriefs in Mathematics; Springer International Publishing: Cham, Switzerland, 2020. [CrossRef]

*Article*

# Chattering Free Sliding Mode Control and State Dependent Kalman Filter Design for Underground Gasification Energy Conversion Process

Sohail Ahmad [1], Ali Arshad Uppal [1,*], Muhammad Rizwan Azam [1] and Jamshed Iqbal [2,*]

[1] Department of Electrical and Computer Engineering, COMSATS University Islamabad, Islamabad 45550, Pakistan
[2] School of Computer Science, Faculty of Science and Engineering, University of Hull, Hull HU6 7RX, UK
* Correspondence: ali_arshad@comsats.edu.pk (A.A.U.); j.iqbal@hull.ac.uk (J.I.);
Tel.: +92-331-3666163 (A.A.U.); +44-1482-462187 (J.I.)

**Abstract:** The fluctuations in the heating value of an underground coal gasification (UCG) process limit its application in electricity generation, where a desired composition of the combustible gases is required to operate gas turbines efficiently. This shortcoming can be addressed by designing a robust control scheme for the process. In the current research work, a model-based, chattering-free sliding mode control (CFSMC) algorithm is developed to maintain a desired heating value trajectory of the syngas mixture. Besides robustness, CFSMC yields reduced chattering due to continuous control law, and the tracking error also converges in finite time. To estimate the unmeasurable states required for the controller synthesis, a state-dependent Kalman filter (SDKF) based on the quasi-linear decomposition of the nonlinear model is employed. The simulation results demonstrate that despite the external disturbance and measurement noise, the control methodology yields good tracking performance. A comparative analysis is also made between CFSMC, a conventional SMC, and an already designed dynamic integral SMC (DISMC), which shows that CFSMC yields 71.2% and 69.9% improvement in the root mean squared tracking error with respect to SMC and DISMC, respectively. Moreover, CFSMC consumes 97% and 23.2% less control energy as compared to SMC and DISMC, respectively.

**Keywords:** underground coal gasification (UCG); state-dependent Kalman filter (SDKF); chattering-free sliding mode control (CFSMC); energy conversion systems

**Citation:** Ahmad, S.; Uppal, A.A.; Azam, M.R.; Iqbal, J. Chattering Free Sliding Mode Control and State Dependent Kalman Filter Design for Underground Gasification Energy Conversion Process. *Electronics* **2023**, *12*, 876. https://doi.org/10.3390/electronics12040876

Academic Editor: Sung Jin Yoo

Received: 14 January 2023
Revised: 7 February 2023
Accepted: 7 February 2023
Published: 9 February 2023

## 1. Introduction

Coal plays a pivotal role in global power generation owing to its affordability and ubiquitous presence worldwide. However, the greenhouse gas emissions due to the combustion of fossil fuels negatively impact the environment and contribute significantly to global warming [1]. These environmental concerns are addressed by employing various clean coal energy technologies, for example, the underground coal gasification (UCG) process, that allows the removal of harmful elements at various stages of the UCG process [2]. UCG process involves drilling two wells from the surface of the earth to the coal beds. The injection well is utilized to inject the oxidants like air and steam into the coal bed. These oxidants then react with the ignited coal, and syngas is produced at the production well [3]. Syngas is a flammable gas comprising of higher hydrocarbons, CO, $H_2$, and $CH_4$, that can be utilized in many applications such as the production of liquid fuel and electric power generation. In comparison to conventional mining and surface gasifiers, UCG provides an efficient and cost-effective solution to produce decarbonized gas without putting human labor in danger underground [4].

Industrial applications like integrated gasification combined cycle turbines (IGCC) efficiently operate on a constant heating value/calorific value of syngas. The desired heating value can be achieved by controlling the flow rate or molar flux of the injected oxidants [5]. The presence of parametric uncertainties, modeling inaccuracies, and external disturbances constitutes a challenging control problem that has recently become an emerging field of research [4].

### 1.1. Related Work

Both model-free and model-based controllers are designed for a UCG system in the literature, to achieve the desired syngas properties. In [6,7], a conventional Proportional Integral (PI) controller is designed to control the temperature, concentration, and heating value of syngas for a lab scale setup for UCG. This model-free controller solely relies on the output measurements. The authors extended their research work to include the effect of uncertainties in the measurements in [8]. Later, in [9], an experimental study is investigated to solve a real-time optimization problem for the UCG process. The optimal control design technique is used to maximize the CO concentration in syngas. The authors in [10] conducted an experimental study to achieve the desired calorific value of the syngas by employing an adaptive model predictive control (MPC). In [11], an Internet-of-Things (IoT) based monitoring system is utilized to assist a deep learning-based optimal control technique. In [12], the authors conducted an experimental study to optimize oxygen flow, airflow, and syngas exhaust to maximize the heating value of syngas.

The controllers designed in the literature presented above are either model-free or utilize data-driven modeling techniques. However, model-based control strategies are proven to be more accurate. In this regard, a multitude of research is conducted on various model-based control approaches for the UCG process. In [5], the authors developed a 1-D control-oriented mathematical model of the Thar coal gasifier, which is utilized to design various sliding mode control (SMC) techniques for maintaining a desired level of the heating value. In [13], a time domain UCG process model is developed. The authors also design a conventional SMC for heating value regulation with the assumption that all the state variables are measurable. To remove this discrepancy, [14] employs a gain-scheduled modified Utkin observer (GSMUO) to estimate the unmeasurable states required for designing dynamic SMC. Moreover, a time-varying reference is used to test the performance of the control techniques. Considering the required computational complexity and implementation resources for nonlinear techniques, several linear control techniques are also investigated for the UCG system.

### 1.2. Gap Analysis

The controllers based on linear models can only work near a particular operating point. To cover the whole operating range between no-load and full load, a robust nonlinear controller is required. The tracking error of the heating value does not converge in finite time for all the aforementioned SMC techniques. However, to cater for the abrupt changes in the demand for electric power, the calorific value of the UCG process also needs to change quickly. Therefore, in an IGCC power plant, a robust and finite time convergent SMC, cf. [15] is required to meet the sudden changes in the electricity demand.

### 1.3. Major Contributions

In the current research article, a model-based, chattering-free SMC (CFSMC), cf. [16], is developed for the UCG process model given in [13]. Apart from robustness, CFSMC exhibits less chattering due to continuous control law, and by the virtue of nonlinearity in the sliding surface, the tracking error also converges in finite time. To reconstruct the unmeasurable states necessary for controller design, a state-dependent Kalman filter (SDKF) is designed, which is a linear discrete-time Kalman filter based on the quasilinear model of [13]. The water influx from the surrounding aquifers is considered as the input disturbance, cf. [14], to evaluate the robustness of the control scheme. Furthermore, the performance of the SDKF is evaluated by introducing a measurement noise. Simulation results indicate that the designed methodology quickly and accurately tracks the desired heating value trajectory, outperforming conventional SMC and DISMC (cf. [14]). For brevity, the contributions of the paper are listed below:

- A finite-time CFSMC is designed for the UCG process to track the desired heating value trajectory.
- The unmeasured states used to synthesize the model-based controller are reconstructed using SDKF.
- A thorough quantitative and qualitative comparison is made between the designed technique and already developed techniques for tracking the heating value for the UCG process.

The remaining paper is organized as follows: Section 2 outlines the control-oriented mathematical model of the UCG plant. Section 3 and Section 4 respectively discuss the synthesis of CFSMC and SDKF for the UCG plant. The simulation results are presented in Section 5, followed by the conclusion in Section 6 of the manuscript.

## 2. Mathematical Model of UCG Process

The current study utilizes a nonlinear control-oriented mathematical model of the UCG process, derived in one of our earlier works [14], for the control design. The mathematical model consists of two solid components: char and coal, and eight gaseous components: $CO_2$, $O_2$, $CO$, $H_2$, $CH_4$, $N_2$, $H_2O$, and tar. Moreover, the model incorporates the effect of water influx from the nearby aquifers as the disturbance. The following mathematical equations describe the nonlinear control-oriented UCG model

$$\left.\begin{aligned}
\dot{\rho}_{Coal} &= -M_1 R_1, \\
\dot{\rho}_{Char} &= M_2(0.766R_1 - R_2 - R_3), \\
\dot{T}_S &= \frac{1}{C_s}(h_t(T - T_s) - \Delta q_2 R_2 - \Delta q_3 R_3), \\
\dot{C}_{CO} &= 0.008R_1 + R_3 - \beta C_{CO}, \\
\dot{C}_{CO_2} &= 0.058R_1 + R_2 - \beta C_{CO_2}, \\
\dot{C}_{H_2} &= 0.083R_1 + R_3 - \beta C_{H_2}, \\
\dot{C}_{CH_4} &= 0.044R_1 - \beta C_{CH_4}, \\
\dot{C}_{Tar} &= 0.0138R_1 - \beta C_{Tar}, \\
\dot{C}_{H_2O} &= 0.055R_1 + 0.075R_2 - 0.925R_3 - \beta C_{H_2O} + \frac{\alpha}{L}u + \frac{1}{L}\phi, \\
\dot{C}_{O_2} &= -1.02R_2 - \beta C_{O_2} + \frac{\delta}{L}u, \\
\dot{C}_{N_2} &= -\beta C_{N_2} + \frac{\gamma}{L}u.
\end{aligned}\right\} \tag{1}$$

The UCG model includes a number of parameters and variables that are listed in Tables 1 and 2 provides the nominal values of the parameters.

Pyrolysis of coal, oxidation of char, and gasification of steam are dominant chemical reactions of the current model of the UCG process and are expressed in Table 3. Molecular formulas of coal, char and tar are respectively $CH_{0.912}O_{0.194}$, $CH_{0.15}O_{0.02}$, and $(CH_{2.782})_9$. The reaction rates of chemical reactions : $R_1$, $R_2$ and $R_3$ are expressed in (2).

**Table 1.** Parameters and states.

| Symbol | Description | Unit |
| --- | --- | --- |
| $M_i$ | Molecular Weight of solids | $\text{mol}^{-1}\,\text{g}$ |
| $T$ | Temperature of gas | K |
| $h_t$ | Transfer coefficient of heat | $\text{cal}\,\text{s}^{-1}\,\text{K}^{-1}\,\text{cm}^{-3}$ |
| $C_s$ | Solids specific heat capacity | $\text{cal}\,\text{K}^{-1}\,\text{g}^{-1}$ |
| $R$ | Chemical reaction rate | $\text{mol}\,\text{cm}^{-3}\,\text{s}^{-1}$ |
| $\Delta q_i$ | Char oxidation heat and steam gasification heat | $\text{cal}\,\text{mol}^{-1}$ |
| $L$ | Reactor's length | cm |
| $\beta C_i$ | Spatial derivative's approximation [13] | $\text{mol}\,\text{cm}^{-3}\,\text{s}^{-1}$ |
| $u$ | Injected gases flow rate | $\text{mol}\,\text{cm}^{-2}\,\text{s}^{-1}$ |
| $\alpha, \delta, \gamma$ | Amount (%) of $H_2O$, $O_2$ and $N_2$ in $u$ | - |
| $\phi$ | Input disturbance: steam's flow rate, generated from water influx from nearby aquifers | $\text{mol}\,\text{cm}^{-2}\,\text{s}^{-1}$ |

**Table 2.** Nominal parameters value.

| $C_s$ | $\beta$ | $P$ | $h_t$ | $L$ | $\Delta q_2$ | $\Delta q_3$ |
| --- | --- | --- | --- | --- | --- | --- |
| 7.3920 | $7 \times 10^{-6}$ | 4.83 | 0.001 | 100 | $-93{,}929$ | 31,309.7 |

**Table 3.** Chemical reactions in UCG.

| Sr. | Chemical Equations |
| --- | --- |

1. **Coal pyrolysis:**

$$CH_{0.912}O_{0.194} \xrightarrow{R_1} 0.766CH_{0.15}O_{0.02} + 0.008CO + 0.055H_2O + 0.083H_2 + 0.044CH_4 + 0.058CO_2 + 0.0138(CH_{2.782})_9$$

2. **Char oxidation:**

$$CH_{0.15}O_{0.02} + 1.02O_2 \xrightarrow{R_2} CO_2 + 0.075H_2O$$

3. **Steam gasification:**

$$CH_{0.15}O_{0.02} + 1.02O_2 \xrightarrow{R_3} CO + H_2$$

$$\left.\begin{aligned}
R_1 &= 5\frac{\rho_{coal}}{M_1}exp\left(\frac{-6039}{T_s}\right), \\[2mm]
R_2 &= \frac{1}{\dfrac{1}{R_{c_2}} + \dfrac{1}{R_{m_2}}}, \\[2mm]
R_3 &= \frac{1}{\dfrac{1}{R_{c_3}} + \dfrac{1}{R_{m_3}}},
\end{aligned}\right\} \tag{2}$$

where,

$$R_{m_2} = \frac{1}{10}h_t m_{O_2},$$

$$R_{c_2} = \frac{1}{M_2}\left(9.55 \times 10^8 \rho_{char} m_{O_2} P\, exp\left(\frac{-22142}{T_s}\right) T_s^{-0.5}\right),$$

$$R_{c_3} = \frac{\rho_{char} m_{H_2O}^2 P^2 exp\left(5.052 - \dfrac{12908}{T_s}\right)}{M_2\left(m_{H_2O}P + exp\left(-22.216 + \dfrac{24880}{T_s}\right)\right)^2},$$

$$R_{m_3} = \frac{1}{10}h_t m_{H_2O},$$

the molar fractions $m_{O_2}$ and $m_{H_2O}$ are expressed as,

$$m_{O_2} = \frac{C_{O_2}}{C_T + C_{H_2O}},$$

$$m_{H_2O} = \frac{C_{H_2O}}{C_T + C_{H_2O}},$$

where

$$C_T = C_{CO} + C_{CO_2} + C_{H_2} + C_{CH_4} + C_{Tar} + C_{O_2} + C_{N_2}. \tag{3}$$

The nonlinear control-oriented model (1) can be expressed in a control-affine form as

$$\dot{x} = f(x) + g_1 u + g_2 \phi, \tag{4}$$

where $x \in R^{11}$ is the state vector, $u$ is the control input, $f, g_1, g_2 \in R^{11}$ are smooth vector fields, and $\phi$ is considered as the external disturbance. The vector of states $x$ is chosen as

$$x = [\rho_{coal}\ \rho_{char}\ T_s\ C_{CO}\ C_{CO_2}\ C_{H_2}\ C_{CH_4}\ C_{Tar}\ C_{H_2O}\ C_{O_2}\ C_{N_2}]^T. \tag{5}$$

Vector fields $f(x)$, $g_1$, and $g_2$ are given by (6)–(8), respectively

$$f(x) = \begin{bmatrix} -M_1 R_1(x), \\ M_2(0.766 R_1(x) - R_2(x) - R_3(x)), \\ \frac{1}{C_s}(h_t(T - x_3) - \Delta q_2 R_2(x) - \Delta q_3 R_3(x)) \\ 0.008 R_1(x) + R_3(x) - \beta x_4, \\ 0.058 R_1(x) + R_2(x) - \beta x_5, \\ 0.083 R_1(x) + R_3(x) - \beta x_6 \\ 0.044 R_1(x) - \beta x_7 \\ 0.0138 R_1(x) - \beta x_8 \\ 0.055 R_1(x) + 0.075 R_2(x) - 0.925 R_3(x) - \beta x_9 \\ -1.02 R_2(x) - \beta x_{10} \\ -\beta x_{11} \end{bmatrix}, \tag{6}$$

$$g_1 = \begin{bmatrix} 0 & 0 & 0 & 0 & 0 & 0 & 0 & 0 & \frac{\alpha}{L} & \frac{\delta}{L} & \frac{\gamma}{L} \end{bmatrix}^T, \tag{7}$$

$$g_2 = \begin{bmatrix} 0 & 0 & 0 & 0 & 0 & 0 & 0 & 0 & \frac{1}{L} & 0 & 0 \end{bmatrix}^T. \tag{8}$$

The measurement vector $y_m$ represents the concentration of the gases measured from the gas analyzer

$$y_m = [x_4\ x_5\ x_6\ x_7\ x_8\ x_{10}\ x_{11}]^T, \tag{9}$$

The heating value $H_v$ of the syngas is the variable to be controlled, characterized as

$$H_v = \frac{H_4 x_4 + H_6 x_6 + H_7 x_7}{C_T}, \tag{10}$$

where $C_T = x_4 + x_5 + x_6 + x_7 + x_8 + x_{10} + x_{11}$, and $H_4$, $H_6$, and $H_7$ represent heat of combustion (kJ/mol) of CO, $H_2$ and $CH_4$ respectively.

The objective of the control design is to keep the heating value at the desired level based on the operating conditions, such as the usage of char and coal in the UCG bed. Hence, the next section will focus on the design of the controller.

### 3. Chattering Free Sliding Mode Control Design

This section presents the design of a CFSMC with the aim of following the desired syngas heating value ($H_{vd}$). Despite being a promising solution, the implementation of the sliding mode-based control law may result in high-frequency oscillations, known as chattering, because of modeling errors, external disturbances, and discretization. Chattering is a common issue with sliding mode controllers; however, by the virtue of smoothening terms in the control law and the sliding surface, CFSMC can effectively resolve the issue of chattering [17]. Moreover, a carefully designed sliding surface ensures the finite time convergence of the tracking error. Furthermore, conventional SMC often has the disadvantage of producing discontinuous control inputs that do not meet the requirement of the current system. Therefore, the above-mentioned issues associated with the conventional SMC are addressed in this paper by a systematic design of CFSMC [16].

The output to be controlled is the difference between the actual $H_v$ and its desired trajectory $H_{vd}$, $e = H_v - H_{vd}$. The control input $u$ appears after differentiating $e$ once, which inferred that the relative degree of the tracking error $e$ is 1 with respect to the control input $u$. Therefore, differentiating the error $e$ with respect to time results in the following error dynamics

$$\dot{e} = \psi(x,t) + d(x,\phi,t) + b(x,t)u, \tag{11}$$

where $\psi(x,t)$, $b(x,t)$, and $d(x,t)$ are smooth and nonlinear functions of states.

It is pertinent to mention here that $\psi(x,t)$ and $b(x,t)$ are the nominal parts of the system, whereas $d(x,\phi,t)$ includes the perturbed part of the system influenced by a smooth and bounded input disturbance $\phi(t) : |\phi(t)| \leq \phi_0$. Considering the control-oriented model given in (4) and reaction rate expressions in (2), it can be observed that the input disturbance affects $R_2$ and $R_3$ by changing the concentration of steam. The functions $\psi(x,t)$, $b(x,t)$ and $d(x,t)$ are defined as

$$\psi(x,t) = \frac{C_T(H_4\dot{x}_4 + H_6\dot{x}_6 + H_7\dot{x}_7) - N(\beta(x_{10} + x_{11}) + \dot{x}_4 + \dot{x}_5 + \dot{x}_6 + \dot{x}_7 + \dot{x}_8)}{C_T^2} - \dot{H}_{vd},$$

$$d(x,\phi,t) = \frac{(C_T H_4 + C_T H_6 - 2N)\tilde{R}_3(x,\phi) - (1 + 1.02)N\tilde{R}_2(x,\phi)}{C_T^2},$$

$$b(x,t) = \frac{-N(\delta + \gamma)}{LC_T^2}, \tag{12}$$

where $\tilde{R}_2$ and $\tilde{R}_3$ are errors between perturbed and nominal reaction rates, and $N = H_4 x_4 + H_6 x_6 + H_7 x_7$.

Now, to mitigate error within a finite time, a terminal sliding manifold is chosen for the system in (11) as

$$S = \dot{e} + c\,sgn(e)|e|^\lambda, \tag{13}$$

where $c$ and $\lambda$ are design parameters. The value of $c$ is selected such that the polynomial $\rho + c$, which corresponds to the system (13), is Hurwitz, meaning its eigenvalues are in the open left half of the complex plane, with $\rho$ as a Laplace operator. By selecting appropriate values of $c$ and $\lambda$, the ideal sliding mode $S = 0$ for the error dynamics (11) can be achieved, resulting in the system converging to its equilibrium point, $e = 0$, from any initial condition along the sliding surface $S = 0$ in a finite amount of time [16]. Hence, the control law is chosen as

$$u = -\frac{1}{b(x,t)}(u_{eq} + u_{dis}), \tag{14}$$

where

$$u_{eq} = -\psi(x,t) - c\,sgn(e)|e|^\lambda,$$
$$\dot{u}_{dis} = -(k_d + k_T + \eta)\,sgn(S) - T\,u_{dis},$$

where $T, k_T, \eta \in \Re^+$ are gains of the controller. The constants $T$ and $k_T$ are selected such that $k_T \geq T d_0$, where $||d(x, \phi, t)|| \leq d_0 \in \Re^+$.

The validity of the sliding mode, i.e., whether the trajectories converge to the manifold $S = 0$ is proved in the subsequent subsection.

*Existence of Sliding Mode*

By using (11) and (14), Equation (13) can be re-expressed as

$$S = d(x, \phi, t) + u_{dis}. \tag{15}$$

Now, to prove that the above surface is attractive, a positive definite candidate Lyapunov function is chosen

$$V(x, t) = \frac{1}{2} S^2 > 0, \tag{16}$$

whose time derivative determined as

$$\begin{aligned}
\dot{V}(x, t) &= S\dot{S}, \\
&= S(\dot{d}(x, \phi, t) + \dot{u}_{dis}), \\
&= S(\dot{d}(x, \phi, t) - (k_d + k_T + \eta)sgn(S) - Tu_{dis}), \\
&\leq (D_0 - k_d)|S| - \eta|S| - |S|(k_T + Tu_{dis}),
\end{aligned} \tag{17}$$

where $\dot{d}(x, \phi, t)$ is the time derivative of $d(x, \phi, t)$, and it is smooth and bounded: $||\dot{d}(x, \phi, t)|| \leq D_0 \in \Re^+$.

Now by selecting $k_d > D_0$ and $k_T \geq T|u_{dis}|$ we can write (17) as

$$\dot{V}(x, t) \leq -\eta|S|, \tag{18}$$

which shows that the system in (11) will reach $S = 0$ in finite time [16].

## 4. State Dependent Kalman Filter Design

This section focuses on designing a state-dependent Kalman filter (SDKF) for the UCG plant to estimate the unmeasurable states needed for the CFSMC controller synthesis. The design of SDKF is based upon the quasi-linear approximation of the UCG mathematical model (1), which is discussed in the following subsection.

*4.1. Quasi-Linear Decomposition*

SDKF design is based upon the quasi-linear decomposition, cf. [18] of the UCG model (1). The conventional Taylor series expansion method can be inadequate in case of arbitrary operating points. To overcome the limitation, a constrained minimization problem is introduced to develop a quasi-linear model that accurately approximates the behavior of the nonlinear model (1) close to any operating point. This decomposition portrays the actual nonlinear dynamics of (1), i.e.,

$$\dot{x} = f(x) + g_1 u + g_2 \phi = F(x)x + g_1 u + g_2 \phi. \tag{19}$$

The expressions for $g_1$ and $g_2$ are already defined in (7) and (8), respectively. The matrix $F(x) \in \Re^{11 \times 11}$ can be expressed as

$$F(x) = \begin{bmatrix} | & | & | & | & & | \\ a_1 & a_2 & a_3 & a_4 & \cdots & a_{11} \\ | & | & | & | & & | \end{bmatrix}, \tag{20}$$

where $a_i$ represents $i$th column of $F(x)$, given as

$$a_i = \nabla f_i(x) + \frac{f_i(x) - x^T \nabla f_i(x)}{\|x\|^2} x, \qquad x \neq 0, \tag{21}$$

where $\nabla f_i(x)$ represents the gradient of $i^{th}$ element of the vector field $f(x)$ in the direction of $x$.

*4.2. SDKF Design Procedure*

SDKF employs the algorithm of the discrete-time linear Kalman filter; however, due to the state-dependent matrices obtained using the quasi-linear decomposition of the nonlinear system, SDKF can estimate the states of a nonlinear system. The Kalman filter holds a prominent place in stochastic estimation theory as it involves filtering of both process (dynamic modeling noise) and measurement noise (sensor noise) and also estimates the states by minimizing the estimated error covariance [19]. To employ SDKF, the quasi-linear UCG model is discretized with sampling time $\Delta t$ to obtain

$$x_k = A_{k-1} x_{k-1} + B_{k-1} u_{k-1} + w_{k-1}, \tag{22}$$

where $x_k$ is the state vector defined in (5), $A_{k-1} = F_{k-1}\Delta t + I_{11}$, $B = g_1 \Delta t$, and $w_{k-1}$ is the zero-mean Gaussian process noise, with known covariance matrix $Q$ defined by [20]

$$Q = E(w_k w_k^T). \tag{23}$$

The measurement equation (9) can be expressed in discrete form as

$$z_k = H x_k + v_k, \tag{24}$$

where $z_k$ is the measurement vector defined in (9), $H \in \Re^{7 \times 11}$ is the output matrix defined by

$$H = \begin{bmatrix} 0 & 0 & 0 & 1 & 0 & 0 & 0 & 0 & 0 & 0 & 0 \\ 0 & 0 & 0 & 0 & 1 & 0 & 0 & 0 & 0 & 0 & 0 \\ 0 & 0 & 0 & 0 & 0 & 1 & 0 & 0 & 0 & 0 & 0 \\ 0 & 0 & 0 & 0 & 0 & 0 & 1 & 0 & 0 & 0 & 0 \\ 0 & 0 & 0 & 0 & 0 & 0 & 0 & 1 & 0 & 0 & 0 \\ 0 & 0 & 0 & 0 & 0 & 0 & 0 & 0 & 0 & 1 & 0 \\ 0 & 0 & 0 & 0 & 0 & 0 & 0 & 0 & 0 & 0 & 1 \end{bmatrix}, \tag{25}$$

and $v_k$ is the zero-mean Gaussian measurement noise, with known covariance matrix $R$ defined by

$$R = E(v_k v_k^T). \tag{26}$$

The system's dynamic model uncertainty and measurement imprecision are used to choose the Q and R matrices. Selection of R is relatively easy as it considers the consequences of unmodeled dynamics, while choosing $Q$ is typically challenging. Given this, $Q$ and $R$ are often viewed as design parameters and are calculated by using the trial-and-error method [21].

The objective of SDKF is to make $E(\hat{x}_k) = E(x_k)$, where $\hat{x}_k$ is the a-posteriori estimated error, and also to make a-posteriori estimated error covariance $P_k$ as small as possible. SDKF is designed in two stages; (i) the predictor stage and (ii) the corrector stage. The predictor stage forecasts a state based on the previous state and is defined as

$$\bar{x}_k = A_{k-1}\hat{x}_{k-1} + Bu_k, \tag{27}$$

$$\bar{P}_k = A_{k-1}P_{k-1}A_{k-1}^T + Q, \tag{28}$$

where $\bar{x}_k$ is the a-priori estimated state and $\bar{P}_k$ is the a-priori estimated error covariance. The estimated state and associated error covariance are then carried forward to the step of the measurement update. Whereas the corrector stage adjusts the anticipated state depending on the most recent output measurements, i.e.,

$$K = \bar{P}_k H^T (H\bar{P}_k H^T + R)^{-1} \tag{29}$$

$$\hat{x}_k = \bar{x}_k + K(z_k - H\bar{x}_k) \tag{30}$$

$$P_k = (I - KH)\bar{P}_k \tag{31}$$

where $K$ is the Kalman gain, $\hat{x}_k$ is the a-posteriori estimated state and $P_k$ is the a-posteriori estimated error covariance. Figure 1 describes the complete working scheme of SDKF.

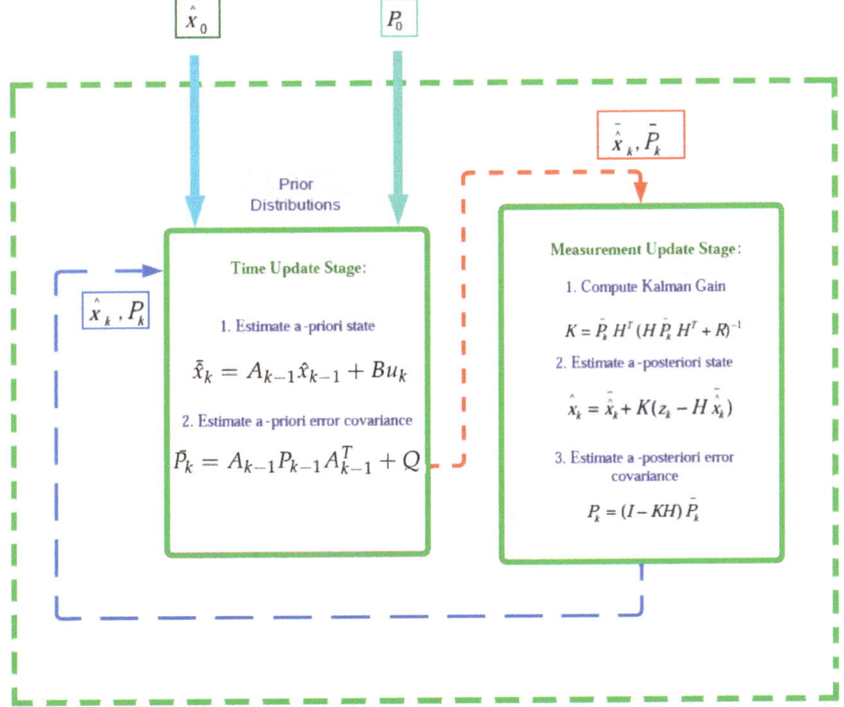


**Figure 1.** SDKF algorithm implementation.

## 5. Results and Discussions

This section presents simulation results of the UCG process along with CFSMC and SDKF. MATLAB/Simulink (Version: R2018a, running on Laptop: Lenovo i3, 3rd generation with 4GB Ram ) is utilized to perform the simulations. A qualitative and quantitative comparison is carried out between the conventional SMC, dynamic integral SMC (DISMC) [14], and the proposed CFSMC for desired trajectory tracking of the calorific value. A modified gain-scheduled Utkin observer (GSMUO) estimates the unmeasurable states for DISMC; however, SDKF is used to reconstruct states for SMC and CFSMC. To replicate the real-time conditions, a comprehensive simulation study is performed, taking into account the practical considerations listed below:

- A white Gaussian noise with a zero mean and a variance of $0.02^2$ is added to $y_m$. This variance is selected based on the typical accuracy of the gas analyzer used for taking measurements in the UCG process [5].
- The values of $\alpha$, $\delta$, and $\gamma$ are chosen to be 0.77, 0.154, and 0.076, respectively in (1).
- The desired heating value trajectory is expressed in Figure 2, which represents a sudden change in the demand for electricity generation.
- The total simulation time is 9 h. To ensure that the heating value reaches the appropriate set point, the UCG system is run in an open loop for initial 5.5 h. The flow rate of gases is maintained at $2 \times 10^{-4}$ moles cm$^{-2}$ s$^{-1}$ during this time. Afterward, the UCG system is run in a close loop configuration for $5.5 \leq t \leq 9$ h. Therefore, for better visualization, the simulation results for evaluating the performance of the controllers are only shown for the closed-loop operation.
- SDKF works for the complete simulation, i.e., $0 \leq t \leq 9$ h.
- The gains of CFSMC in (13) and (14) are: $c = 69$, $\lambda = 0.1$, $T = 0.1$, $k_d = 0.5$, $k_T = -2$, and $\eta = 1$.

The tracking performance of the selected controllers is shown in Figure 2. The reference trajectory shows a sudden change from a higher to a lower level of the heating value. To track the abrupt change in the reference trajectory, the gains of the controllers are kept on the higher side, which results in poor performance of SMC and DISMC as compared to CFSMC. Figure 3 depicts the tracking error of different control schemes.

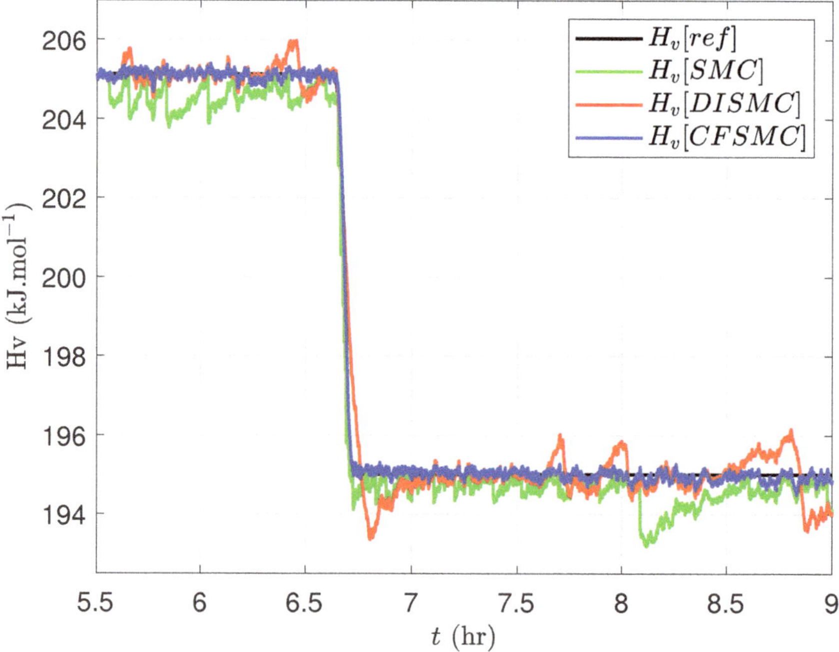


**Figure 2.** Syngas Heating value in the closed-loop operation with time.

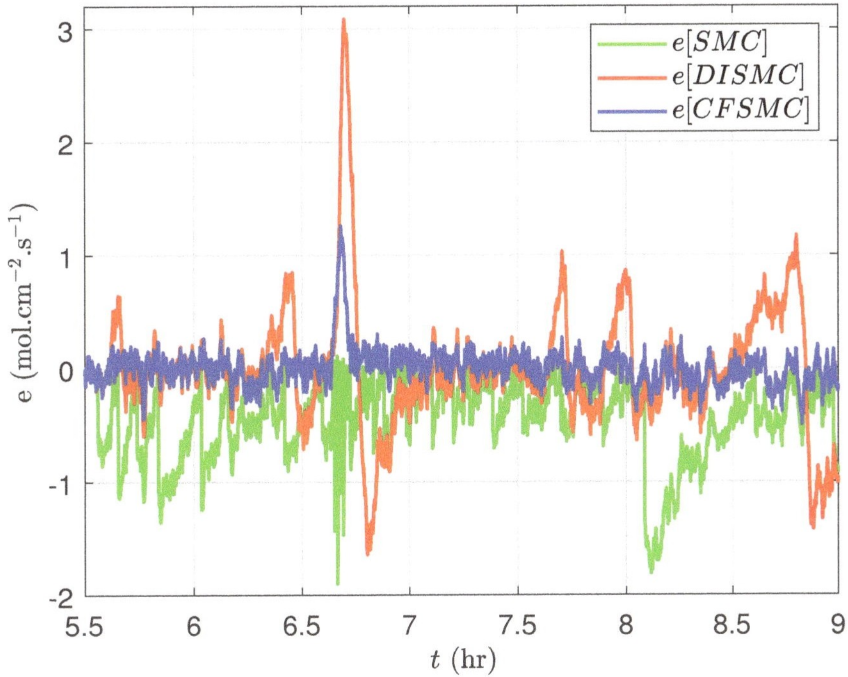

**Figure 3.** Tracking -error with various controllers during closed-loop operation with time.

The manipulated flow rate of the inlet gases for different control schemes is presented in Figure 4. It is evident from the figure that the tracking performance of CFSMC is the most superior as compared to DISMC and SMC.

A sufficient amount of steam is necessary to ensure the smooth operation of the UCG reactor. Generally, there are two sources of steam: a mixture of the inlet gas (which in the current case contains 77% steam) and water ingress from surrounding aquifers, which is considered as an external disturbance, and its time profile is depicted in Figure 5. It is evident from Figure 4 that all the controllers manipulate $u$ to mitigate the effect of $\phi$. However, the results in Figures 2 and 3 demonstrate that CFSMC exhibits more robustness to compensate $\phi$ as compared to DISMC and SMC.

Figure 6 shows the sliding variable designed for CFSMC, which is given by (13). Despite the disturbance and measurement noise, the system trajectories are confined to the manifold $S = 0$.

A comprehensive quantitative analysis has been performed to assess the performance of CFSMC, DISMC, and SMC. The attributes considered for performance evaluation are root mean squared error $e_{rms}$ and the average power $P_{avg}$ of the control input $u$. These performance indices are characterized as

$$e_{rms} = \sqrt{\frac{1}{N}\sum_{i=1}^{N}(H_v[i] - H_{vd}[i])^2},$$

$$P_{avg} = \frac{1}{N}\sum_{i=1}^{N}(u_i)^2, \tag{32}$$

where $N$ represents the number of samples.

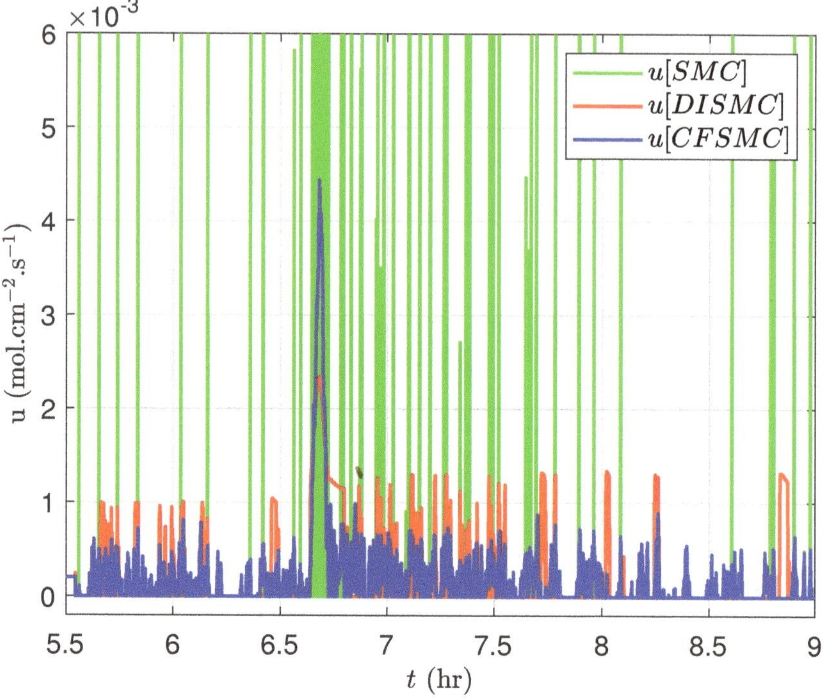

**Figure 4.** Control input for different controllers with time.

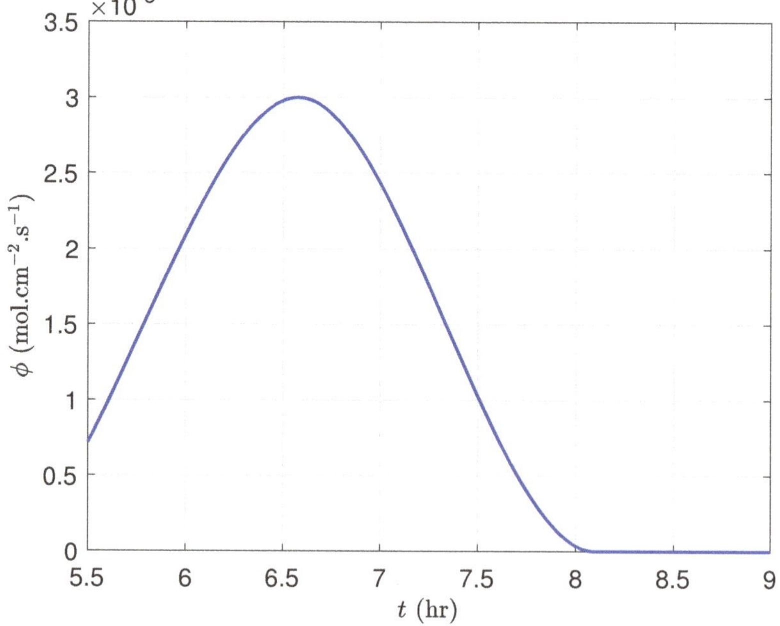

**Figure 5.** Steam flow rate generated by water influx from nearby aquifers with time.

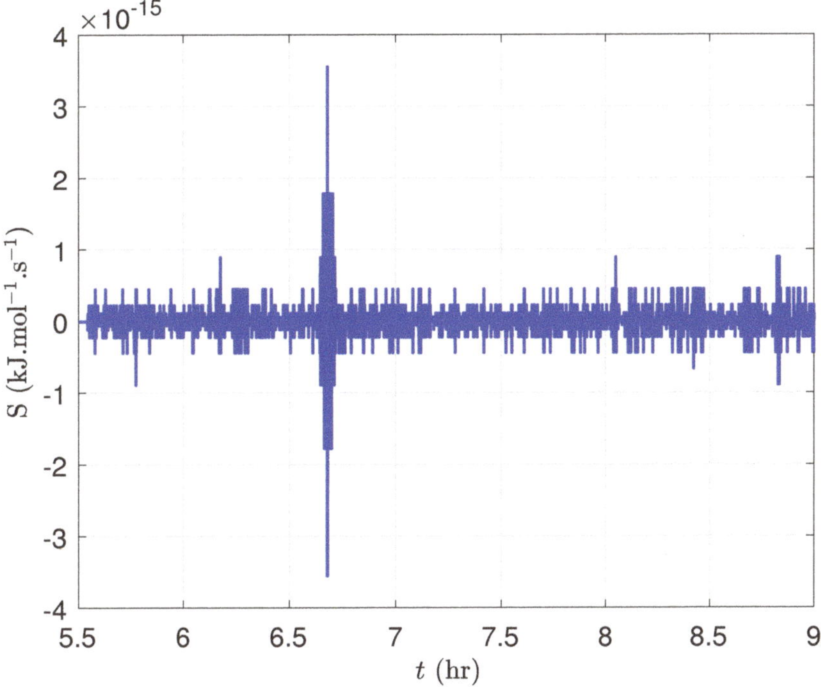

**Figure 6.** Sliding manifold for CFSMC with time.

The values of the performance indices for CFSMC, DISMC, and SMC are provided in Table 4. Here, it can be seen that all the controllers consume the same control energy; however, the performance of CFSMC is superior to its counterparts. In fact, CFSMC shows 57.1% and 58.9% better tracking performance than SMC and DISMC, respectively.

**Table 4.** Comparison of SMC, DISMC, and CFSMC.

| Controller | $e_{rms}$ | $P_{avg}$ |
|---|---|---|
| SMC | 0.5736 | $5.0364 \times 10^{-6}$ |
| DISMC | 0.5479 | $1.9422 \times 10^{-7}$ |
| CFSMC | 0.1653 | $1.4981 \times 10^{-7}$ |

The remaining part of the section discusses the performance of SDKF. To evaluate the estimation performance of SDKF; it is essential to start the UCG plant and GSMUO with contrasting initial conditions. The initial state vectors for the UCG process model and SDKF are selected as

$$x^T(0) = \begin{bmatrix} 0.5 & 0 & 497 & 0 & 0 & 0 & 0 & 0 & 0 & 4.2 \times 10^{-4} & 1.6 \times 10^{-3} \end{bmatrix},$$

$$\hat{x}^T(0) = \begin{bmatrix} 0.48 & 0 & 480 & 0 & 0 & 0 & 0 & 0 & 0 & 4.2 \times 10^{-4} & 1.6 \times 10^{-3} \end{bmatrix}.$$

It is pertinent to mention that some values in $x(0)$ and $\hat{x}(0)$ are equal or very close to each other; this is because the initial values of some states are known.

Seven out of eleven states are measurable, but a full state SDKF is designed so that the measurement noise can be filtered. The results in Figures 7 and 8 demonstrate how SDKF filters out the measurement noise from the measured states. The estimated states are then used for the synthesis of the controller.

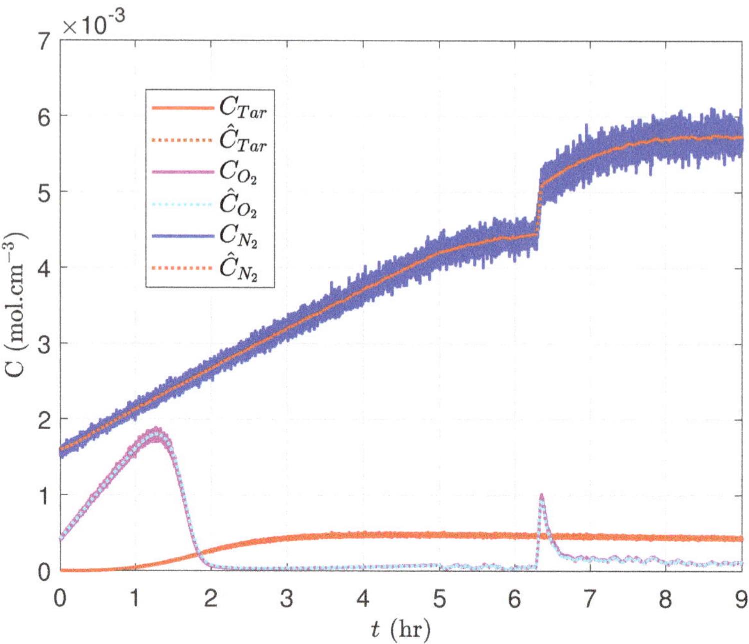

**Figure 7.** Measured and estimated concentrations of gases with time.

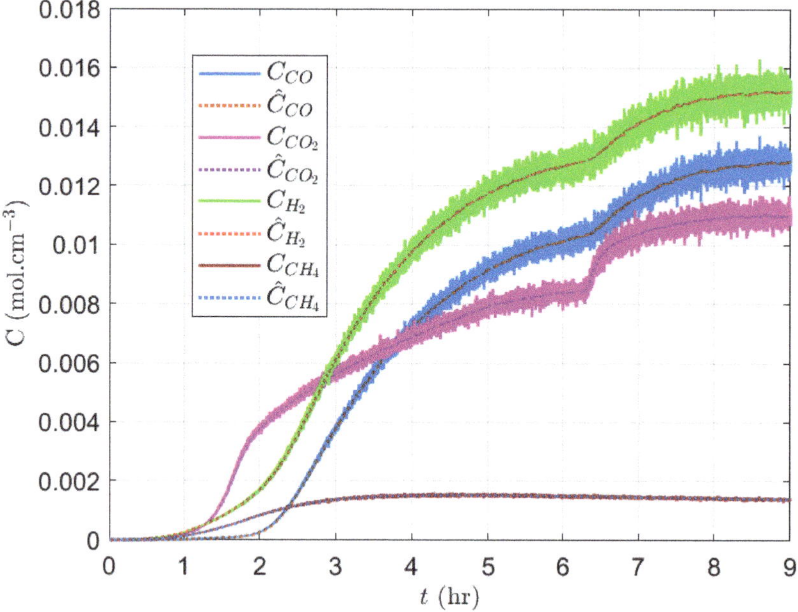

**Figure 8.** Measured and estimated concentrations of gases with time.

Figures 9–11 depict the estimation of the unmeasurable states of the UCG model. It is evident from the simulation results that the estimated and actual states are well aligned. The difference in the true and estimated values of the steam concentration is due to the disturbance $\phi$.

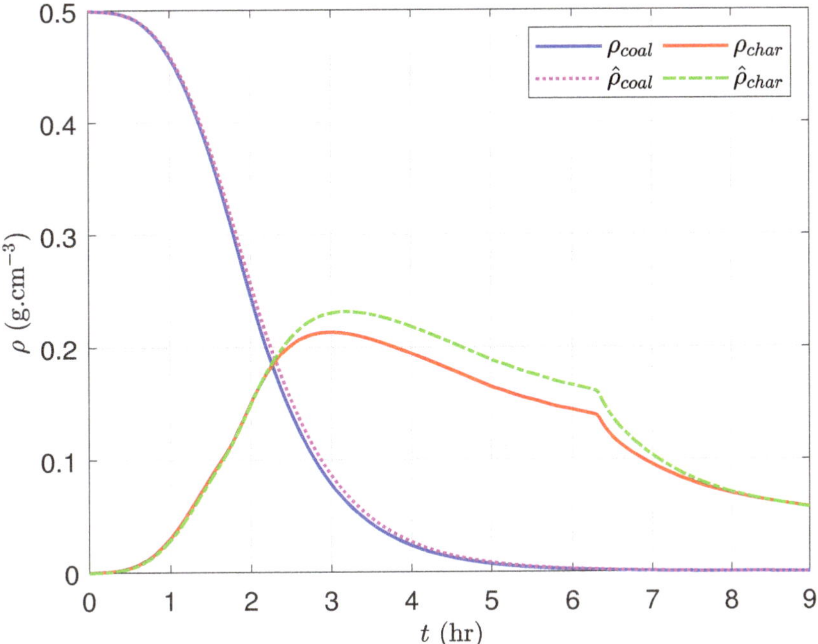

**Figure 9.** True and reconstructed solid densities with time.

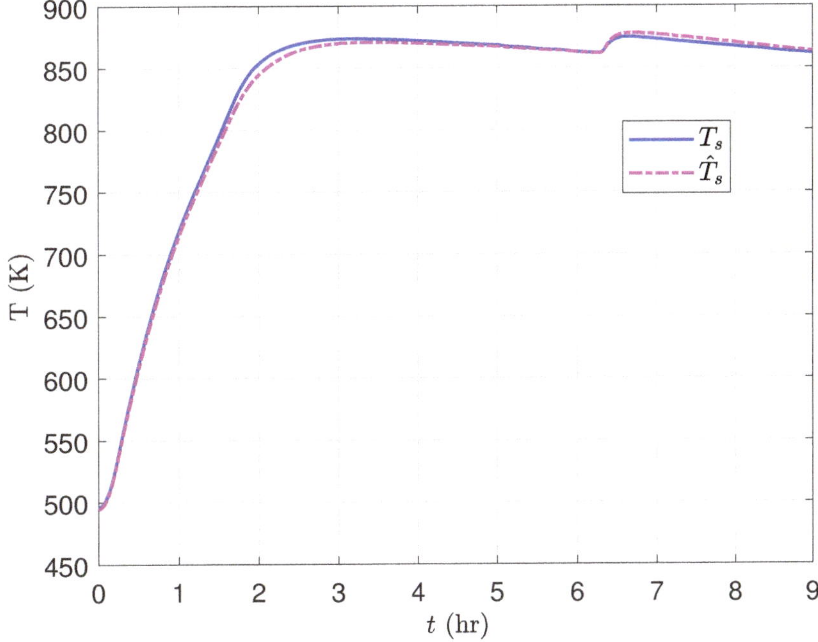

**Figure 10.** True and reconstructed solid temperature with time.

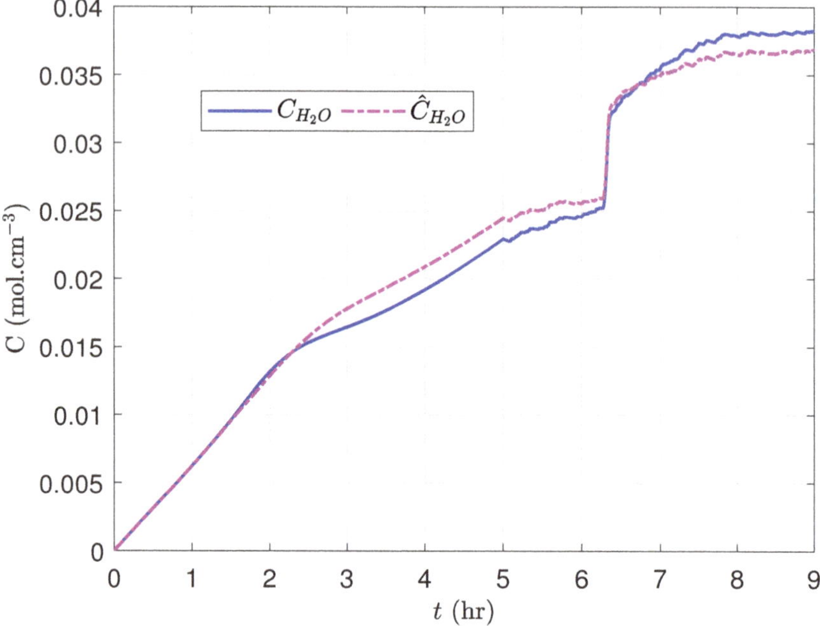

**Figure 11.** True and estimated steam concentration with time.

From the simulation results and comparative analysis of different control techniques, it can be seen that the performance of SDKF is slightly better than GSMUO designed in [14]. However, the design of GSMUO is quite complex as compared to SDKF, which is simply a linear discrete-time Kalman filter implemented using the quasi-linear UCG model.

## 6. Conclusions

In this study, a model-based CFSMC is proposed for tracking the desired heating value trajectory of the UCG process in the presence of external disturbance and measurement noise. A formal analysis of the stability of the controller is also presented. To enable the feedback control design, the unmeasurable UCG process states are estimated by utilizing SDKF, which is based on the quasi-linearization of the nonlinear UCG model. A detailed qualitative and quantitative comparison is also made between CFSMC-SDKF, SMC-SDKF, and DISMC-GSMUO techniques. The comparison demonstrates that the CFSMC-SDKF technique outperforms its counterparts for a rapidly changing reference trajectory.

A possible future extension of this work is the cascade control of the IGCC power plant, which is comprised of the UCG plant and a combined cycle turbine. The efficacy of this technique can be analyzed for tracking the sudden variations in the demand for electrical power.

**Author Contributions:** S.A.: Methodology, formal analysis, software, original draft writing. A.A.U.: conceptualization, visualization, supervision, formal analysis, validation, review & editing. M.R.A.: methodology, formal analysis, writing, review & editing. J.I.: conceptualization, supervision, validation, review & editing. All authors have read and agreed to the published version of the manuscript.

**Funding:** This research received no external funding.

**Data Availability Statement:** We don't have any data to share.

**Conflicts of Interest:** The authors declare no conflict of interest.

**Abbreviations**

The manuscript uses the following abbreviations:

| | |
|---|---|
| CFSMC | Chattering-free sliding mode control |
| DISMC | Dynamic integral sliding mode control |
| GSMUO | Gain-scheduled modified Utkin Observer |
| IGCC | Integrated gasification combined cycle |
| SDKF | State dependent Kalman Filter |
| UCG | Underground coal gassification |

## References


1. Blinderman, M.; Klimenko, A. Underground Coal Gasification and Combustion. In *Underground Coal Gasification and Combustion*; Woodhead Publishing: Sawston, UK, 2017; pp. 1–650. [CrossRef]
2. Perkins, G. Underground coal gasification—Part II: Fundamental phenomena and modeling. *Prog. Energy Combust. Sci.* **2018**, *67*, 234–274. [CrossRef]
3. Perkins, G. Underground coal gasification—Part I: Field demonstrations and process performance. *Prog. Energy Combust. Sci.* **2018**, *67*, 158–187. [CrossRef]
4. Green, M. Recent developments and current position of underground coal gasification. *Proc. Inst. Mech. Eng. Part A J. Power Energy* **2017**, *232*, 095765091771877. [CrossRef]
5. Uppal, A.A.; Bhatti, A.; Aamir, E.; Samar, R.; Khan, S. Control oriented modeling and optimization of one dimensional packed bed model of underground coal gasification. *J. Process Control* **2014**, *24*, 269–277. [CrossRef]
6. Karol, K.; Ján, K. The monitoring and control of underground coal gasification in laboratory conditions. *Acta Montan. Slovaca* **2008**, *13*, 111–117.
7. Karol, K.; Ján, K. Development of control and monitoring system of UCG by Promotic. In Proceedings of the 12th International Carpathian Control Conference (ICCC), Velke Karlovice, Czech Republic, 25–28 May 2011; pp. 215–219. [CrossRef]
8. Kačur, J.; Durdán, M.; Laciak, M.; Flegner, P. A Comparative Study of Data-Driven Modeling Methods for Soft-Sensing in Underground Coal Gasification. *Acta Polytech.* **2019**, *59*.
9. Karol, K.; Ján, K. Extremum Seeking Control of Carbon Monoxide Concentration in Underground Coal Gasification. *IFAC-PapersOnLine* **2017**, *50*, 13772–13777. [CrossRef]
10. Kacur, J.; Flegner, P.; Durdán, M.; Laciak, M. Model predictive control of UCG: An experiment and simulation study. *Inf. Technol. Control* **2019**, *48*, 557–578. [CrossRef]
11. Xiao, Y.; Yin, J.; Hu, Y.; Wang, J.; Yin, H.; Qi, H. Monitoring and control in underground coal gasification: Current research status and future perspective. *Sustainability* **2019**, *11*, 217.
12. Kačur, J.; Laciak, M.; Durdán, M.; Flegner, P. Model-free control of UCG based on continual optimization of operating variables: An experimental study. *Energies* **2021**, *14*, 4323.
13. Uppal, A.A.; Bhatti, A.I.; Samar, R.; Ahmed, Q.; Aamir, E. Model development of UCG and calorific value maintenance via sliding mode control. In Proceedings of the 2012 International Conference on Emerging Technologies, Islamabad, Pakistan, 8–9 October 2012; pp. 1–6. [CrossRef]
14. Uppal, A.A.; Butt, S.S.; Khan, Q.; Aschemann, H. Robust tracking of the heating value in an underground coal gasification process using dynamic integral sliding mode control and a gain-scheduled modified Utkin observer. *J. Process Control* **2019**, *73*, 113–122. [CrossRef]
15. YU, X.; MAN, Z. Model reference adaptive control systems with terminal sliding modes. *Int. J. Control* **1996**, *64*, 1165–1176. [CrossRef]
16. Feng, Y.; Fengling, H.; Yu, X. Chattering free full-order sliding-mode control. *Automatica* **2014**, *50*, 1310–1314. [CrossRef]
17. Yu, X.; Kaynak, O. Sliding-Mode Control With Soft Computing: A Survey. *IEEE Trans. Ind. Electron.* **2009**, *56*, 3275–3285. [CrossRef]
18. Teixeira, M.; Zak, S. Stabilizing Controller Design for Uncertain Nonlinear Systems Using Fuzzy Models. *IEEE Trans. Fuzzy Syst.* **1999**, *7*, 133–142. [CrossRef]
19. Li, Q.; Li, R.; Ji, K.; Dai, W. Kalman Filter and Its Application. In Proceedings of the 8th International Conference on Intelligent Networks and Intelligent Systems (ICINIS), Tianjin, China, 1–3 November 2015; pp. 74–77. [CrossRef]
20. Kim, Y.; Bang, H. Introduction to Kalman Filter and Its Applications. In *Introduction to Kalman Filter and Its Applications*; IntechOpen: London, UK, 2018; pp. 1–16. [CrossRef]
21. Chaudhry, A.M.; Uppal, A.A.; Bram, S. Model Predictive Control and Adaptive Kalman Filter Design for an Underground Coal Gasification Process. *IEEE Access* **2021**, *9*, 130737–130750. [CrossRef]

 *electronics*

*Article*

# Design of Predefined Time Convergent Sliding Mode Control for a Nonlinear PMLM Position System

**Saleem Riaz [1], Chun-Wu Yin [2], Rong Qi [1], Bingqiang Li [1,*], Sadia Ali [3] and Khurram Shehzad [4,*]**

[1]  School of Automation, Northwestern Polytechnical University, Xi'an 710072, China
[2]  School of Information and Control Engineering, Xi'an University of Architecture and Technology, Xi'an 710055, China
[3]  Department of Electrical and Computer Engineering, International Islamic University, Islamabad 04436, Pakistan
[4]  Department of Electrical and Computer Engineering, COMSATS University, Islamabad 22060, Pakistan
*  Correspondence: libingqiang@nwpu.edu.cn (B.L.); khurram.shehzad@comsats.edu.pk (K.S.)

**Abstract:** The significant role for a contemporary control algorithm in the position control of a permanent magnet linear motor (PMLM) system is highlighted by the rigorous standards for accuracy in many modern industrial and robotics applications. A robust predefined time convergent sliding mode controller (PreDSMC) is designed for the high precision position tracking of a permanent magnet linear motor (PMLM) system with external disturbance, and its convergence time is independent of the system's initial value and model parameters. We verified theoretically that the performance function conditions are satisfied, the motor speed is steady and constrained, and the motor position tracking error converges to zero within the prescribed time. First, we designed a sliding mode (SM) surface with predetermined time convergence, which mathematically demonstrates that the tracking error converges to zero within the predefined time and shows that the position tracking accuracy is higher. Secondly, we developed a PreDSMC law that is independent of initial state and based on the predefined time convergence Lyapunov stability criterion. Finally, to prove the accuracy and higher precision of the proposed PreDSMC, comparative numerical simulations are performed for PMLM with compound disturbances. Simulation findings show that the suggested robust predefined control method considerably reduces the impacts of friction and external disturbances; consequently, it may increase the control performance when compared to the typical proportional integral derivative (PID) controller, the nonsingular fast terminal SMC, and the linear SMC.

**Keywords:** nonlinear control; robust sliding mode control; predefined time sliding mode control; permanent magnet linear motor system

**Citation:** Riaz, S.; Yin, C.-W.; Qi, R.; Li, B.; Ali, S.; Shehzad, K. Design of Predefined Time Convergent Sliding Mode Control for a Nonlinear PMLM Position System. *Electronics* **2023**, *12*, 813. https://doi.org/10.3390/electronics12040813

Academic Editor: Ahmed Abu-Siada

Received: 5 January 2023
Revised: 30 January 2023
Accepted: 2 February 2023
Published: 6 February 2023



## 1. Introduction

Today, permanent magnet linear motors (PMLM) have a significant role in civil, industrial, military, and other high-precision applications [1,2]. Especially, PMLM is widely used in the high-precision manufacturing industry because of its higher thrust density, higher acceleration, speed, and very high precision. As a result, PMLM research has recently received more prominent attention from researchers. However, the performance of these motors degrades due to external disturbances, force ripple, and friction effects [3]. Concerning high-precision motion control, it is always very critical to attain a fast-dynamic response, especially when increasing the tracking accuracy is a vital goal of the PMLM.

Numerous control algorithms for PMLM analysis and design focusing on various applications have been presented in the literature. For instance, a robust adaptive algorithm was developed in [4] for the compensation and control of PMLM displacement in the presence of force ripple and friction. The authors of [5] presented a water-cooled PMLM system that was designed for studying the temperature behavior under various work tasks. The performance of PMLM rate control based on flow-oriented primary control was studied in [6]. Sliding mode control (SMC) is a preferred method for managing control problems in nonlinear systems. It is widely used in practical applications for its insensitivity to parameter changes and perturbations [7–13]. Design of an adaptive 2-SMC method for a class of unknown nonlinear discrete systems with multiple inputs and multiple outputs (MIMO) was described in [14]. Stable pressure control for coke oven gas distributors was achieved by combining SMC and data-driven control (DDC) methods [15]. The traditional SMC ensures the asymptotic stability of the closed loop control only in the sliding mode phase, i.e., the system to the equilibrium point indefinitely. However, the finite-time terminal SMC-based PMLM system realizes uncertainties by balancing control performance and robustness. An efficient tactic that not only guarantees control quality but can also improve robustness is predictive interference compensation, presented in the literature [16–18]. The observer-based design method is proficient for estimating perturbations such as SMC [19–21] and the finite-time approach [22–24]. Owing to the supremacy of finite-time convergence [25–27], terminal sliding mode control (TSMC) has been presented in [28–30].

Nonlinear systems have become the topic of extensive study during the past few decades because such systems provide a more critical perspective of the available information on the qualities of memory and legacy [31–33]. Some examples of physical systems involving nonlinearities include electrical circuits containing supercapacitive elements which are vital in the design of batteries and fuel cells [34–37], mechanical systems containing viscoelastic materials [38,39], and biological systems [40,41]. Moreover, sophisticated control design can benefit from the reliability of fractional-order approaches. Sliding mode approaches [42–44] are amongst the most often used methods because of their effectiveness in reducing the impact of uncertainty in the control of nonlinear systems.

Dynamic characteristics involve enforcing sliding motion with finite and fixed-time convergence [45–47]. Recent publications have investigated sliding mode and fixed-time tools, elaborating on the remarkable qualities of these uncertain systems [48]. Moreover, the time at which the sliding action is imposed (the settling-time function) is unbounded with respect to the initial condition. Nevertheless, sliding mode methods ensure finite-time convergence of the sliding function.

Recent research has established the concept of fixed-time stability to address the issue of an unbounded settling-time function. When starting from this state, the work in question enforces a sliding mode within a uniformly defined time. Designing appropriate controllers for nonlinear systems has been studied using the predefined time-stability approach [49–53]. Therefore, in the case of a second-order nonlinear system based on the Lyapunov stability criterion, predefined time SMCs have been proposed. These works differ from prior research as they suggest auxiliary controllers through a suitable dynamic extension. This research was influenced by adaptive predefined time stability.

The robustness of perturbations is rarely explored for predefined settling-time strategies, especially for autonomous systems [54–56]. The controller proposed in [57] could only mitigate the disturbances if the system's initial states are known. In [53] and [55], complementary random time convergence is accomplished by changing the sliding phase solely; moreover, simulations presume that the reaching time of SMC is predictable. Additionally, [56] used a time-changing piecewise controller to discard matching instabilities and attain preset time by altering system parameters. The authors in [58,59] addressed this problem by developing unique terminal sliding surface-based control in a combined predetermined settling period for the category of nonlinear second-order systems containing coupled disturbances.

In [60–62], nonsingular terminal SMC is used to evaluate the overall effectiveness of a fixed-time control mechanism with mismatched disturbances. It is pertinent to mention that the constraints on the settling time were too conservative and required a significant amount of control effort. In [63], the authors extended this work and demonstrated the ultimate boundedness with predetermined temporal stability. Moreover, it is believed that the predetermined time in [59,63,64] is a factor of the system's starting states and characteristics.

Research in [65] proposed a back-stepping strategy to establish fixed-time stability for higher-order systems with rigorous feedback. In [66], predetermined time stability is attained for dynamical systems with distributed order in the context of uncertainty. In addition, a recent study [44] designed a unique SMC that leads to preset time convergence. Further, nonsingular terminal SMC is being proposed to avoid singularities and it has been implemented in some practical control systems [67–71]. Despite this, inner discontinuities appear as never before in previous terminal SMC schematics. As a result, some effective techniques have been used to overcome the chattering problem resulting from internal discontinuities. In particular, the finite-time terminal SMC can achieve fast state convergence to improve tracking accuracy and limit the jitter problem of phase. In this article, motivated by the advantages of the finite-time terminal SMC, we designed a predefined time control method for the PMLM system.

A permanent magnet linear motor (PML) not only has advantages of high force density, low heat loss, and high precision and accuracy, but also has simple mechanical structure [72], so it is the first choice for motor motion control systems involving high strength, speed, and high precision [3]. Today, PML has been widely used in the precision manufacturing industry. Compared with traditional rotary motors [73–76], linear motors do not require indirect coupling mechanisms such as gear boxes, chains, and screw couplings, which greatly reduces contact-type nonlinearity and mechanical interference such as friction and recoiling [77–79]. As a result, PML can meet the growing demand for high performance servo system applications and has been successfully used in machine tools, semiconductor manufacturing systems, etc. Since PML is not equipped with a transport mechanism, the realizable performance of PML, such as the ability to reduce uncertainty and the influence of external interference, is inevitably lost to some extent. Therefore, it is very important to reduce these interference effects through appropriate mechanical design or control scheme. With new methods of mechanical design, the effect can be maintained at the allowable level, but this method is more expensive. The controller design based on a PML system is an economical and feasible method to suppress these effects.

The available literature not only ignores the impact of predefined stability and disturbances, but assumes that the predefined settling time is a factor of the initial states. Moreover, the convergence rate conventions are extra cautious and require a significant control effort. To the best of our knowledge, predefined time convergence for nonlinear systems containing uncertainties and unknown initial conditions remains an unsolved topic. The contributions of our paper are summarized below:

- A predefined time sliding surface of the tracking error is designed, and we also theoretically prove that the tracking error found at the sliding surface converges to the equilibrium point within a prescribed time. Furthermore, its derivative is bounded, which gives the tracking error superior information. A Lyapunov-based stability criterion with predefined time convergence is elaborated and proved theoretically. The results depict the predefined convergence time, which is independent of the initial value and the parameters of the controlled system.
- A robust predefined SMC law is designed and proven theoretically. We utilize the predefined sliding function to capture the control input variables, which simplifies the complexity of the designed controller. It is shown that the trajectory tracking error of the closed-loop system has a convergent nature within a predefined time. Moreover, the trajectory satisfies the performance, and it is worth noting that its derivative is bounded.
- We have designed a predefined robust control law to approximate the uncertain part of the PMLM, i.e., friction and ripple forces. The presented scheme in this paper has a wide range of applications. Our proposed PreDSMC only needs the trajectory tracking error of the PMLM and its derivative information to realize the high-precision

trajectory tracking control of the system. It is independent of the initial conditions and model parameters of the system, which is why it is called a robust controller.

This study is organized into the following sections: Section 2 presents the mathematical model and control objective of the PMLM. Section 3 presents the linearization method of nonlinear parts and the variation method for the tracking error performance based on the predefined criterion of the sliding function. Moreover, the sliding mode surface with a predefined time for tracking error convergence and the Lyapunov stability criterion for convergence are presented in the same section. Section 4 presents the proposed PreDSMC control law. To verify the robustness of the proposed PreDSMC control algorithm, simulation results for various cases are presented in Section 5. Lastly, our study is concluded in Section 6.

## 2. Problem Formulation

In general, the second-order nonlinear dynamic system can be mathematically expressed as

$$\begin{cases} \dot{x}_1(t) = x_2(t) \\ \dot{x}_2(t) = f(\theta, x(t), t) + Nu(t) + d(t) \\ y(t) = x_1(t) \end{cases} \tag{1}$$

where $x(t) = [x_1(t), x_2(t)]^T$ and $\theta\,(t)$ are the internal parameter variables. Mass and $f(\theta, x(t), t)$ are smooth continuous functions and $N$ includes some nonlinearities such as Stribeck friction and ripple forces. Nomenclature is given in Table 1.

**Table 1.** Parameters of the PMLM model.

| Parameter | Symbol | Value | Unit |
| --- | --- | --- | --- |
| Motor mass | $m$ | 5.4 | Kg |
| Resistance | $R$ | 16.8 | $\Omega$ |
| Force constant | $k_f$ | 130 | |
| Back electromotive force | $k_e$ | 123 | V/(rad/s) |
| | $f_c$ | 10 | |
| Friction force | $f_s$ | 20 | N |
| | $f_v$ | 10 | |
| | $x_{2s}$ | 0.1 | |
| | $A_1$ | 8.5 | |
| | $\omega_1$ | 314 | |
| Ripple force | $A_2$ | 4.25 | N |
| | $\omega_2$ | 314 | |
| | $A_3$ | 2 | |
| | $\omega_3$ | 314 | |

*PMLM Model Representation*

Although a second-order system may frequently be used to illustrate the dynamics of a PMLM [27], owing to the low value of stator inductance relative to the resistance, we neglect it in this study along with loads and slight perturbations. A typical PMLM mathematical model is characterized as follows:

$$\begin{cases} \dot{x}_1(t) = x_2(t) \\ \dot{x}_2(t) = -\dfrac{k_f k_e}{Rm} x_2(t) + \dfrac{k_f}{Rm} u(t) - \dfrac{d(t)}{m} \\ y(t) = x_1(t) \end{cases} \tag{2}$$

where the parameters are $M = \frac{k_f k_e}{Rm}, N = \frac{k_f}{Rm},$ and $F = \frac{d(t)}{m}$. Assuming the reference signal is denoted as $x_r$ and its first and second order derivatives, i.e., $\dot{x}_r$ and $\ddot{x}_r$ are sequential and bounded, respectively. Consequently, position tracking error and speed tracking errors can be expressed as $e_1(t) = x_r(t) - x_1(t), e_2(t) = \dot{x}_r(t) - x_2(t)$.

According to the model described above, the state-variable formulation of a PMLM is easily derivable. Additionally, the error dynamics equation can be described as:

$$\begin{cases} \dot{e}_1(t) = e_2(t) \\ \dot{e}_2(t) = -Me_2(t) - Nu(t) + F + M\dot{x}_r + \ddot{x}_r \\ y(t) = x_1(t) \end{cases} \tag{3}$$

The control goal of this article is to develop a robust controller that can accurately follow the reference position curve in the predefined time. The Stribeck curve in Figure 1 is a well-known tool for modeling the friction. It essentially indicates the relationship between the friction force and angular velocity for different values of friction. The Stribeck friction model can be expressed as in Equation (2). When the static friction is $\left| \dot{\theta}(t) \right| < \alpha$, with $\alpha$ threshold, the dynamic friction $F_{friction}(t)$ can be given by the following piecewise expression:

$$F_{\text{friction}}(t) = \begin{cases} f_m f(t) > f_m \\ f(t) - f_m < f < f_m \\ -f_m f(t) < -f_m \end{cases}$$

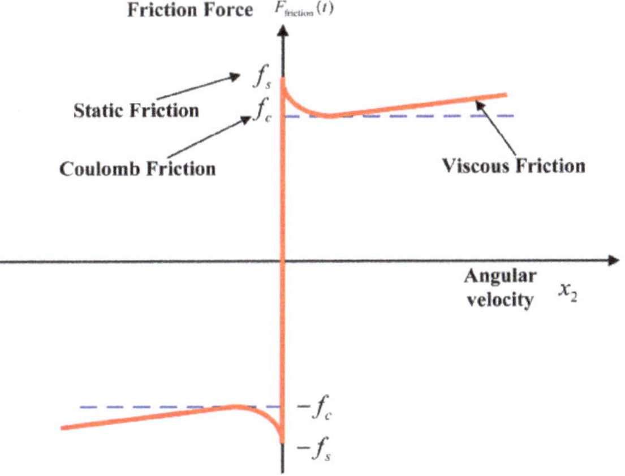


**Figure 1.** Stribeck friction curve.

The Stribeck friction curve is a well-known model of the friction. It shows the relationship between the friction force and angular velocity for different values of the friction. A friction model can be expressed as in Equation (4). When the static friction is $|\theta'(t)| < \alpha$, with threshold, the dynamic friction can be given by (4). One can see that the static friction is $|\theta'(t)| > \alpha$, and can be given as above in which Fs denotes the driving force, Fc represents the maximum static friction force, and it is a small scaling factor.

$$F_{\text{friction}}(t) = \left[ f_c + (f_s - f_c)e^{-(x_2/x_{2s})^2} + f_v x_2 \right] \text{sgn}(x_2) \tag{4}$$

where $f_v$ is the viscous friction coefficient, $x_{2s}$ is the assumed measured lubrication parameter, $f_c$ is the minimum value of the Coulomb friction, and $f_s$ is the static friction. According to the cogging effect in the structure of the PMLM, the pulsating force $F_{\text{ripple}}(t)$ generated by the reluctance is modeled as

$$F_{\text{ripple}}(t) = \sum_{i=1}^{n} A_i \sin(\omega_i x_1) \tag{5}$$

where $\omega_i$ is the angular velocity and $A_i$ is the amplitude. The load disturbance torque $F_L(t)$ is a constant value. The overall system is described as follows:

$$\begin{cases} \dot{x}_1(t) = x_2(t) \\ \dot{x}_2(t) = Mx_2(t) + Nu(t) + d(t) \\ y(t) = x_1(t) \end{cases} \tag{6}$$

## 3. Predefined Time Sliding Surface Design

Definition 1: For the nonlinear system (2), if there is a predefined constant $T_s > 0$ such that the condition is satisfied for any $t \in [0, \infty]$:

(1) At the time $t \to T_s$, $\lim\limits_{t \to T_s} y(t) = 0$;

(2) When $t \geq T_s$, and has $y(t) \equiv 0$, then the nonlinear system (2) is globally predefined time stable.

To take the full advantage of strong robustness of the sliding mode controller and ensure that the trajectory tracking error $e(t)$ can converge to the equilibrium point within the predefined time $T_{s1}$, the sliding mode surface with the predefined time convergence is designed as follows:

$$S = e_2(t) + \frac{\pi}{2p_1 T_{s1}\sqrt{a_1 b_1}}\left(a_1 e_1^{1-p_1}(t) + b_1 e_1^{1+p_1}(t)\right) \tag{7}$$

where the parameter satisfies $0\langle p_1\langle 1, a_1\rangle 0, b_1\rangle 0$; $T_{s1}$ is the predefined time. For the sliding mode surface (7), Theorem 1 is given below.

*3.1. Predefined Time Error Convergence Criterion*

**Theorem 1.** *For any predefined time $T_{s1} > 0$ and parameter $0 < p_1\langle 1$, $a_1\rangle 0$, $b_1 > 0$, when the sliding mode surface (3) satisfies $S = 0$, the following holds true:*

(1) *If the initial value of the tracking error is $e_1(0) \neq 0$, then $e_1(t)$ will converge to zero within the predefined time $T_s$, and the convergence time*

$$t_s = \frac{2T_{s1}}{\pi}\arctan\left(\sqrt{\frac{b_1}{a_1}}e_1^{p_1}(0)\right) < T_{s1} \tag{8}$$

(2) *If the initial value of the tracking error is $e_1(0) = 0$, then at that the time $e_1(t)$ that converges to zero is $t_s = 0$, that is, when $S = 0$ is the trajectory tracking error $e_1(t) \equiv 0$.*

**Proof of Theorem 1.**
When $S = 0$, there is:

$$e_2 = \frac{de_1}{dt} = -\frac{\pi}{2p_1 T_{s1}\sqrt{a_1 b_1}}a_1 e_1^{1-p_1}\left(1 + \frac{b_1}{a_1}e_1^{2p_1}\right) \tag{9}$$

After transforming (8), we obtain

$$\frac{p_1 e_1^{p_1-1}de_1}{1 + \left(\sqrt{\frac{b_1}{a_1}}e_1^{p_1}\right)^2} = -\frac{\pi\sqrt{a_1}}{2T_{s1}\sqrt{b_1}}dt \tag{10}$$

The following measures were taken to ease the integration:

$$\frac{d\left(\sqrt{\frac{b_1}{a_1}}e_1^{p_1}\right)}{1 + \left(\sqrt{\frac{b_1}{a_1}}e_1^{p_1}\right)^2} = -\sqrt{\frac{b_1}{a_1}}\frac{\pi}{2T_{s1}}\sqrt{\frac{a_1}{b_1}}dt = -\frac{\pi}{2T_{s1}}dt \tag{11}$$

Assuming that the tracking error $e_1(t)$ converges to zero at time $t_s$, i.e., $e_1(t_s) = 0$, integrating both sides of Equation (10) on $[0, t_s]$ gives

$$\int_0^{t_s} \frac{d\left(\sqrt{\frac{b_1}{a_1}} e_1^{p_1}\right)}{1 + \left(\sqrt{\frac{b_1}{a_1}} e_1^{p_1}\right)^2} = \int_0^{t_s} -\frac{\pi}{2T_{s1}} dt \tag{12}$$

$$arctan\left(\sqrt{\frac{b_1}{a_1}} e_1^{p_1}\right) \Big|_0^{t_s} = -\frac{\pi t}{2T_{s1}} \Big|_0^{t_s} \tag{13}$$

Simplifying with the help of the steps given in Equations (12) and (13), we obtain

$$t_s = \frac{2T_{s1}}{\pi} arctan\left(\sqrt{\frac{b_1}{a_1}} e_1^{p_1}(0)\right) \le \frac{2T_{s1}}{\pi}\frac{\pi}{2} = T_{s1} \tag{14}$$

When $e_1(0) = 0$ and $t_s = \frac{2T_{s1}}{\pi} arctan\left(\sqrt{\frac{b_1}{a_1}} e_1^{p_1}(0)\right) = 0$ represent constant $e_1(t) \equiv 0$, it means that when the sliding surface $S = 0$, the tracking error $e_1(t)$ is also equal to zero and hence the theorem is proved. $\square$

*3.2. Predefined Time Convergence Stability*

**Theorem 2.** *For nonlinear system $\dot{x}(t) = f(x), f(0) = 0, x(0) = x_0$, for any predefined time $T_s > 0$, and parameter $0\langle p\langle 1, a\rangle 0, b\rangle 0$, if there is a radially unbounded and positive definite Lyapunov function $V(t)$ that satisfies*

$$\dot{V}(t) \le -\frac{\pi}{2pT_s\sqrt{ab}}\left(aV^{1-p}(t) + bV^{1+p}(t)\right) \tag{15}$$

*then*

(1)   *If $V(0) \neq 0$, the system is stable at the global predefined time and converges to the equilibrium point.*

$$t_s = \frac{2T_s}{\pi} arctan\left(\sqrt{\frac{b}{a}} V^p(0)\right) < T_s$$

(2)   *If $V(0) = 0$, then $V(t) \equiv 0$ and it indicates that the system state is always at the equilibrium point.*

**Proof of Theorem 2.**
Let $\dot{V}(t) = -\frac{\pi}{2pT_s\sqrt{ab}}\left(aV^{1-p}(t) + bV^{1+p}(t)\right) + \Delta$, then

$$\frac{dV}{dt} = -\frac{\pi}{2pT_s\sqrt{ab}} aV^{1-p}\left(1 + \frac{b}{a}V^{2p} - \frac{2pT_s\sqrt{ab}}{\pi a V^{1-p}}\Delta\right) \tag{16}$$

After transforming (15) and taking the differential, we have

$$\frac{\pi}{2T_s} dt = -\frac{1}{1 + \left(\sqrt{\frac{b}{a}}V^p\right)^2 - \frac{2pT_s\sqrt{ab}}{\pi a V^{1-p}}\Delta} d\left(\sqrt{\frac{b}{a}}V^p\right) \tag{17}$$

Let us assume that at time $t_s$, $V(t_s) = 0$. Next, integrate both sides of Equation (16). Since, $V \geq 0$, $\Delta \geq 0$ then $\frac{2pT_s\sqrt{ab}}{\pi a V^{1-p}}\Delta \geq 0$, we obtain:

$$\int_0^{t_s} \frac{\pi}{2T_s}\,dt = -\int_{V(0)}^{V(t_s)} \frac{d\left(\sqrt{\frac{b}{a}}V^p\right)}{1+\left(\sqrt{\frac{b}{a}}V^p\right)^2 - \frac{2pT_s\sqrt{ab}}{\pi a V^{1-p}}\Delta} \leq -\int_{V(0)}^{V(t_s)} \frac{d\left(\sqrt{\frac{b}{a}}V^p\right)}{1+\left(\sqrt{\frac{b}{a}}V^p\right)^2}$$

After simplification, we have

$$\frac{\pi}{2T_s}t_s \leq arctan\left(\sqrt{\frac{b}{a}}V^p(0)\right) - arctan\left(\sqrt{\frac{b}{a}}V^p(t_s)\right)$$

$$\begin{cases} V(t) \leq \left(\sqrt{\frac{a}{b}}\tan\left(arctan\left(\sqrt{\frac{b}{a}}V^p(0)\right) - \frac{\pi}{2T_s}t\right)\right)^{\frac{1}{p}} \\ t_s = \frac{2T_s}{\pi}arctan\left(\sqrt{\frac{b}{a}}V^p(0)\right) \leq T_{s2} \end{cases}$$

$\square$

## 4. Design of Predefined Time Control Law

First, take the Lyapunov function as $V_1 = 0.5S^TS$, and derive it with respect to $t$:

$$\dot{V}_1 = S^T\dot{S} = S^T\left(\dot{e}_2 + \frac{\pi}{2p_1 T_{s1}.\sqrt{a_1 b_1}}(a_1(1-p_1)e_1^{-p_1} + b_1(1+p_1)e_1^{p_1})e_2\right)$$

After simplification, we obtain

$$\dot{V}_1 = S^T\left(-Me_2(t) - Nu(t) + F + M\dot{x}_r + \ddot{x}_r + \frac{\pi}{2p_1 T_{s1}\sqrt{a_1 b_1}}(a_1(1-p_1)e_1^{-p_1} + b_1(1+p_1)e_1^{p_1})e_2\right) \tag{18}$$

The controller is designed as: $u_{total} = u_a + u_b$ where

$$\begin{aligned} u_a &= \tfrac{1}{N}\left(-Me_2(t) + F + M\dot{x}_r + \ddot{x}_r + \frac{\pi}{2p_1 T_{s1}\sqrt{a_1 b_1}}(a_1(1-p_1)e_1^{-p_1} + b_1(1+p_1)e_1^{p_1})e_2\right) \\ u_b &= \frac{\pi}{2Np_2 T_{s2}\sqrt{a_2 b_2}}(a_2 S^{1-p_2} + b_2 S^{1+p_2}) \end{aligned} \tag{19}$$

then

$$\begin{aligned} -Me_2(t) - Nu(t) + F + M\dot{x}_r + \ddot{x}_r + &\frac{\pi}{2p_1 T_{s1}\sqrt{a_1 b_1}}(a_1(1-p_1)e_1^{-p_1} + b_1(1+p_1)e_1^{p_1})e_2 \\ &= -\frac{\pi}{2p_2 T_{s2}\sqrt{a_2 b_2}}(a_2 S^{1-p_2} + b_2 S^{1+p_2}) \end{aligned} \tag{20}$$

Therefore,

$$\begin{aligned} \dot{V}_1 &= S\left(-\frac{\pi}{2p_2 T_{s2}\sqrt{a_2 b_2}}(a_2 S^{1-p_2} + b_2 S^{1+p_2})\right) \\ &= -\frac{\pi}{2p_2 T_{s2}\sqrt{a_2 b_2}}(a_2 S^{2-p_2} + b_2 S^{2+p_2}) \\ &= -\frac{\pi}{2\bar{p}T_{s2}\sqrt{\bar{a}\bar{b}}}(\bar{a}V_1^{1-\bar{p}} + \bar{b}V_1^{1+\bar{p}}) \end{aligned} \tag{21}$$

Among them $\bar{a} = a_2 2^{1-0.5p_2}, \bar{b} = b_2 2^{1+0.5p_2}, \bar{p} = 0.5p_2$, according to Theorem 2, it can be known that system (2) is globally stable at predefined time, and the sliding mode surface $S$ will converge to zero within the predefined time; the convergence time is $t_s = \frac{2T_{s2}}{\pi}arctan\left(\sqrt{\frac{b_2}{a_2}}V_1^{p_2}(0)\right) < T_{s2}$.

**Remarks.** *In SMC, the usage of the signum function often results in a chattering phenomenon. Therefore, we must define a special function in order to mitigate this chattering effect. We can now define a continuous function, i.e., sat (s), for the compensation of chattering in the linear SMC (LSMC) control signal such that:*

$$sat(s) = \begin{cases} \frac{s}{\zeta} & \text{for } |s| < \zeta \\ sgn(s) & \text{for } |s| \geq \zeta \end{cases} \tag{22}$$

*where $\zeta$ is taken as a positive constant. Its value can be chosen so that the control action and chattering effect are not compromised. For practical applications, any random but positive value can be set.*

$$\begin{cases} \dot{x}_1 = x_2 \\ \dot{x}_2 = -Mx_2 + Nu - F + d_x \\ y = x_1 \end{cases}$$

Here, the system states are $[x_1, x_2]^T$; $x_1$ denotes the displacement and $x_2$ represents the velocity of the PMLM. From this equation, we can set the sliding surface as: $s = e_2 + \lambda e_1$. In accumulation, the LSMC is taken as:

$$u_{\text{LSMC}} = \frac{1}{N}\left[\ddot{x}_r + Mx_2 + \mu e_2 + \phi_1 sat(s)\right] \tag{23}$$

The renowned PID control law is given as: $u^{\text{PID}} = \Gamma_p e_1 + \Gamma_i \int e_1 dt + \Gamma_d e_2$.

To compare with the finite control method, we can select the nonsingular fast terminal SMC (NFTSMC), which is given as follows:

$$\begin{cases} u_{total} = u_{main} + u_{comp} \\ u_{main} = \frac{1}{N}\left[-Me_2 + M\dot{x}_r + \ddot{x}_r + F + \frac{1}{\beta_1 v_1}|e_2|^{2-v_1}sgn(e_2) + \frac{\beta_2}{\beta_1}|e_2|^{2-v_1}|e_1|^{v_1-1}\right] \\ u_{comp} = \frac{1}{N}[k_1 s + k_2 sgn(s)] \\ s = e_1 + \beta_1|e_2|^{v_1}sgn(e_2) + \beta_2|e_1|^{v_1}sgn(e_1) \end{cases} \tag{24}$$

In this formula, $u_{total}$ is the main control law. For nonsingularity, there is a compensation technique that provides a control law to compensate for disturbances. The following control is our proposed PreDSMC input control law:

$$\begin{cases} u_a = \frac{1}{N}\left(-Me_2(t) + F + M\dot{x}_r + \ddot{x}_r + \frac{\pi}{2p_1 T_{s1}\sqrt{a_1 b_1}}(a_1(1-p_1)e_1^{-p_1} + b_1(1+p_1)e_1^{p_1})e_2\right) \\ u_b = \frac{\pi}{2Np_2 T_{s2}\sqrt{a_2 b_2}}(a_2 S^{1-p_2} + b_2 S^{1+p_2}) \end{cases} \tag{25}$$

According to Theorem 2, based on the proposed PreDSMC in Equation (25), a closed-loop system is able to converge to a defined surface within a predefined time $T_{s2}$. When the system state reaches and stabilizes in the sliding mode surface, (please refer to Equation (21), i.e., $S = 0$), then according to Theorem 1, a tracking error will always converge to the equilibrium point within a predefined time $T_{s1}$. Thus, the system also converges to its origin to the respective predefined time $T_s = T_{s1} + T_{s2}$. According to the conclusion of Theorem 1, the system tracking error and its derivative are bounded.

## 5. Numerical Simulation Analysis

### 5.1. PMLM Simulation

Taking the PMLM mathematical model as the simulation target [80–82], the relevant parameters of the PMLM model are set as given in Table 1. Additionally, the overall proposed robust control structure is shown in Figure 2.

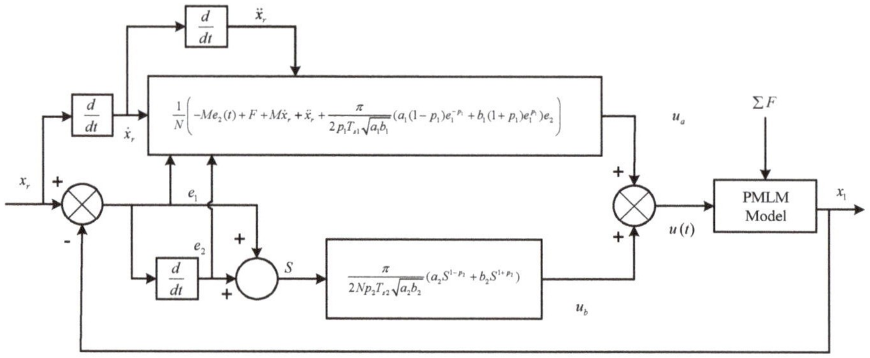

**Figure 2.** Structure of the proposed PreDSMC.

PreDSMC is a robust controller that is particularly designed for PMLM position tracking in this study. The main goal is to track the displacement within a predefined time with a higher accuracy for the PMLM model. Two cases have been considered to check the robust performance of the proposed control scheme. The first is step tracking, and the second is sinusoid tracking. The control parameters are chosen according to the PMLM system for higher performance and accuracy. Multiple control schemes are considered for the purpose of comparison analysis with the PreDSMC. The control gains are chosen accordingly to achieve higher accuracy and precision in both the step and sinusoid tracking. The numerical values of these parameters are listed in Table 2.

**Table 2.** Parameters of the controllers.

| Control Algorithm | Step Tracking Control Parameters | Sinusoid Tracking Control Parameters |
|---|---|---|
| PID [82,83] | $\Gamma_p = 800$, $\Gamma_d = 50$, $\Gamma_i = 2$ | $\Gamma_p = 23,000$, $\Gamma_d = 5000$, $\Gamma_i = 2$ |
| LSMC [82,83] | $\mu = 100$, $\phi_1 = 200$, $\zeta = 0.5$ | $\mu = 100$, $\phi_1 = 200$, $\zeta = 0.5$ |
| NFTSMC [82–84] | $k_1 = 700$, $k_2 = 10$, $\beta_1 = 0.1$, $\beta_2 = 0.1$, $v_1 = 1$ | $k_1 = 5000$, $k_2 = 7000$, $\beta_1 = 0.1$, $\beta_2 = 0.1$, $v_1 = 1.1$ |
| Proposed PreDSMC | $T_{s1} = 0.5$, $p_1 = 0.8$, $a_1 = 100$, $b_1 = 150$ <br> $T_{s2} = 0.2$, $p_2 = 0.4$, $a_2 = 10$, $b_2 = 1$ | $T_{s1} = 0.5$, $p_1 = 0.7$, $a_1 = 100$, $b_1 = 150$ <br> $T_{s2} = 0.5$, $p_2 = 0.3$, $a_2 = 10$, $b_2 = 1$ |

*5.2. Step Tracking Response of PMLM Displacement*

We chose the step signal as a reference to simulate the displacement of the PMLM, and the load is set to $d_{\text{load}} = 0$ N. The step response of NFTSMC, LSMC, PID, and our proposed PreDSMC control is shown in Figure 3a, and the respective error curve is given in Figure 3b. The predefined time for our proposed controller is set to $T_s = 0.5$ s. The convergence of PID, LSMC, and NFTSMC is very slow as compared to our proposed PreDSMC. Owing to the steady state error, the convergence time of the other three controllers is about 1 s. Thus, the proposed control strategy has a good convergence time, which is settled before 0.5 s. The error tracking results in Figure 3b depict that our proposed control strategy has a good convergence as compared to NFTSMC, LSMC, and PID control.

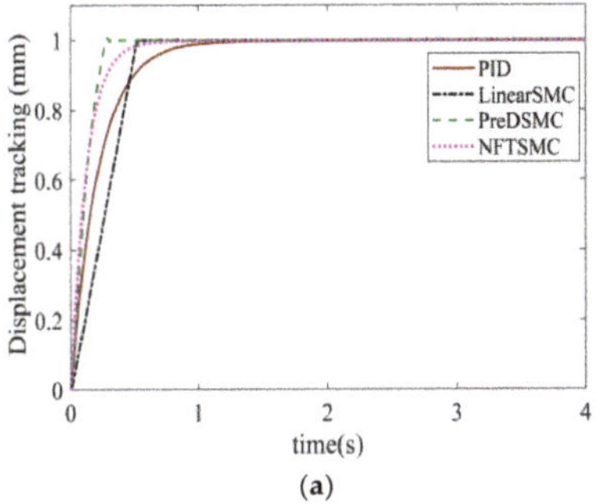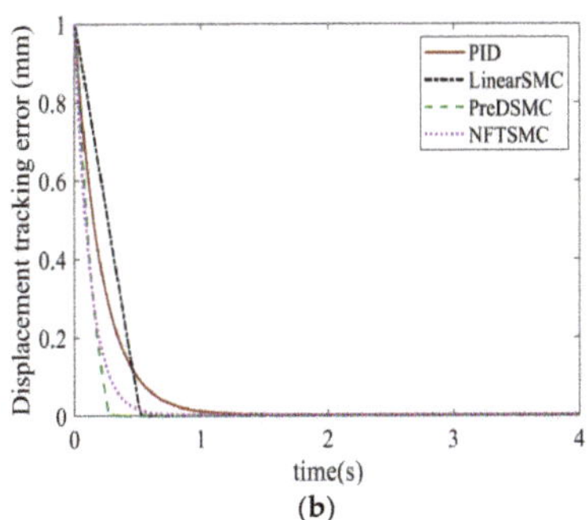

**Figure 3.** (**a**) Comparative step tracking performance of displacement for all control laws. (**b**) The corresponding step tracking error of displacement.

To analyze the effectiveness of our proposed PreDSMC controller, we can compare the results of the derivative of the error signal as well as the control input. Figure 4a shows that the derivative error of our proposed controller is much less than that of LSMC, PID, and NFTSMC. Figure 4b depicts that our proposed controller needs less control effort as compared to NFTSMC, while NFTSMC has a chattering effect in the control input. It can clearly be seen that our proposed control does not have any chattering effect. The overall error convergence time and control response shows that the PreDSMC has priority over the PID, LSMC, and NFTSMC. To this end, we can summarize that the proposed PreDSMC control law has less steady state error, and the control effect is very impressive. It can precisely and accurately track the desired step signal, so we can conclude that the proposed controller is more robust as compared to the other three control algorithms.

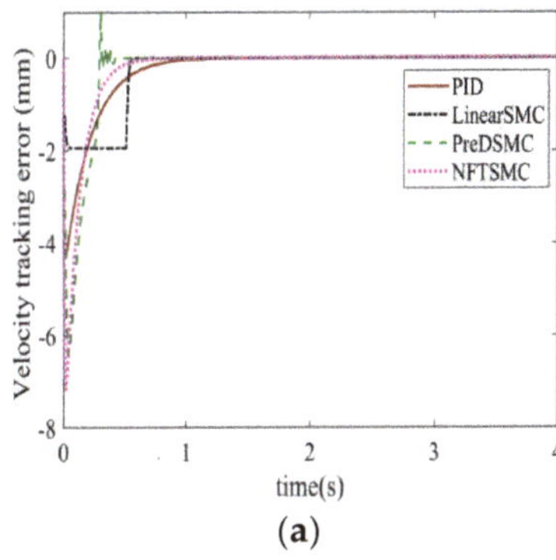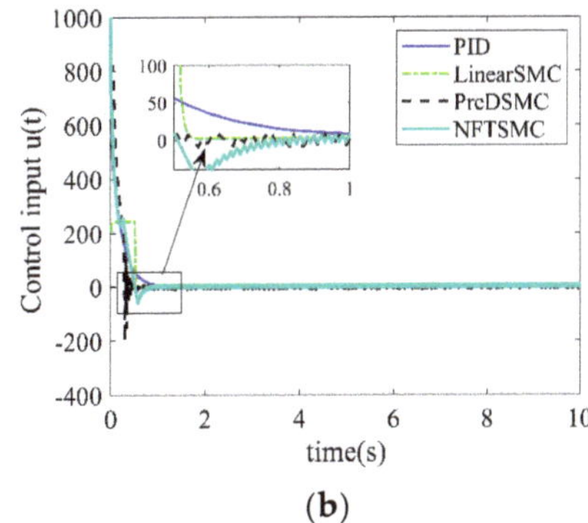

**Figure 4.** (**a**) Comparison of velocity tracking error for the case of step response of displacement for all control laws. (**b**) Velocity tracking error for the corresponding step tracking of displacement.

### 5.3. Sinusoidal Tracking Response of PMLM Displacement

A sinusoidal displacement signal is taken as a reference, i.e., $x_r = 5\sin(t)$, with an amplitude of 5 mm and frequency of $\pi$ rad/s. Like the previous case of step response, $d_{\text{load}}$ is taken as 0 N. In the same manner, the predefined settling time PreDSMC in the sinusoidal case for the proposed controller is set to $T_s = 0.5$ s. It means we want the controller to settle before 0.5 s. It can be clearly observed from Figure 5a that our proposed controller convergence is very fast as compared to PID, LSMC, and NFTSMC. Moreover, the tracking accuracy of PreDSMC in the case of predefined time control is very high. The respective steady state error of the four controllers can be seen in Figure 5b. The steady state error in the case of PID is very large, about $-0.3$ to $0.3$ mm, while for LSMC and NFTSMC it is almost $-0.2$ to $0.2$ mm. Our proposed controller has a steady state error of almost

0.01 mm. It depicts that our proposed PreDSMC is more robust as compared to the simple PID, LSMC, and NFTSMC.

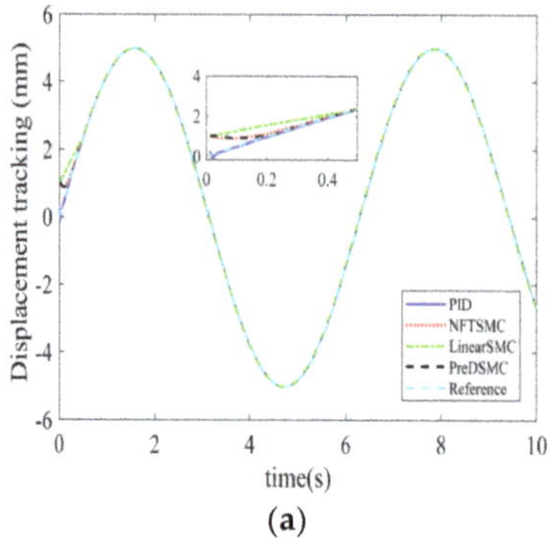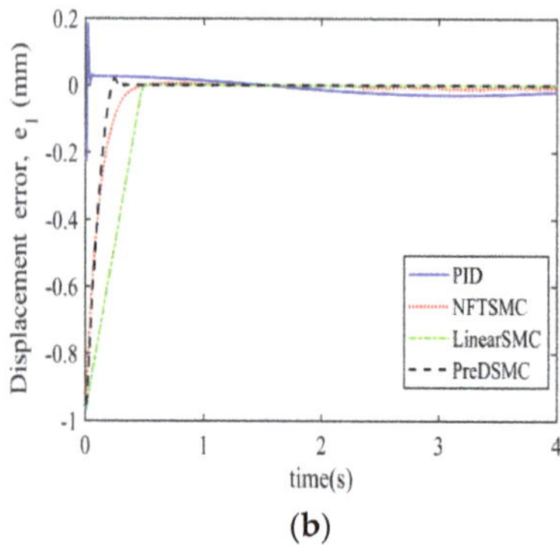

**Figure 5.** (**a**) Sinusoidal tracking performance of displacement for all control laws. (**b**) The comparison of sinusoidal tracking error of displacement.

Figure 6a,b depict the derivative of the tracking error displacement and the control effort of the four control algorithms. It is clear from Figure 6a that the tracking error derivative in the case of PID and NFTSMC is much larger as compared to LSMC and the proposed PreDSMC control strategy. Moreover, we can analyze that the tracking error in the case of the proposed PreDSMC is less as compared to the other control laws. Figure 6b shows the control effort of these four control algorithms. The control effort of our proposed controller is much higher, and it is quite astonishing that PID has almost the same control effort but with a large steady state error.

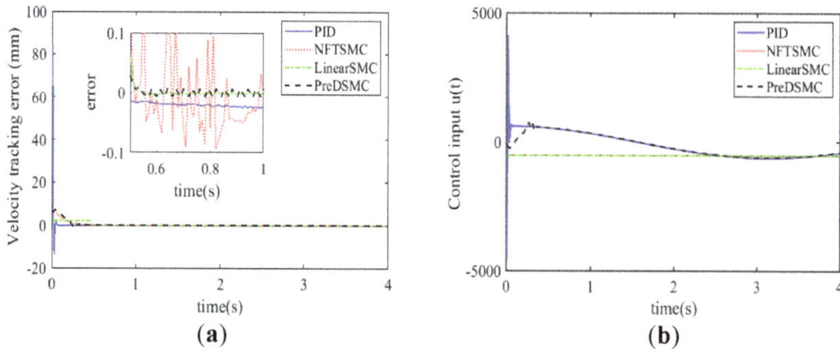

**Figure 6.** (**a**) Comparison of sinusoid velocity tracking error displacement for all controllers. (**b**) Control input for different control algorithms.

To recapitulate, the proposed PreDSMC is robust and has the capability of high precision displacement tracking of PMLM. Furthermore, the proposed model can converge the error within the prescribed time. It is obvious from the simulation results given above that the proposed PreDSMC method is robust and able to obtain the fastest convergence within the predefined time and has smallest tracking errors. It also possesses preferable control features when compared with the PID, LSMC, and NFTSMC control laws.

## 6. Conclusions

In this article, a robust predefined time SMC control algorithm has been designed which has predefined convergent time characteristics. Predefined feature means its convergence time can be chosen in advance, which is especially designed for a PMLM with external compound disturbances, bounded state, and control saturation constraints. The designed predefined time SMC algorithm described in this paper not only ensures the position tracking error convergence of the PMLM within the predefined time, but also verifies that the velocity tracking error is bounded, and the control input meets the predefined boundedness requirements. Furthermore, the motor position can be tracked with high accuracy within the predefined performance function. The desired position is tracked with $10^{-4}$ order-of-magnitude accuracy within the chosen time to achieve a balance between the motor tracking accuracy and the tracking velocity. To recapitulate from the robustness point of view and the scope of application, it can also be used in future work for high-precision trajectory tracking control of nonlinear systems such as piezoelectric positioning, robotic arms, and other second-order nonlinear models. Future study may take into account other nonlinearities as PMLM servo system parameter uncertainty, gear backlash, and coupled frictional nonlinearity with external load variations.

**Author Contributions:** The following contributions reflect the role of individual authors in this study and manuscript. Conceptualization, S.R. and B.L.; methodology, S.R.; software, S.R.; validation, C.-W.Y., S.R. and B.L.; formal analysis, K.S.; investigation, R.Q.; resources, C.-W.Y.; data curation, C.-W.Y.; writing—original draft preparation, S.R.; writing—review and editing, B.L., K.S. and S.A.; visualization, R.Q.; supervision, R.Q. and B.L.; project administration, B.L.; funding acquisition, B.L. All authors have read and agreed to the published version of the manuscript.

**Funding:** This research has been funded by the following agencies: (1) Science Foundation of Shaanxi Province for Outstanding Youth, 2022JC-32, and (2) Joint Key Project of Shaanxi Key R & D Program, 2021GXLH-01-14.

**Data Availability Statement:** Not applicable.

**Conflicts of Interest:** The funders had no role in the design of the study; in the collection, analyses, or interpretation of data; in the writing of the manuscript; or in the decision to publish the results.

## Nomenclature

| Notation | Description |
| --- | --- |
| $x_1(t), x_2(t)$ | Position and speed of PMLM |
| $y(t)$ | Output of system |
| $u(t)$ | Input voltage |
| $R$ | Resistance |
| $m$ | Motor mass |
| $k_f$ | Force constant |
| $k_e$ | Back electromotive force |
| $F_{\text{friction}}(t)$ | Friction force |
| $F_{\text{ripple}}(t)$ | Ripple force |
| $F_L(t)$ | Load |
| $f_v$ | Viscous friction coefficient |
| $x_{2s}$ | Lubrication parameters |
| $f_c$ | Minimum value of Coulomb friction |
| $f_s$ | Static friction |
| $\omega_i, A_i$ | Angular velocity and amplitude of ripple force |
| $M, N, F$ | Intermediate variable of system |
| $y_d(t)$ | Reference trajectory |
| $e_1(t), e_2(t)$ | Position error and speed error |
| $S$ | Sliding mode surface |
| $T_{s1}, T_{s2}$ | Predefined time |
| $q$ | Parameter of sliding mode surface |
| $t_s$ | Convergence time |
| $V(t)$ | Lyapunov function |
| $p, a, b$ | Parameters of Lyapunov function |

# References


1. Kim, J.; Choi, S.; Cho, K.; Nam, K. Position estimation using linear hall sensors for permanent magnet linear motor systems. *IEEE Trans. Ind. Electron.* **2016**, *63*, 7644–7652. [CrossRef]
2. Liu, X.; Wu, Q.; Zhen, S.; Zhao, H.; Li, C.; Chen, Y.-H. Robust constraint-following control for permanent magnet linear motor with optimal design: A fuzzy approach. *Inf. Sci.* **2022**, *600*, 362–376. [CrossRef]
3. Chen, S.-L.; Tan, K.K.; Huang, S.; Teo, C.S. Modeling and compensation of ripples and friction in permanent-magnet linear motor using a hysteretic relay. *IEEE/ASME Trans. Mechatron.* **2009**, *15*, 586–594. [CrossRef]
4. Tan, K.; Huang, S.; Lee, T. Robust adaptive numerical compensation for friction and force ripple in permanent-magnet linear motors. *IEEE Trans. Magn.* **2002**, *38*, 221–228. [CrossRef]
5. Lu, Q.; Zhang, X.; Chen, Y.; Huang, X.; Ye, Y.; Zhu, Z. Modeling and investigation of thermal characteristics of a water-cooled permanent-magnet linear motor. *IEEE Trans. Ind. Appl.* **2014**, *51*, 2086–2096. [CrossRef]
6. Cao, R.; Cheng, M.; Zhang, B. Speed control of complementary and modular linear flux-switching permanent-magnet motor. *IEEE Trans. Ind. Electron.* **2015**, *62*, 4056–4064. [CrossRef]
7. Utkin, V.I. *Sliding Modes in Control and Optimization*; Springer Science & Business Media: Berlin/Heidelberg, Germany, 2013.
8. Edwards, C.; Colet, E.F.; Fridman, L.; Colet, E.F.; Fridman, L.M. *Advances in Variable Structure and Sliding Mode Control*; Springer: Berlin/Heidelberg, Germany, 2006; Volume 334.
9. Mou, F.; Wu, D.; Dong, Y. Disturbance rejection sliding mode control for robots and learning design. *Intell. Serv. Robot.* **2021**, *14*, 251–269. [CrossRef]
10. Dong, H.; Yang, X.; Basin, M.V. Practical Tracking of Permanent Magnet Linear Motor Via Logarithmic Sliding Mode Control. *IEEE/ASME Trans. Mechatron.* **2022**, *27*, 4112–4121. [CrossRef]
11. Cao, W.; Lan, Q.; Ma, T. Second-Order Sliding Mode Controller Design for Motion Control of PMLM System. In *Advances in Guidance, Navigation and Control*; Springer: Singapore, 2022; pp. 2097–2107.
12. Sepestanaki, M.A.; Barhaghtalab, M.H.; Mobayen, S.; Jalilvand, A.; Fekih, A.; Skruch, P. Chattering-Free Terminal Sliding Mode Control Based on Adaptive Barrier Function for Chaotic Systems With Unknown Uncertainties. *IEEE Access* **2022**, *10*, 103469–103484. [CrossRef]
13. Wang, H.; Hu, Y.; Ye, M.; Zhang, J.; Cao, Z.; Zheng, J.; Man, Z. Real-Time Control Systems with Applications in Mechatronics. In *Handbook of Real-Time Computing*; Springer: Singapore, 2022; pp. 605–640.
14. Hou, M.; Wang, Y.; Han, Y. Data-driven discrete terminal sliding mode decoupling control method with prescribed performance. *J. Frankl. Inst.* **2021**, *358*, 6612–6633. [CrossRef]
15. Weng, Y.; Gao, X. Data-driven robust output tracking control for gas collector pressure system of coke ovens. *IEEE Trans. Ind. Electron.* **2016**, *64*, 4187–4198. [CrossRef]
16. Chen, W.-H.; Yang, J.; Guo, L.; Li, S. Disturbance-observer-based control and related methods—An overview. *IEEE Trans. Ind. Electron.* **2015**, *63*, 1083–1095. [CrossRef]
17. Liu, S.; Liu, Y.; Wang, N. Nonlinear disturbance observer-based backstepping finite-time sliding mode tracking control of underwater vehicles with system uncertainties and external disturbances. *Nonlinear Dyn.* **2017**, *88*, 465–476. [CrossRef]
18. Sun, H.; Li, S.; Sun, C. Finite time integral sliding mode control of hypersonic vehicles. *Nonlinear Dyn.* **2013**, *73*, 229–244. [CrossRef]
19. Yang, Z.; Ding, Q.; Sun, X.; Lu, C.; Zhu, H. Speed sensorless control of a bearingless induction motor based on sliding mode observer and phase-locked loop. *ISA Trans.* **2022**, *123*, 346–356. [CrossRef]
20. Zhou, M.; Cheng, S.; Xu, W.; Feng, Y.; Su, H. Control and Observation of Induction Motors Based on Full-Order Terminal Sliding-Mode Technique. In *Advances in Control Techniques for Smart Grid Applications*; Springer: Singapore, 2022; pp. 327–363.
21. Liao, K.; Xu, W.; Bai, L.; Gong, Y.; Ismail, M.M.; Boldea, I. Improved Position Sensorless Piston Stroke Control Method for Linear Oscillatory Machine via an Hybrid Terminal Sliding-Mode Observer. *IEEE Trans. Power Electron.* **2022**, *37*, 14186–14197. [CrossRef]
22. Du, H.; Qian, C.; Yang, S.; Li, S. Recursive design of finite-time convergent observers for a class of time-varying nonlinear systems. *Automatica* **2013**, *49*, 601–609. [CrossRef]
23. Chen, L.; Liu, Z.; Dang, Q.; Zhao, W.; Wang, G. Robust trajectory tracking control for a quadrotor using recursive sliding mode control and nonlinear extended state observer. *Aerosp. Sci. Technol.* **2022**, *128*, 107749. [CrossRef]
24. Zhang, L.; Yang, J. Continuous nonsingular terminal sliding mode control for nonlinear systems subject to mismatched terms. *Asian J. Control* **2022**, *24*, 885–894. [CrossRef]
25. Chen, W.; Du, H.; Chen, C.-C. Stability and Robustness Analysis of Finite-Time Consensus Algorithm for Second-Order Multiagent Systems Under Sampled-Data Control. *IEEE Trans. Syst. Man Cybern. Syst.* **2022**, 1–8. [CrossRef]
26. Du, H.; Jiang, C.; Wen, G.; Zhu, W.; Cheng, Y. Current sharing control for parallel DC–DC buck converters based on finite-time control technique. *IEEE Trans. Ind. Inform.* **2018**, *15*, 2186–2198. [CrossRef]
27. Du, H.; Zhai, J.; Chen, M.Z.; Zhu, W. Robustness analysis of a continuous higher order finite-time control system under sampled-data control. *IEEE Trans. Autom. Control* **2018**, *64*, 2488–2494. [CrossRef]
28. Jin, M.; Lee, J.; Chang, P.H.; Choi, C. Practical nonsingular terminal sliding-mode control of robot manipulators for high-accuracy tracking control. *IEEE Trans. Ind. Electron.* **2009**, *56*, 3593–3601.
29. Mobayen, S.; Alattas, K.A.; Assawinchaichote, W. Adaptive continuous barrier function terminal sliding mode control technique for disturbed robotic manipulator. *IEEE Trans. Circuits Syst. I Regul. Pap.* **2021**, *68*, 4403–4412. [CrossRef]

30. Sai, H.; Xu, Z.; Xia, C.; Sun, X. Approximate continuous fixed-time terminal sliding mode control with prescribed performance for uncertain robotic manipulators. *Nonlinear Dyn.* **2022**, *110*, 431–448. [CrossRef]

31. Kaufmann, E.; Mboumi, E. Positive solutions of a boundary value problem for a nonlinear fractional differential equation. *Electron. J. Qual. Theory Differ. Equ.* **2008**, *2008*, 1–11. [CrossRef]

32. Khan, A.; Nigar, U. Sliding mode disturbance observer control based on adaptive hybrid projective compound combination synchronization in fractional-order chaotic systems. *J. Control Autom. Electr. Syst.* **2020**, *31*, 885–899. [CrossRef]

33. Liu, S.; Zhang, L.; Niu, B.; Zhao, X.; Ahmad, A.M. Adaptive neural finite-time hierarchical sliding mode control of uncertain under-actuated switched nonlinear systems with backlash-like hysteresis. *Inf. Sci.* **2022**, *599*, 147–169. [CrossRef]

34. Bertrand, N.; Sabatier, J.; Briat, O.; Vinassa, J.-M. Embedded fractional nonlinear supercapacitor model and its parametric estimation method. *IEEE Trans. Ind. Electron.* **2010**, *57*, 3991–4000.

35. Martynyuk, V.; Ortigueira, M. Fractional model of an electrochemical capacitor. *Signal Process.* **2015**, *107*, 355–360. [CrossRef]

36. Islam, M.M.; Siffat, S.A.; Ahmad, I.; Liaquat, M.; Khan, S.A. Adaptive nonlinear control of unified model of fuel cell, battery, ultracapacitor and induction motor based hybrid electric vehicles. *IEEE Access* **2021**, *9*, 57486–57509. [CrossRef]

37. Thounthong, P.; Tricoli, P.; Davat, B. Performance investigation of linear and nonlinear controls for a fuel cell/supercapacitor hybrid power plant. *Int. J. Electr. Power Energy Syst.* **2014**, *54*, 454–464. [CrossRef]

38. Penas, R.; Balmes, E.; Gaudin, A. A unified non-linear system model view of hyperelasticity, viscoelasticity and hysteresis exhibited by rubber. *Mech. Syst. Signal Process.* **2022**, *170*, 108793. [CrossRef]

39. Şengül, Y. Nonlinear viscoelasticity of strain rate type: An overview. *Proc. R. Soc. A* **2021**, *477*, 20200715. [CrossRef]

40. Bezerra, J.I.; Molter, A.; Rafikov, M.; Frighetto, D.F. Biological control of the chaotic sugarcane borer-parasitoid agroecosystem. *Ecol. Model.* **2021**, *450*, 109564. [CrossRef]

41. Sabir, Z.; Baleanu, D.; Ali, M.R.; Sadat, R. A novel computing stochastic algorithm to solve the nonlinear singular periodic boundary value problems. *Int. J. Comput. Math.* **2022**, *99*, 2091–2104. [CrossRef]

42. Aslmostafa, E.; Mirzaei, M.J.; Asadollahi, M.; Badamchizadeh, M.A. Stabilization problem for a class of nonlinear MIMO systems based on prescribed-time sliding mode control. *Arab. J. Sci. Eng.* **2022**, *47*, 15083–15094. [CrossRef]

43. Liang, C.-D.; Ge, M.-F.; Liu, Z.-W.; Zhan, X.-S.; Park, J.H. Predefined-time stabilization of TS fuzzy systems: A novel integral sliding mode based approach. *IEEE Trans. Fuzzy Syst.* **2022**, *30*, 4423–4433. [CrossRef]

44. Liang, C.-D.; Ge, M.-F.; Liu, Z.-W.; Ling, G.; Zhao, X.-W. A novel sliding surface design for predefined-time stabilization of Euler–Lagrange systems. *Nonlinear Dyn.* **2021**, *106*, 445–458. [CrossRef]

45. Meng, X.; Gao, C.; Jiang, B.; Wu, Z. Finite-time Synchronization of Variable-order Fractional Uncertain Coupled Systems via Adaptive Sliding Mode Control. *Int. J. Control Autom. Syst.* **2022**, *20*, 1535–1543. [CrossRef]

46. Aslmostafa, E.; Mirzaei, M.J.; Asadollahi, M.; Badamchizadeh, M.A. Synchronization problem for a class of multi-input multi-output systems with terminal sliding mode control based on finite-time disturbance observer: Application to chameleon chaotic system. *Chaos Solitons Fractals* **2021**, *150*, 111191. [CrossRef]

47. Mobayen, S.; Tchier, F. Nonsingular fast terminal sliding-mode stabilizer for a class of uncertain nonlinear systems based on disturbance observer. *Sci. Iranica. Trans. D Comput. Sci. Eng. Electr.* **2017**, *24*, 1410–1418. [CrossRef]

48. Roy, P.; Roy, B.K. Sliding mode control versus fractional-order sliding mode control: Applied to a magnetic levitation system. *J. Control Autom. Electr. Syst.* **2020**, *31*, 597–606. [CrossRef]

49. Ni, J.; Liu, L.; Liu, C.; Hu, X. Fractional order fixed-time nonsingular terminal sliding mode synchronization and control of fractional order chaotic systems. *Nonlinear Dyn.* **2017**, *89*, 2065–2083. [CrossRef]

50. Tian, Y.; Cai, Y.; Deng, Y. A fast nonsingular terminal sliding mode control method for nonlinear systems with fixed-time stability guarantees. *IEEE Access* **2020**, *8*, 60444–60454. [CrossRef]

51. Xie, S.; Chen, Q.; He, X. Predefined-time approximation-free attitude constraint control of rigid spacecraft. *IEEE Trans. Aerosp. Electron. Syst.* **2022**, 1–11. [CrossRef]

52. Pan, Y.; Ji, W.; Liang, H. Adaptive Predefined-time Control for Lü Chaotic Systems via Backstepping Approach. *IEEE Trans. Circuits Syst. II Express Briefs* **2022**, *69*, 5064–5068. [CrossRef]

53. Mazhar, N.; Malik, F.M.; Raza, A.; Khan, R. Predefined-time control of nonlinear systems: A sigmoid function based sliding manifold design approach. *Alex. Eng. J.* **2022**, *61*, 6831–6841. [CrossRef]

54. Abdel Hameed, M.; Hassaballah, M.; Hosney, M.E.; Alqahtani, A. An AI-Enabled Internet of Things Based Autism Care System for Improving Cognitive Ability of Children with Autism Spectrum Disorders. *Comput. Intell. Neurosci.* **2022**, *2022*, 2247675. [CrossRef]

55. Xiong, X.; Pal, A.K.; Liu, Z.; Kamal, S.; Huang, R.; Lou, Y. Discrete-time adaptive super-twisting observer with predefined arbitrary convergence time. *IEEE Trans. Circuits Syst. II Express Briefs* **2020**, *68*, 2057–2061. [CrossRef]

56. Gómez-Gutiérrez, D. On the design of nonautonomous fixed-time controllers with a predefined upper bound of the settling time. *Int. J. Robust Nonlinear Control* **2020**, *30*, 3871–3885. [CrossRef]

57. Pal, A.K.; Kamal, S.; Nagar, S.K.; Bandyopadhyay, B.; Fridman, L. Design of controllers with arbitrary convergence time. *Automatica* **2020**, *112*, 108710. [CrossRef]

58. Sánchez-Torres, J.D.; Gómez-Gutiérrez, D.; López, E.; Loukianov, A.G. A class of predefined-time stable dynamical systems. *IMA J. Math. Control Inf.* **2018**, *35*, i1–i29. [CrossRef]

59. Sánchez-Torres, J.D.; Muñoz-Vázquez, A.J.; Defoort, M.; Jiménez-Rodríguez, E.; Loukianov, A.G. A class of predefined-time controllers for uncertain second-order systems. *Eur. J. Control* **2020**, *53*, 52–58. [CrossRef]
60. Moulay, E.; Léchappé, V.; Bernuau, E.; Defoort, M.; Plestan, F. Fixed-time sliding mode control with mismatched disturbances. *Automatica* **2022**, *136*, 110009. [CrossRef]
61. Nian, W.; Chen, H.; Ding, D. A New Non-Singular Terminal Sliding Mode Control and Its Application to Chaos Suppression in Interconnected Power System. *Int. J. Adv. Netw. Monit. Control.* **2018**, *3*, 5–12. [CrossRef]
62. Wang, L.; Liu, X.; Cao, J.; Hu, X. Fixed-time containment control for nonlinear multi-agent systems with external disturbances. *IEEE Trans. Circuits Syst. II Express Briefs* **2021**, *69*, 459–463. [CrossRef]
63. Jiménez-Rodríguez, E.; Muñoz-Vázquez, A.J.; Sánchez-Torres, J.D.; Defoort, M.; Loukianov, A.G. A Lyapunov-like characterization of predefined-time stability. *IEEE Trans. Autom. Control* **2020**, *65*, 4922–4927. [CrossRef]
64. Sánchez-Torres, J.D.; Defoort, M.; Muñoz-Vázquez, A.J. Predefined-time stabilisation of a class of nonholonomic systems. *Int. J. Control* **2020**, *93*, 2941–2948. [CrossRef]
65. Liu, B.; Hou, M.; Wu, C.; Wang, W.; Wu, Z.; Huang, B. Predefined-time backstepping control for a nonlinear strict-feedback system. *Int. J. Robust Nonlinear Control* **2021**, *31*, 3354–3372. [CrossRef]
66. Muñoz-Vázquez, A.J.; Fernández-Anaya, G.; Sánchez-Torres, J.D.; Meléndez-Vázquez, F. Predefined-time control of distributed-order systems. *Nonlinear Dyn.* **2021**, *103*, 2689–2700. [CrossRef]
67. Feng, Y.; Yu, X.; Man, Z. Non-singular terminal sliding mode control of rigid manipulators. *Automatica* **2002**, *38*, 2159–2167. [CrossRef]
68. Jie, W.; Yudong, Z.; Yulong, B.; Kim, H.H.; Lee, M.C. Trajectory tracking control using fractional-order terminal sliding mode control with sliding perturbation observer for a 7-DOF robot manipulator. *IEEE/ASME Trans. Mechatron.* **2020**, *25*, 1886–1893. [CrossRef]
69. Cruz-Ortiz, D.; Chairez, I.; Poznyak, A. Non-singular terminal sliding-mode control for a manipulator robot using a barrier Lyapunov function. *ISA Trans.* **2022**, *121*, 268–283. [CrossRef]
70. Sun, G.; Wu, L.; Kuang, Z.; Ma, Z.; Liu, J. Practical tracking control of linear motor via fractional-order sliding mode. *Automatica* **2018**, *94*, 221–235. [CrossRef]
71. Sun, G.; Ma, Z.; Yu, J. Discrete-time fractional order terminal sliding mode tracking control for linear motor. *IEEE Trans. Ind. Electron.* **2017**, *65*, 3386–3394. [CrossRef]
72. Yang, C.; Che, Z.; Zhou, L. Composite feedforward compensation for force ripple in permanent magnet linear synchronous motors. *J. Shanghai Jiaotong Univ.* **2019**, *24*, 782–788. [CrossRef]
73. Liu, W.; Chen, S.; Huang, H. Adaptive nonsingular fast terminal sliding mode control for permanent magnet synchronous motor based on disturbance observer. *IEEE Access* **2019**, *7*, 153791–153798. [CrossRef]
74. Shengquan, L.; Juan, L.; Yongwei, T.; Yanqiu, S.; Wei, C. Model-based model predictive control for a direct-driven permanent magnet synchronous generator with internal and external disturbances. *Trans. Inst. Meas. Control* **2020**, *42*, 586–597. [CrossRef]
75. Ting, C.S.; Chang, Y.N.; Shi, B.W.; Lieu, J.F. Adaptive backstepping control for permanent magnet linear synchronous motor servo drive. *IET Electr. Power Appl.* **2015**, *9*, 265–279. [CrossRef]
76. Hui, Y.; Chi, R.; Huang, B.; Hou, Z. Extended state observer-based data-driven iterative learning control for permanent magnet linear motor with initial shifts and disturbances. *IEEE Trans. Syst. Man Cybern. Syst.* **2019**, *51*, 1881–1891. [CrossRef]
77. Zhu, J.; Pan, Y.; Huang, T.; Luan, C.; Li, K. Research on Controlled Characteristic of the Novel Electromagnetic Recuperator. In Proceedings of the 2019 IEEE 3rd Information Technology, Networking, Electronic and Automation Control Conference (ITNEC), Chengdu, China, 15–17 March 2019; 2019; pp. 1980–1984.
78. Zhao, Y.; Li, Y.; Lu, Q. An Accurate No-Load Analytical Model of Flat Linear Permanent Magnet Synchronous Machine Accounting for End Effects. *IEEE Trans. Magn.* **2022**, *59*, 8100111. [CrossRef]
79. Du, H.; Chen, X.; Wen, G.; Yu, X.; Lü, J. Discrete-time fast terminal sliding mode control for permanent magnet linear motor. *IEEE Trans. Ind. Electron.* **2018**, *65*, 9916–9927. [CrossRef]
80. Fu, D.; Zhao, X.; Zhu, J. A novel robust super-twisting nonsingular terminal sliding mode controller for permanent magnet linear synchronous motors. *IEEE Trans. Power Electron.* **2021**, *37*, 2936–2945. [CrossRef]
81. Wang, L.; Zhao, J.; Zheng, Z. Robust Speed Tracking Control of Permanent Magnet Synchronous Linear Motor Based on a Discrete-Time Sliding Mode Load Thrust Observer. *IEEE Trans. Ind. Appl.* **2022**, *58*, 4758–4767. [CrossRef]
82. Li, J.; Du, H.; Cheng, Y.; Wen, G.; Chen, X.; Jiang, C. Position tracking control for permanent magnet linear motor via fast nonsingular terminal sliding mode control. *Nonlinear Dyn.* **2019**, *97*, 2595–2605. [CrossRef]
83. Gao, W.; Chen, X.; Du, H.; Bai, S. Position tracking control for permanent magnet linear motor via continuous-time fast terminal sliding mode control. *J. Control. Sci. Eng.* **2018**, *2018*, 3813624. [CrossRef]
84. Zhang, J.; Wang, H.; Cao, Z.; Zheng, J.; Yu, M.; Yazdani, A.; Shahnia, F. Fast nonsingular terminal sliding mode control for permanent-magnet linear motor via ELM. *Neural Comput. Appl.* **2020**, *32*, 14447–14457. [CrossRef]

 *electronics*

*Article*

# Hybrid Backstepping-Super Twisting Algorithm for Robust Speed Control of a Three-Phase Induction Motor

Sadia Ali [1,*], Alvaro Prado [2] and Mahmood Pervaiz [3]

1   Department of Electrical and Computer Engineering, International Islamic University, Islamabad 44000, Pakistan
2   Departamento de Ingeniería de Sistemas y Computación (DISC), Universidad Católica del Norte (UCN), Antofagasta 1270709, Chile
3   Departments of Electrical and Computer Engineering, COMSATS University, Islamabad 45550, Pakistan
*   Correspondence: sadia.ali@iiu.edu.pk

**Abstract:** This paper proposes a Hybrid Backstepping Super Twisting Algorithm for robust speed control of a three-phase Induction Motor in the presence of load torque uncertainties. First of all, a three-phase squirrel cage Induction Motor is modeled in MATLAB/Simulink. This is then followed by the design of different non-linear controllers, such as sliding mode control (SMC), super twisting SMC, and backstepping control. Furthermore, a novel controller is designed by the synergy of two methods, such as backstepping and super twisting SMC (Back-STC), to obtain the benefits of both techniques and, thereby, improve robustness. The sigmoid function is used with an exact differentiator to minimize the high-speed discontinuities present in the input channel. The efficacy of this novel design and its performance were evidenced in comparison with other methods, carried out by simulations in MATLAB/Simulink. Regression parameters, such as ISE (Integral Square error), IAE (Integral Absolute error) and ITAE (Integral Time Absolute error), were calculated in three different modes of operation: SSM (Start-Stop Mode), NOM (Normal Operation Mode) and DRM (Disturbance Rejection Mode). In the end, the numerical values of the regression parameters were quantitatively analyzed to draw conclusions regarding the tracking performance and robustness of the implemented non-linear control techniques.

**Keywords:** backstepping control; super twisting control; sliding mode control; 3-phase induction motor; uncertainties

**Citation:** Ali, S.; Prado, A.; Pervaiz, M. Hybrid Backstepping-Super Twisting Algorithm for Robust Speed Control of a Three-Phase Induction Motor. *Electronics* **2023**, *12*, 681. https://doi.org/10.3390/ electronics12030681

Academic Editor: Akshya Swain

Received: 28 December 2022
Revised: 19 January 2023
Accepted: 24 January 2023
Published: 29 January 2023

## 1. Introduction

Early electricity generation through a three-phase induction machine was deemed a revolutionary advance in the power industry. An induction motor is the most widely utilized electrical machine in the energy industry. Nearly 80% of the energy utilized in industries is produced by three-phase induction motors. A three-phase induction motor has a wide number of applications in areas such as Electric Vehicles, energy saving, and monitoring systems [1–3]. This article centered on the nonlinear speed control of a three-phase induction motor. Even though the linear control techniques are simpler and computationally inexpensive, these control techniques cannot handle disturbance rejection and model uncertainties [4]. Moreover, due to the discontinuous nature of the nonlinearities, the linear approximation becomes an issue. These "hard nonlinearities" comprise of saturation, Coulomb friction, dead-zones hysteresis, and backlash. They are frequently found in control system engineering and their properties cannot be derived from linear procedures. In [5], the performance of a surface mounted permanent magnet synchronous motor (SPMSM) was analyzed by comparing the results of a conventional Proportional Integral (PI) controller with the proposed tracking differentiator–proportional integral and derivative (TD–PID) controller. From the published results, it is evident that the conventional linear PI controller falls short in terms of peak overshoot, chattering and settling time.

Many nonlinear control techniques have been utilized to achieve optimum speed control of this multivariable machinery. Researchers have introduced different nonlinear control techniques, such as sliding mode control [6], input-output linearization control [7], direct torque control [8], backstepping control [9] and so on to achieve high-performance control for induction motors [10]. The novel idea presented in this paper has significant weight as it combines two different non-linear controllers, namely, backstepping and super twisting algorithm. This combination results in a controller that has the advantages of both techniques for robust and efficient speed tracking.

The rotor's speed can be controlled by the variable supply provided to the stator. Nowadays, most electricity generation is done using a three-phase induction machine. Efficient speed control of such highly nonlinear dynamic machinery is a challenging task. Load uncertainties and additional nonlinear disturbances can further complicate the task of designing a controller [11].

Backstepping control calls for the division of entire systems into subsystems making it easier to derive and compute the desired control input. It is a recursive process extending outwards to consecutive subsystems until the final optimal control is reached. In [12], integral and classical backstepping approaches based on IFOC (Indirect Field Orientation Control) were applied for robust speed control of a squirrel cage three-phase induction motor. The integral approach provided global system stability and increased robustness in the presence of model uncertainties. However, classical Backstepping approaches result in fluctuations in the armature current and a slight steady state error between the actual and desired rotor speed. A simple Backstepping control might not be able to reject the disturbances effectively. Different robust backstepping techniques have been implemented in literature. In [13], a robust adaptive backstepping technique was applied to mobile robotic manipulators. The simulations demonstrated better tracking performance and robustness in comparison with a conventional PID controller. In [14], a sliding mode observer for estimation of flux components and actual speed was utilized in conjunction with a Backstepping controller to improve robustness. The motor was run in speed inversion mode as well. The results depicted no significant changes in the speed, currents, or voltages of the induction motor.

The sliding mode control (SMC) is a robust design technique that is useful for compensating model uncertainties. It provides very effective tracking control. An SMC combined with input–output feedback linearization for two quasi-induction motor drives was presented in [15]. The motors were connected in two configurations: series and parallel. The results in speed start-up and speed reversal modes depicted a small tracking error but the chattering effect was reduced by replacing the signum function with the saturation function. A robust variable step perturb-and-observe sliding mode controller was designed in [16] for a permanent magnet synchronous generator. The results demonstrated an increase in efficiency and enhanced settling time as opposed to a simple variable step perturb-and-observe controller. In [17], SMC in conjunction with a type-2 neuro-fuzzy controller was applied to an induction motor. The speed response depicted satisfactory behavior with small peaks occurring at fast transitions. The results were depicted in terms of the amount of overshoot and learning features. SMC is a very robust technique; however, due to the discontinuous nature of the control input, the system experiences the chattering effect. In [18], a classical SMC was compared to a fuzzy SMC approach for robust speed control of a doubly-fed induction motor. The speed tracking results showed a profound chattering in classical SMC in comparison with a fuzzy SMC.

The chattering effect can usually be avoided by using higher-order SMC approaches [19]. The super twisting algorithm is a sub-branch of the higher order sliding mode (HOSM) control. In [20], a linearized block control, in conjunction with a super twisting algorithm, was applied to a squirrel cage induction motor. This technique provided reduced chattering, along with disturbance rejection, in the presence of variable load torque. A computerized tuning method for the parameters of the Super Twisting controller technique could further minimize chattering [21] and minimize core losses in a three-phase induction motor [22]. An adaptive Backstepping super twisting SMC was designed and compared with different techniques in [23]. The proposed design showed superior cyclic path tracking and disturbance rejection qualities.

The SMC techniques have been modified and extended, and their effects have been further enhanced over the years to improve the performance of many nonlinear systems. All non-linear controllers have their merits and demerits. For instance, SMC offers satisfactory tracking performance and disturbance rejection; however, it lacks optimum performance due to the chattering effect [24]. The Backstepping controller does not have to deal with the chattering effect; however, the disturbance rejection of load torque is not as effective as in the case of SMC [25]. The super twisting control reduces the chattering effect in the actual speed of the rotor which is dependent on the exact parameter tuning of the controller [21].

Even though multiple non-linear controllers have been designed for the speed control of a three-phase Induction motor, very few have addressed the high-speed discontinuities that emerge, due to the differentiation of approximated step changes in input. Moreover, an extensive performance comparison is due. Through an extensive performance comparison, different non-linear control techniques could be analyzed and the most appropriate technique could be selected for a particular application.

Owing to the above mentioned facts, this article has the following contributions:

- A Novel Backstepping super twisting SMC with exact differentiation and signum approximation (Back-STC-EA) was designed for the robust speed control of a three-phase Induction Motor.
- This controller not only reduces the chattering effect, as opposed to a basic SMC, but also improves the disturbance rejection capability, in comparison with the classical Backstepping controller.
- The exact differentiation and signum approximation reduces the overall effect of high-speed discontinuities present in the desired speed response.
- An extensive performance comparison was carried out between the conventional Backstepping controller, SMC, Back–SMC, Back–STC and the novel Back–STC–EA controller.
- A quantitative and graphical analysis was performed in terms of regression parameters (ISE, IAE, ITAE) and simulation results. This analysis is performed under three different modes of operation: SSM, NOM and DRM.

The rest of this article is organized as follows. The problem is formulated in Section 2. The mathematical model and the uncompensated simulation results for the squirrel cage three-phase induction motor are presented in Section 3. The design methodologies of the nonlinear control techniques, such as Backstepping control, SMC, Back–SMC, Back–STC and Back–STC–EA are proposed in Section 4. The simulation results, numerical comparison and evaluation are presented in Section 5. Section 6 presents conclusions and future work. Appendices are present at the end of the article.

## 2. Problem Formulation

The proposed problem is stated as follows:

### 2.1. System Description

The non-linear system considered in this paper can be represented by the following equations:

$$\dot{\varphi} = f(\varphi(t), \omega(t), c(t)) \tag{1}$$

$$\dot{\omega}(t) = g(\varphi(t), \eta(t)) \tag{2}$$

where $\varphi(t)$ are the state variables, $\omega(t)$ is the output, $c(t)$ is the control input and $\eta(t)$ is the disturbance in load. However, the following conditions apply:

- In Start Stop Mode (SSM) and Normal Operation Mode (NOM) disturbance in load $\eta(t)$ is not considered. (i.e., $\eta(t) = 0$).
- In Disturbance rejection mode (DRM), a disturbance in load ($\eta(t)$) is introduced.

The following assumption is made:

- The functions $f(\cdot)$ and $g(\cdot)$ are continuously differentiable, or are made continuously differentiable, by using exact differentiation and signum approximation.

The illustration diagram for the formulated problem is provided in Figure 1.

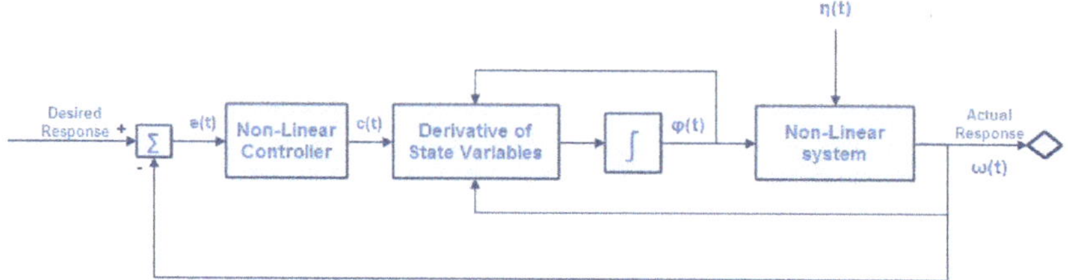


**Figure 1.** Illustration diagram for Problem Formulation.

*2.2. Problem Statement*

Design a control input $c(t)$ using different non-linear control techniques for the system (1) and (2), such that a robust and stable output is achieved in the presence of uncertainties. Afterwards, numerically compare and analyze the results of the non-linear control techniques under consideration.

### 3. Mathematical Model

A mathematical model, when consciously selected, can reduce the amount of work and produce more accurate results. A mathematical model of a three-phase nonlinear induction motor was selected and implemented in MATLAB/Simulink [26–28]. For simplification of the model, park transformation was used. The park transformation rotated the *abc* reference frame to a *dq* (direct-quadrature) reference frame. The park transformation is utilized very often in MATLAB/Simulink with three-phase induction motors, due to the perfect alignment of the rotor flux with the d-axis, which implies that the q-axis component of the rotor flux can be taken as zero. These reference voltages are further utilized to compute the flux linkages, which, in turn, compute the rotor and stator current. These currents are then used to derive the final equations for speed and electromagnetic torque [27]. The overall block diagram is presented in Figure 2. The final equations for the flux linkage variables are given as follows (Symbols are given in Abbreviations):

$$\frac{dF_{sq}}{dt} = \omega_b\left[V_{sq} - \frac{\omega_e F_{sd}}{\omega_b} + \frac{R_s}{X_{ls}}\left\{\frac{F_{rq}X_m}{X_{lr}} + F_{sq}\left(\frac{X_m}{X_{ls}} - 1\right)\right\}\right] \tag{3}$$

$$\frac{dF_{sd}}{dt} = \omega_b\left[V_{sq} + \frac{\omega_e F_{sq}}{\omega_b} + \frac{R_s}{X_{ls}}\left\{\frac{F_{rd}X_m}{X_{lr}} + F_{sd}\left(\frac{X_m}{X_{ls}} - 1\right)\right\}\right] \tag{4}$$

$$\frac{dF_{rq}}{dt} = \omega_b\left[\frac{\omega_e - \omega_r}{\omega_b}F_{rd} + \frac{R_r}{X_{lr}}\left\{\frac{F_{sq}X_m}{X_{ls}} + F_{rq}\left(\frac{X_m}{X_{lr}} - 1\right)\right\}\right] \tag{5}$$

$$\frac{dF_{rd}}{dt} = \omega_b\left[\frac{\omega_e - \omega_r}{\omega_b}F_{rq} + \frac{R_r}{X_{lr}}\left\{\frac{F_{sd}X_m}{X_{ls}} + F_{rd}\left(\frac{X_m}{X_{lr}} - 1\right)\right\}\right] \tag{6}$$

The final equations of the speed and electromagnetic torque are as follows:

$$\frac{d\omega_r}{dt} = k_1[T_e - T_L] \tag{7}$$

$$T_e = k_2\left[F_{sd}i_{sq} - F_{sq}i_{sd}\right] \tag{8}$$

where $k_1 = \frac{P}{2j}$, and $k_2 = \frac{3p}{4\omega_b}$. These equations may be further utilized to achieve the optimum control of the induction motor. The field distribution variables include the supply voltage variables, and the stator and rotor currents, which can be used as the control inputs in the design of nonlinear control systems. In (7), the actual speed of the rotor is dependent on the load torque. Hence, when the load varies, the rotor speed varies as well. The effect of uncertainties present in load torque is compensated by automatically adjusting the electromagnetic torque.

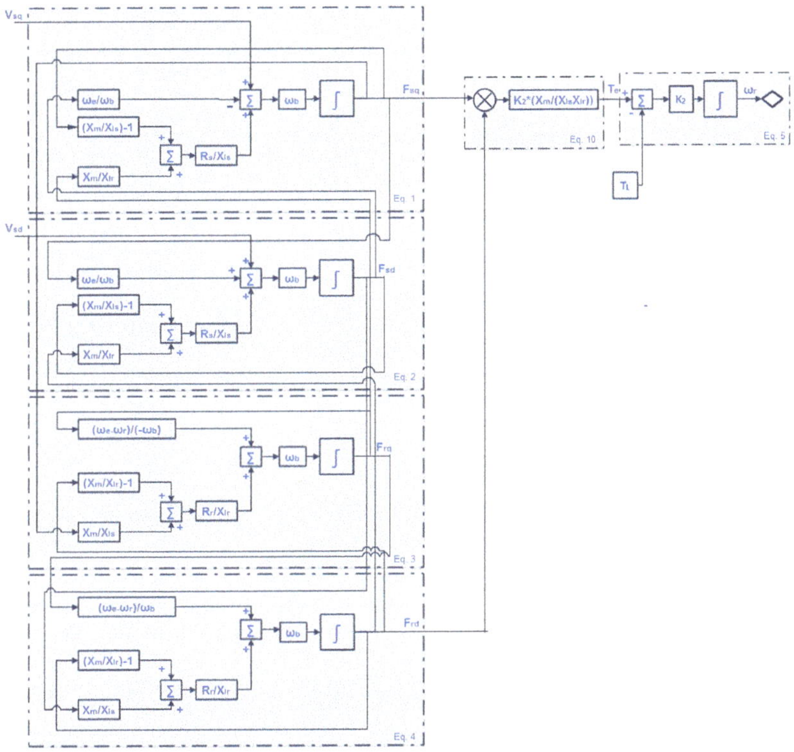

**Figure 2.** Internal dynamics of 3-phase induction motor model.

*3.1. Simulation Results and Findings of Uncompensated 3-Phase Induction Motor*

3.1.1. Motor under Test

A 1.1 KW, 220 V, 50 Hz, 4 poles Squirrel cage induction motor has been selected as a plant. The parameters are given in Table 1.

**Table 1.** Parameters of the induction motor under test.

| Induction Motor Parameters | Symbol | Numerical Value | Unit |
|---|---|---|---|
| Stator Resistance | $R_s$ | 0.19 | $\Omega$ |
| Rotor Resistance | $R_r$ | 0.39 | $\Omega$ |
| Leakage Stator Inductance | $L_{ls}$ | $0.21 \times 10^{-3}$ | H |
| Leakage Rotor Inductance | $L_{lr}$ | $0.60 \times 10^{-3}$ | H |
| Magnetizing Inductance | $L_m$ | $4 \times 10^{-3}$ | H |
| Nominal Current | $I_m$ | 10 | Amps |
| Rotor's Inertia | $J$ | 0.0226 | kg m$^2$ |
| Base Speed | $\omega_e$ | 314.159 | rad/s |

### 3.1.2. Simulation Study

The d–q model of the 3-phase induction motor was implemented using Simulink. The obtained actual rotor speed is illustrated in Figure 3. The frequency of supplied voltage to the stator was 50 Hz. The base speed of the motor was produced by the Rotational Magnetic Field. In Figure 3, the actual rotor-speed is plotted with the base-speed (reference speed) for the 3-phase induction motor. It can be seen that the actual speed tracked the base speed. However, as soon as the load torque of 10 Nm was applied at 0.875 s, the rotor's speed dropped, which showed that the machine was not invariant to load disturbance. There were undesired oscillations as well at the start. To solve these issues related to robustness and speed tracking, a number of different nonlinear controllers were designed, as described in the next Section.

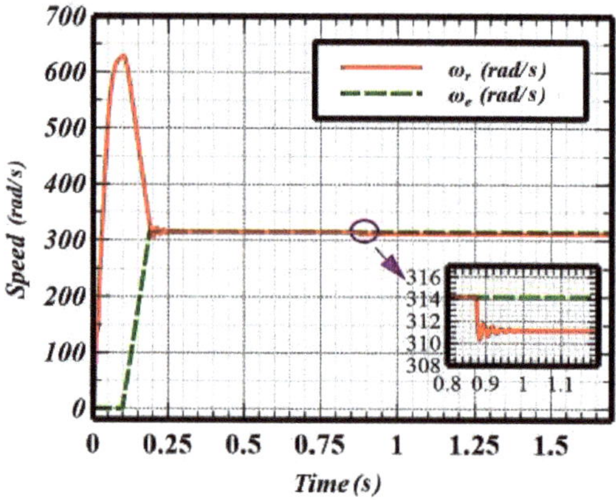


**Figure 3.** Base Speed ($\omega_e$) and actual rotor speed ($\omega_r$).

## 4. Design and Simulation of Nonlinear Controllers

### 4.1. Backstepping Controller

The backstepping control technique is widely utilized for achieving the control of numerous nonlinear systems [29]. It has applications in robotics, military and biomedical engineering services [30]. Referring to (3) to (6) it can be seen that the flux variables are taken as the state variables. The overall block diagram of a backstepping controller with a 3-phase induction motor is presented in Figure 4.

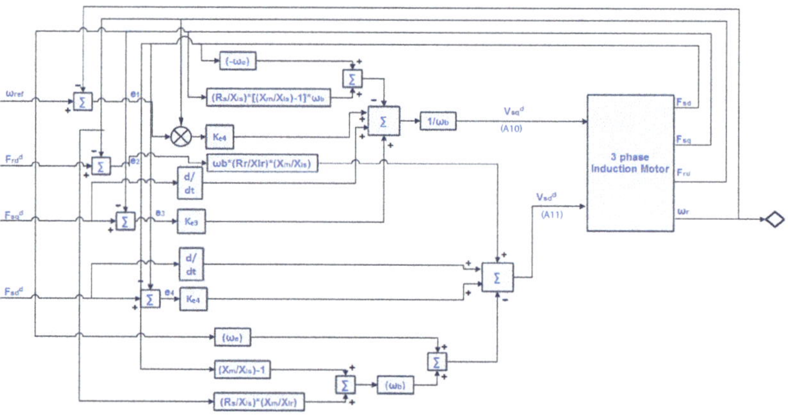

**Figure 4.** Internal Dynamics of backstepping control for 3-phase induction motor model.

For ease of computation, Equation (8) is rewritten in terms of only the flux variables. This is done by substituting stator current variables in the electromagnetic torque equation, resulting as follows [27]:

$$i_{sq} = F_{sq}\left[\frac{X_{ls} - X_m}{X_{ls}^2}\right] - \frac{X_m}{X_{ls}X_{lr}}F_{rq} \tag{9}$$

$$i_{sd} = F_{sd}\left[\frac{X_{ls} - X_m}{X_{ls}^2}\right] - \frac{X_m}{X_{ls}X_{lr}}F_{rd} \tag{10}$$

where $X_{ls}$ and $X_m$ are stator leakage and magnetizing reactance. The above equations ensure that current variables are now only dependent on the flux linkage variables. Substituting Equations (9) and (10) into (8) yields the following equation of the electromagnetic flux:

$$T_e = k_2\left(\frac{X_m}{X_{ls}X_{lr}}\right)\left[F_{rd}F_{sq} - F_{sd}F_{rq}\right] \tag{11}$$

The park transformation ensures that the rotor flux is only directed along the d-axis, hence flux across the quadrature axis can be assumed to be zero ($F_{rq} = 0$). Substituting the value of $F_{rq}$ into (11), the electric torque could be obtained according to the following simplified expression:

$$T_e \approx k_2\left(\frac{X_m}{X_{ls}X_{lr}}\right)F_{rd}F_{sq} \tag{12}$$

The value of electromagnetic torque in (12) is substituted into the equation of the rotor's speed in (7), which yields the following:

$$\frac{d\omega_r}{dt} = k_4 F_{rd}F_{sq} - k_1 T_L \tag{13}$$

where $k_4 = \frac{p}{2j}\frac{3p}{4\omega_b}\left(\frac{X_m}{X_{ls}X_{lr}}\right)$.

For designing the backstepping control, the mathematical model of the induction motor is divided into two subsystems.

**Subsystem 1:** This subsystem consists of the state space equations of the rotor's speed and the rotor's flux in the d-axis direction. The model of the rotor's speed and the rotor's flux dynamics can be written as follows:

$$\frac{d\omega_r}{dt} = k_4 F_{rd}F_{sq} - k_1 T_L \tag{14}$$

$$\frac{dF_{rd}}{dt} = \omega_b\left[\frac{R_r}{X_{lr}}\left\{\frac{F_{sd}X_m}{X_{lr}} + F_{rd}\left(\frac{X_m}{X_{lr}} - 1\right)\right\}\right] \tag{15}$$

The stator's fluxes ($F_{sq}$, $F_{sd}$) are taken as the intermediate control inputs, which are designed using composite Lyapunov stability criteria. These control inputs are fed to subsystem 2.

**Subsystem 2:** This subsystem consists of the state space equation of the stator's fluxes in the direction of the dq-axis.

$$\frac{dF_{sq}}{dt} = \omega_b\left[V_{sq} - \frac{\omega_e F_{sd}}{\omega_b} + \frac{R_s}{X_{ls}}\left\{F_{sq}\left(\frac{X_m}{X_{ls}} - 1\right)\right\}\right] \tag{16}$$

$$\frac{dF_{sd}}{dt} = \omega_b\left[V_{sd} + \frac{\omega_e F_{sd}}{\omega_b} + \frac{R_s}{X_{ls}}\left\{\frac{F_{rd}X_m}{X_{lr}} + F_{sd}\left(\frac{X_m}{X_{lr}} - 1\right)\right\}\right] \tag{17}$$

In this subsystem, the supply voltages ($V_{sq}$, $V_{sd}$) are taken as the final control inputs, which are designed using the combined Lyapunov functions for both subsystems.

**Step 1**: To control the speed of an induction motor, the speed tracking error should be zero which implies that the rotor's speed follows the reference speed exactly. The following error signals are generated for subsystem 1:

$$e_1(t) = \omega_{\text{ref}} - \omega_r \tag{18}$$

$$e_2(t) = F_{rd}^d - F_{rd} \tag{19}$$

where $\omega_{\text{ref}}$ and $F_{rd}^d$ are the desired values of speed and flux linkage across the d-axis, respectively. The following Lyapunov stability function is defined to derive the expressions for the intermediate control inputs ($F_{sq}$ and $F_{sd}$):

$$V_{12} = \frac{1}{2}e_1^2 + \frac{1}{2}e_2^2 \tag{20}$$

The proof of Lyapunov stability criteria and the derivation of the intermediate control inputs are presented in Appendix A. The final intermediate control inputs are as follows:

$$F_{sq}^d = \frac{k_{e1}e_1 + \frac{d\omega_{\text{ref}}}{dt} + k_1 T_L}{k_4 F_{rd}} \tag{21}$$

$$F_{sd}^d = \frac{k_{e2}e_2 + \frac{dF_{rd}^d}{dt} - \omega_b \frac{R_r}{X_{lr}}\left(\frac{X_m}{X_{lr}} - 1\right)F_{rd}}{\omega_b \frac{X_m}{X_{ls}} \frac{R_r}{X_{lr}}} \tag{22}$$

**Step 2**: Subsystem 2 takes the control inputs designed by subsystem 1 and, then, using a combined Lyapunov function including all the errors, it designs the final control inputs ($V_{sq}$ and $V_{sd}$). The following error signals are generated for subsystem 2:

$$e_3 = F_{sq}^d - F_{sq} \tag{23}$$
$$e_4 = F_{sd}^d - F_{sd} \tag{24}$$

To derive the final control inputs ($V_{sq}$ and $V_{sd}$), we will substitute the intermediate control inputs $F_{sq}^d$ and $F_{sd}^d$ into (23) and (24), respectively:

$$e_3 = \frac{k_{e1}e_1 + \frac{d\omega_{\text{ref}}}{dt} + k_1 T_L}{k_4 F_{rd}} - F_{sq} \tag{25}$$

$$e_4 = \frac{k_{e2}e_2 + \frac{dF_{rd}^d}{dt} - \omega_b \frac{R_r}{X_{lr}}\left(\frac{X_m}{X_{lr}} - 1\right)F_{rd}}{\omega_b e f} - F_{sd} \tag{26}$$

The Lyapunov stability function for the entire system is as follows:

$$V = \frac{1}{2}\left[e_1^2 + e_2^2 + e_3^2 + e_4^2\right] \tag{27}$$

The proof of Lyapunov stability criteria for the entire system and the derivation of the final control inputs are presented in Appendix B. The final control inputs that satisfy the Lyapunov stability criteria are as follows:

$$V_{sq}^d = \frac{k_{e3}e_3 + \frac{dF_{sq}^d}{dt} - \left(-\omega_e F_{sd} + \omega_b\left[\frac{R_s}{X_{ls}}\left(\frac{X_m}{X_{ls}} - 1\right)F_{sq}\right]\right) + k_4 F_{rd}e_1}{\omega_b} \tag{28}$$

$$V_{sd}^d = \frac{k_{e4}e_4 + \frac{dF_{sd}^d}{dt} - \left(\omega_e F_{sq} + \omega_b\left[\frac{R_s}{X_{ls}}\left(\frac{X_m}{X_{ls}} - 1\right)F_{sd}\right]\right) + \omega_b \frac{R_r}{X_{lr}} \frac{X_m}{X_{ls}}e_2}{\omega_b} \tag{29}$$

### 4.2. Design of Sliding Mode Controller for an Induction Motor

The sliding mode control is a robust design technique that is useful in compensating for model uncertainties and provides very effective tracking control. The sliding mode control starts with the design of a sliding surface [31]. The sliding surface is designed in a way that depicts that the actual parameter tracks the reference value and the system is stable. The basic principle behind sliding mode theory is to design a control algorithm that forces the system to stay on a sliding surface [32]. Sliding mode control has two phases: (i) reaching phase (ii) sliding phase. The model considered in this design has the viscous co-efficient of friction denoted by $Fc$. The overall block diagram of a sliding mode controller with a 3-phase induction motor is presented in Figure 5.

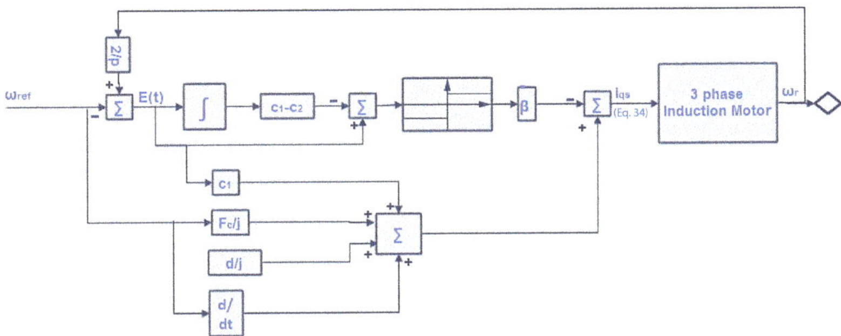


**Figure 5.** Internal Dynamics of SMC for 3-phase induction motor.

For the design of SMC for an induction motor, the electrical speed is first converted to mechanical speed as follows:

$$\omega_m = 2\frac{\omega_r}{p} \tag{30}$$

The mechanical equation of the induction motor then becomes:

$$\frac{d\omega_m}{dt} = -\frac{F_c}{j}\omega_m - \frac{T_L}{j} + \frac{T_e}{j} \tag{31}$$

Note that (31) is very similar to (7), except for the additional nonlinearity introduced through the viscous coefficient of friction. For field-oriented control, (31) becomes:

$$\frac{d\omega_m}{dt} = -\frac{F_c}{j}\omega_m - \frac{T_L}{j} + Ci_{qs}, \text{ with } C = \frac{3pF_{sd}}{4\omega_b j} \tag{32}$$

The $T_L$ (i.e., Load torque) has an uncertain behavior which is compensated by the slidling mode control (SMC) law. The tracking error for speed is given as:

$$E(t) = \omega_m - \omega_{\text{ref}} \tag{33}$$

The derivative of the error signal: (33) is:

$$\frac{dE(t)}{dt} = -\frac{F_c}{j}E(t) + u(t) + d(t) \tag{34}$$

where $u(t)$ is the control law and $d(t)$ denotes the uncertainty due to the load torque. The sliding surface is defined as:

$$S(t) = E(t) - (c_1 - c_2)\int E(\tau)d\tau \tag{35}$$

The control law is selected as follows:

$$u(t) = c_1 E(t) - \beta \text{sign}(S(t)) \tag{36}$$

The following limitations must be met to achieve the desired performance of the SMC:

- The constant gain $c_1$ should be selected such that the term $(c_1 - c_2)$ is strictly negative; therefore, $c_1 < 0$.
- The gain $\beta$ must be greater than the uncertainty, i.e., $\beta > d(t)$

Using Lyapunov stability criteria the final control input $i_{qs}(t)$ is designed as follows:

$$V(S) = \frac{1}{2}S^2 \tag{37}$$

Taking the derivatives of the sliding surface and Lyapunov function, the following is obtained:

$$\frac{dS}{dt} = \frac{dE}{dt} - (c_1 - c_2)E(t) \tag{38}$$

$$\frac{dV}{dt} = S\frac{dS}{dt} \tag{39}$$

Substituting the values of the sliding surface from (35) and the derivative of the sliding surface from (38) into the Lyapunov function, and solving for the negative definite condition, the following is obtained:

$$i_{qs} = c_1 E(t) - \beta \text{sign}(S(t)) + \frac{F_c}{j}\omega_{\text{ref}} + \frac{d(t)}{j} + \frac{d\omega_{\text{ref}}}{dt} \tag{40}$$

This control input in (40) is fed to the induction motor (Figure 5) to achieve the desired result and to compensate for the uncertainty in the load torque.

### 4.3. Design of the Backstepping Sliding Mode Controller for an Induction Motor

The sliding mode controller is cascaded with a backstepping controller to further improve the performance of speed tracking for the induction motor. The backstepping controller offers good tracking performance; however, the uncertainty in load torque is not fully compensated. A sliding mode controller has a very good disturbance rejection quality; however, it experiences the chattering effect. To compensate for these individual demerits, a composite controller was designed to take advantage of the salient features of each controller. The overall block diagram of a backstepping sliding mode controller with a 3-phase induction motor is presented in Figure 6.

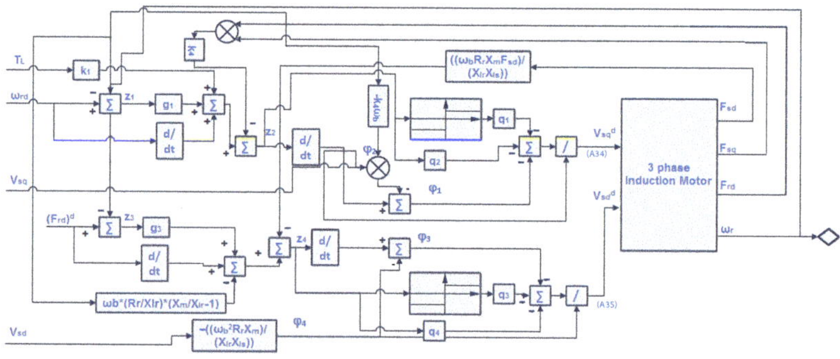

**Figure 6.** Internal Dynamics of backstepping SMC of induction motor.

The two subsystems mentioned in (14)–(17) are considered here. The proof of Lyapunov stability criteria and the derivation of the final control inputs are presented in Appendix C. The Equations (41) and (42) are the final control inputs, which are supplied to the induction motor model to achieve optimum control.

$$V_{sq}^d = \frac{-\varphi_1 - q_1 sign(z_2) - q_2 z_2}{\varphi_2} \tag{41}$$

$$V_{sd}^d = \frac{-\varphi_3 - q_2 sign(z_4) - q_4 z_4}{\varphi_4} \tag{42}$$

### 4.4. Design of the Backstepping Super Twisting Sliding Mode Control (STSMC) Law for an Induction Motor

A traditional SMC has many distinct features but it also has a limitation in terms of the chattering effect. Chattering refers to oscillations with finite amplitude and frequency. Higher order sliding mode provides an additional advantage in terms of eliminating or reducing the chattering phenomena. It also has all the characteristics of a traditional SMC. Hence, in order to eliminate the effect of chattering, a super twisting algorithm was cascaded with a backstepping controller [33]. A super twisting algorithm consists of two parts: (i) a discontinuous function of sliding surface variable (ii) a continuous function of sliding surface variable. The overall block diagram of a Backstepping Super Twisting controller with a 3-phase induction motor is presented in Figure 7.

The control law was designed by adding up the effects of a switching control and the equivalent control of the system. The switching control was implemented by (43) and (44):

$$U_{swc} = -r_1\sqrt{|S|}sign(S) + v \tag{43}$$

$$\frac{dv}{dt} = -k_2 \text{sign}(S) \tag{44}$$

The final control law was as follows:

$$U_{control} = U_{swc} + U_{eq} \tag{45}$$

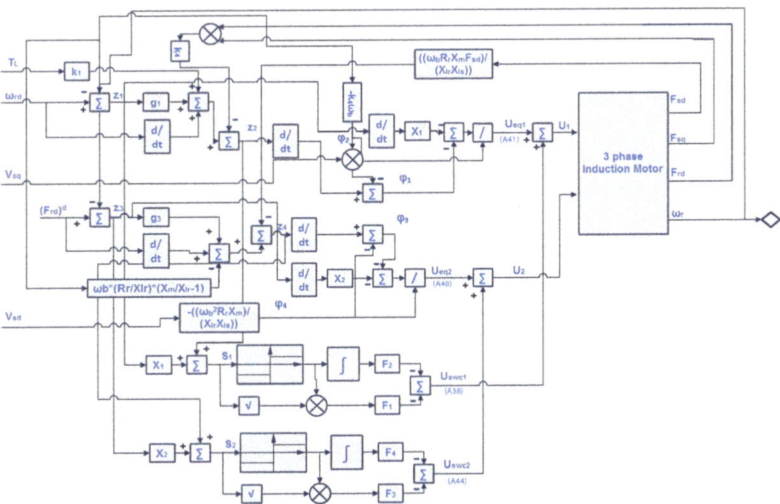


**Figure 7.** Block diagram of the proposed Backstepping Super Twisting algorithm for an induction motor.

For induction motors, the backstepping controller has already been defined in (A53) and (A55). The control law of HOSM (Super twisting control) for the first subsystem is as follows:

$$U_1 = U_{swc1} + U_{eq1} \tag{46}$$

To derive the switching control law, a sliding surface is defined as:

$$S_1 = z_2 + X_1 z_1 \tag{47}$$

where $z_1$ and $z_2$ are given in (A17) and (A24).

Using the super twisting algorithm, the switching control is based on (43) and (44)

$$U_{swc1} = -F_1\sqrt{|S_1|}sign(S_1) - F_2 \int sign(S_1)ds_1 \tag{48}$$

where $F_1$ and $F_2$ are positive constants.

To derive the equivalent control input for subsystem 1, the derivative of the sliding surface was computed first:

$$\frac{dS_1}{dt} = \frac{dz_2}{dt} + X_1\frac{dz_1}{dt} \tag{49}$$

In super twisting algorithm, the derivative of the sliding mode controller is equal to zero ($\dot{s} = 0$). Hence, the equivalent control was designed as follows:

$$\varphi_1 + \varphi_2 V_{sq} + X_1\frac{dz_1}{dt} = 0 \tag{50}$$

$$U_{eq1} = V_{sq}^d = \frac{-\varphi_1 - \frac{X_1 dz_1}{dt}}{\varphi_2} \tag{51}$$

The control law of HOSM for the second subsystem is as follows:

$$U_2 = U_{swc2} + U_{eq2} \tag{52}$$

To derive the switching control law for this subsystem, we define a sliding surface

$$S_2 = z_4 + X_2 z_3 \tag{53}$$

where $z_3$ and $z_4$ have already been defined in (A36) and (A43). Using the super twisting algorithm, the switching control is defined using (43) and (44):

$$U_{swc2} = -F_3\sqrt{|S_2|}sign(S_2) - F_4 \int sign(S_2)ds_2 \tag{54}$$

To derive the equivalent control for the subsystem 2, the derivative of the sliding surface is derived first:

$$\frac{dS_2}{dt} = \frac{dz_4}{dt} + X_2\frac{dz_3}{dt} \tag{55}$$

The equivalent control was designed as follows:

$$\varphi_3 + \varphi_4 V_{sd} + X_2\frac{dz_3}{dt} = 0 \tag{56}$$

$$U_{eq2} = V_{sd}^d = \frac{-\varphi_3 - \frac{X_2 dz_3}{dt}}{\varphi_4} \tag{57}$$

The Equations (48), (50), (54) and (57) represent the final control inputs of the Backstepping Super twisting control law.

Back-STC with Exact Differentiation and Signum Approximation (Back-STC + ea)

The control law defined in (48), (50), (54) and (57) contain derivatives of the reference speed and flux values. Differentiating a step change in an input results in a discontinuity, which leads to an error in the final output. To overcome this problem, an exact differentiator [34] and an approximation of sigmoid function [35] were designed to minimize the effect of discontinuity and to obtain better results. The exact differentiator that was used here was as follows:

$$z_0 = v_0 \tag{58}$$

$$v_0 = -\sigma_0 \mid z_0 - f(t) \mid^{\left(\frac{n}{n+1}\right)} \operatorname{sign}(z_0 - f(t)) + z_1, \text{ and } \dot{z}_1 = v_1 \tag{59}$$

$$v_1 = -\sigma_1 \mid z_1 - v_o \mid^{\left(\frac{n-1}{n}\right)} \operatorname{sign}(z_1 - v_0) + z_2, \text{ and } \dot{z}_2 = v_2, \dots, \tag{60}$$

$$v_{(n-1)} = -\sigma_{n-1} \mid z_{n-1} - v_{n-2} \mid^{\frac{1}{2}} \operatorname{sign}(z_{n-1} - v_{n-2}) + z_n \tag{61}$$

$$\dot{z}_n = -\sigma_{(n)} \operatorname{sign}(z_n - v_{n-1}) \tag{62}$$

The sigmoid function is approximated as follows:

$$S_{app} = -K_{app} sat\left(\frac{s}{\varnothing}\right) \tag{63}$$

where $s$ is the sliding surface for the particular system; $\varnothing$ is the scaling factor for the approximation and $K_{app}$ is a positive constant.

## 5. Numerical Evaluation & Comparison

A comparison study was conducted in terms of Minimization Criteria: Integral Square Error (ISE), Integral Absolute Error (IAE) and Integral Time Absolute Error (ITAE). The induction motor operated under three different modes: Start & stop mode (SSM), Normal Operation Mode (NOM) and Disturbance Rejection Mode (DRM). ISE, IAE and ITAE were statistical parameters used to evaluate the performance of the design system. Furthermore, a graphical analysis was performed to analyze the results more critically.

### 5.1. Start–Stop Mode (SSM)

In this mode, the speed tracking efficiency of the motor under an approximated step reference signal was examined. An approximated step signal was utilized to avoid very large discontinuities. The reference signal ran the motor between 0.1–2 s. This SSM reference speed signal was applied with all the control techniques implemented in Section 4. The results are plotted in Figure 8. Moreover, the statistical parameters were computed for each nonlinear controller as shown in Table 2.

According to the values of evaluation parameters (ITAE, ISE & IAE) stated in Table 2, SMC could be considered the most promising design in comparison with the rest of the controllers, owing to the fact that it had the least values for the statistical parameters (ITAE: 0.0019 rad/s, IAE: 0.0018 rad/s and ISE: 0.000001 rad/s). However, SMC experienced chattering effects, as depicted in Figure 8b, which made it unreliable from a hardware perspective. On the contrary, Back–SMC had the highest values for all the statistical parameters (ITAE: 0.5893 rad/s, IAE: 0.2987 rad/s and ISE: 2.3060) and could be considered to have the most unsatisfactory behavior in comparison with the other designs. Backstepping cont. (ITAE: 0.0061 rad/s, IAE: 0.0071 rad/s and ISE: 0.0004 rad/s) and Back–STC (ITAE: 0.0018 rad/s, IAE: 0.0020 rad/s, and ISE: 0.00009) also provided satisfactory speed tracking capabilities, considering the small values of the regression parameters. Back–STC–EA also had small values for all three parameters (ITAE: 0.0476 rad/s, IAE: 0.0411 rad/s and ISE: 0.0008 rad/s). However, it exhibits a minimal chattering effect at around 0.1 s; although later it smoothed out as depicted in Figure 8e.

It is worth noting that in Figure 8c, Back–SMC had abrupt transitions at $t = 1.007$ and 2.025 s. These anomalies were the primary cause of the increased value for the ISE parameter.

The magnitudes of these peaks were 0.9544 rad/s and 12.3 rad/s, respectively. This further confirmed the unsatisfactory speed tracking behavior of the designed Back–SMC.

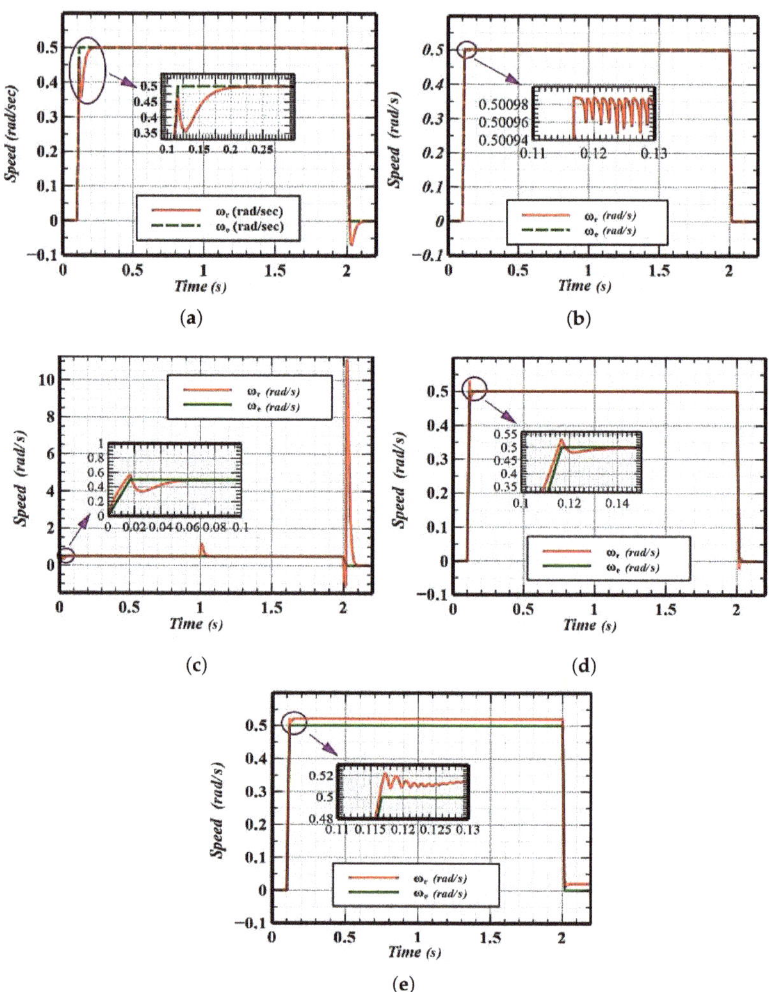

**Figure 8.** Output speed signals for SSM. (**a**) Backstepping control. (**b**) SMC, (**c**) Back-SMC, (**d**) Back-STC, (**e**) Back-STC-EA.

**Table 2.** Numerical evaluation-start-stop mode.

| Control Technique | ITAE (rad/s) | IAE (rad/s) | ISE (rad/s) |
|---|---|---|---|
| Backstepping cont. | 0.0061 | 0.0071 | 0.0004 |
| SMC | 0.0019 | 0.0018 | 0.000001 |
| Back-SMC | 0.5893 | 0.2987 | 2.3060 |
| Back-STC | 0.0018 | 0.0020 | 0.00009 |
| Back-STC-EA | 0.0476 | 0.0411 | 0.0008 |

*5.2. Normal Operation Mode (NOM)*

The second mode that was used to evaluate the performance of the controllers applied to the nonlinear induction motor model was the NOM. In this mode, a regulated reference speed signal with a maximum tolerance of 1.5% was applied to the respective controllers in order to examine their behavior in response to a varying speed signal. In normal mode, the transition period from one speed to another was longer. Hence, the chances of discontinuous behavior were smaller than those from the SSM. For a clearer representation of results from NOM, the actual and desired rotor speeds for each control technique were separately plotted in Figure 9. The results of evaluation parameters for NOM are presented in Table 3.

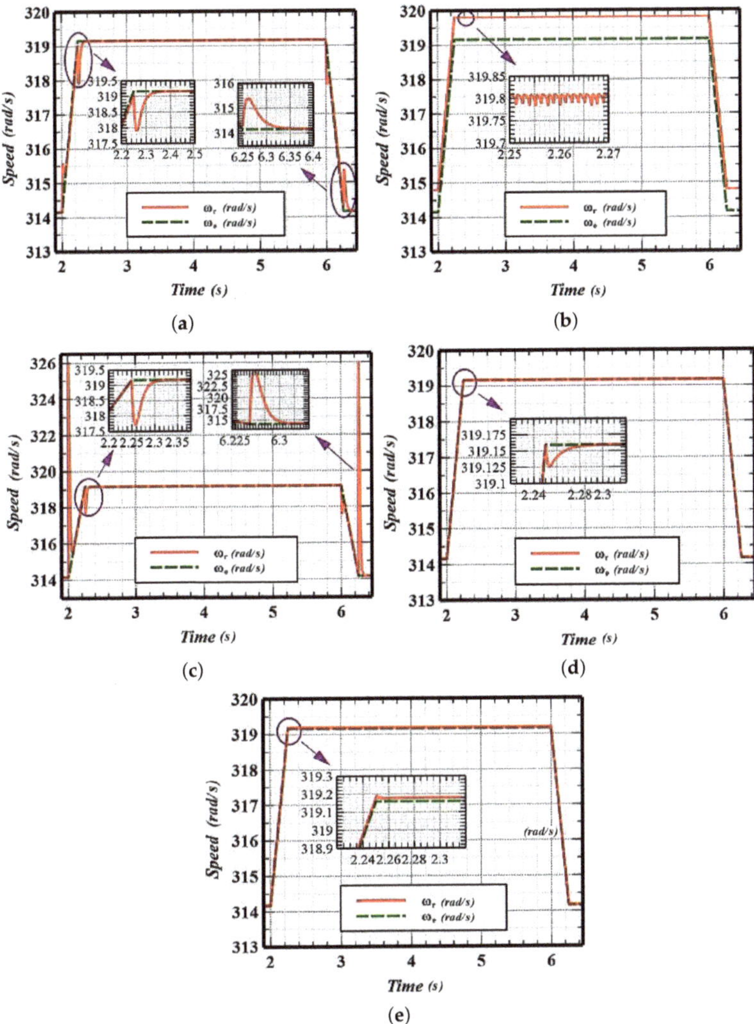

**Figure 9.** Output speed signals for NOM: (**a**) Backstepping control, (**b**) SMC, (**c**) Back-SMC, (**d**) Back-STC, (**e**) Back-STC-EA.

**Table 3.** Numerical Evaluation—NOM.

| Control Technique | ITAE (rad/s) | IAE (rad/s) | ISE (rad/s) |
| --- | --- | --- | --- |
| Backstepping cont. | 2.315 | 1.751 | 24.93 |
| SMC | 13.66 | 4.171 | 2.682 |
| Back-SMC | 2.552 | 2.848 | 147.1 |
| Back-STC | 0.0073 | 0.1011 | 0.2468 |
| Back-STC-EA | 0.4187 | 0.1484 | 0.152 |

The values of ITAE, IAE were the highest for SMC, when compared to the other designs (i.e., 13.66 rad/s and 4.171 rad/s, respectively). Moreover, in Figure 9b, it can be seen that the SMC experienced chattering effects and a steady state error of 0.64 rad/s.

Hence, SMC could be regarded as the most unsatisfactory design for the speed control of a 3-phase induction motor, when compared to the other controllers in NOM. Back–SMC had the highest value of ISE (i.e., 147.1) and it can be seen from Figure 9c that abrupt peaks were present at $t = 2$ and 6.25 s. The magnitudes of these peaks were 325.8 rad/s and 325.7 rad/s, respectively. These abrupt transitions were undesirable from a hardware perspective as they could cause damage to the motor under test. Backstepping cont. also had a considerably high ISE (24.93 rad/s). Figure 9a shows that Backstepping cont. also had abrupt transitions at $t = 2, 2.225, 6.02, 6.27$ s. The magnitudes of these abrupt transitions were 319.45, 315.5, 315.7 and 318.2 rad/s respectively. However, the magnitudes of these transitions were lower than for Back–SMC. Both Back–STC and Back–STC–EA had minimum values for ITAE, ISE, and IAE, as compared to the other three controllers, and depicted good speed tracking characteristics as shown in the Figure 9d,e. From Figure 9d, it can be seen that a slight dip was evident at 2.24 s. While in Figure 9e, this dip no longer existed. Hence, it could be concluded that the use of exact differentiation and signum approximation improved the Back–STC algorithm to some extent.

### 5.3. Normal Operation Mode (NOM) with Bounded Matched Disturbance

To further assess the robustness and efficiency of the non-linear controllers discussed above, a bounded matched disturbance [36] was introduced in the closed loop system for NOM. The bounded disturbance is shown in Figure 10.

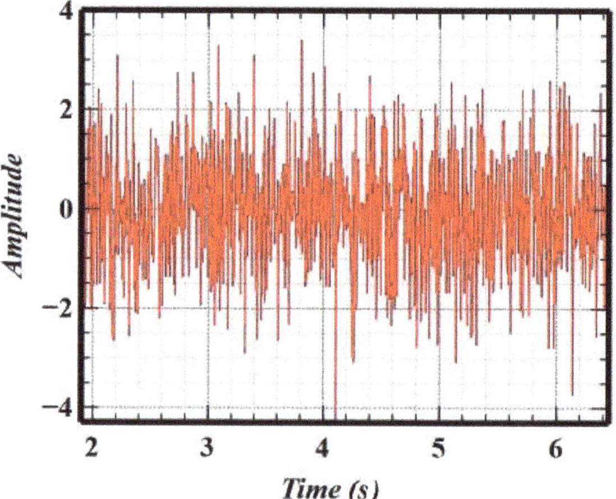

**Figure 10.** Bounded matched disturbance in NOM.

The responses of all the control techniques in the presence of bounded matched disturbance are presented in Figure 11. From Figure 11a, it is evident that the Backstepping controller was not able to fully reject the disturbance, owing to the variations in the actual rotors' speed curve. However, the actual speed of the rest of the controllers, namely SMC (Figure 11b), Back–SMC (Figure 11c), Back–STC (Figure 11d) and Back–STC–EA (Figure 11e) seemed to have rejected the disturbance and were presenting robust results. This could also be seen by comparing the results in Figure 11 with the results given in Figure 9. Comparing Figures 9b and 11b, it is evident that the SMC behaved the same and rejected the bounded matched disturbance; however, the chattering persisted. The speed tracking behavior of the rest of the 3 controllers, namely Back–SMC, Back–STC and Back–STC–EA remained the same, both in disturbance and without disturbance modes. Hence, it could be concluded that these controllers rejected the bounded matched disturbance.

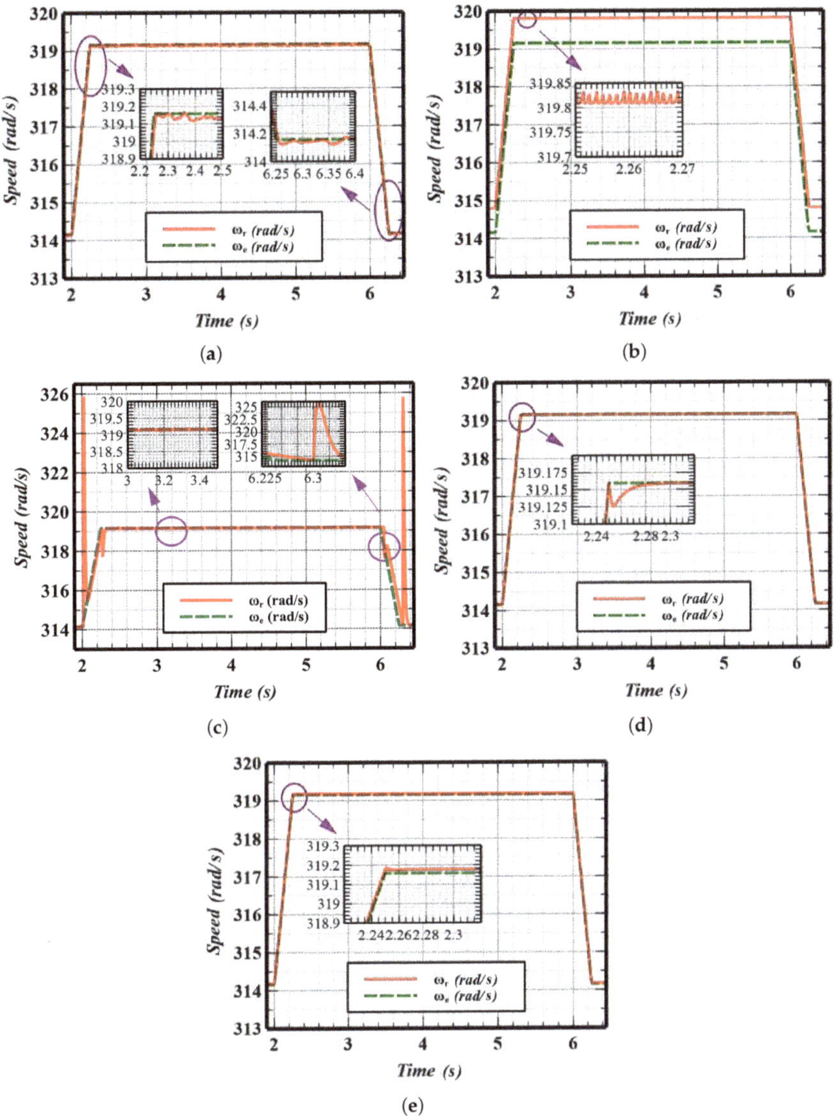

**Figure 11.** Output speed signals for NOM with bounded matched disturbance. (**a**) Backstepping control. (**b**) SMC (**c**) Back-SMC (**d**) Back-STC (**e**) Back-STC-EA.

*5.4. Disturbance Rejection Mode (DRM)*

The third mode is the most important as it concerns itself with the disturbance rejection in load torque of the 3-phase Induction motor. The parametric uncertainty introduced in the load torque should be compensated by a robust controller. In this mode, uncertainty was introduced in the load torque while the motor was running, as shown in Figure 12. The load torque had a sudden abrupt change of 10 Nm at 2 s. The values of the minimizing regression criteria (ITAE, IAE and ISE) are evaluated and presented in Table 4. The actual speeds of all the implemented nonlinear control techniques are plotted in comparison with the desired speed in Figure 13.

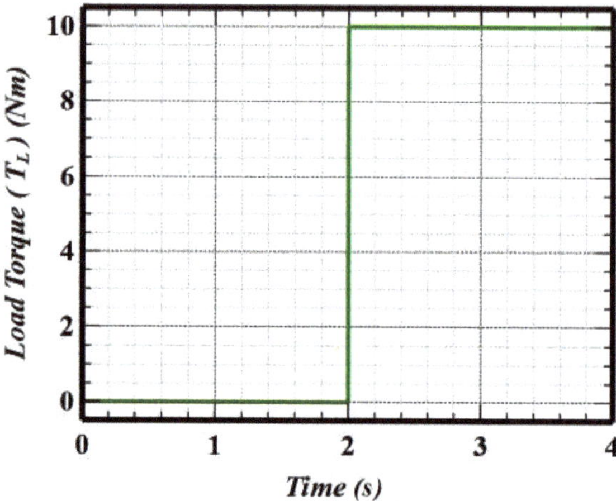

**Figure 12.** Uncertainty in Load Torque (Nm).

The values of ITAE, IAE and ISE were the highest for Backstepping cont. (i.e., 7.217, 4.156 and 8.417 rad/s, respectively). Moreover, it can be seen in Figure 13a, the backstepping cont. did not reject the disturbance as the rotor speed had a steady state error of 1 rad/s after the disturbance in load torque was introduced at $t = 2$ s. Hence, the designed Backstepping cont. did not achieve robust conditions in its performance.

The SMC also had high values for the three regression parameters (i.e., ITAE: 5.065 rad/s, IAE: 2.353 rad/s and ISE: 1.457 rad/s). This could be attributed to the steady state error of 0.64 rad/s between the rotor's desired and actual speeds. However, it can be seen in Figure 13b that the SMC rejected the disturbance introduced in load torque at $t = 2$ s. Hence, by inspection, the designed SMC approached robust characteristics.

Back–SMC had the lowest values for the evaluating parameters (i.e., ITAE: 0.0424 rad/s, IAE: 0.0514 rad/s and ISE: 0.0437 rad/s). It can be seen in Figure 13c that the Back–SMC rejected the disturbance introduced due to the load torque and came back to the reference speed after a slight peak with an % overshoot of 0.42. Hence, the designed Back–SMC was considered robust.

The values of evaluating parameters for both Back–STC and Back–STC–EA were considerably lower, as shown in Table 4. It can be seen in Figure 13d,e that both these controllers rejected the disturbance in load torque and returned to their pre-disturbed rotor speeds. Hence, both of these designed controllers showed robustness in the presence of load torque uncertainties introduced in the 3-phase Induction Motor operation.

A somewhat similar study was conducted in [33]. A disturbance observer-based Back–SMC with super twisting sliding mode observer was implemented for speed control of an Induction motor. An uncertainty in load torque was introduced to monitor the robustness of the proposed design. For the desired speed of 200 rpm and an uncertainty of 2 Nm a speed fluctuation of $-80$ rpm was observed. However, in our proposed design (Back–STC–EA), for the desired speed of 314.1593 rpm and uncertainty of 10 Nm, a speed fluctuation of $-0.1$ rpm was observed, as shown in Figure 13e. This further confirmed the robustness and efficiency of the proposed controller.

Though the work is limited to the MATLAB/Simulink environment, we plan to realize the proposed control strategy in a real 3-phase induction motor in the near future to provide a more rigorous understanding of the motor speed control.

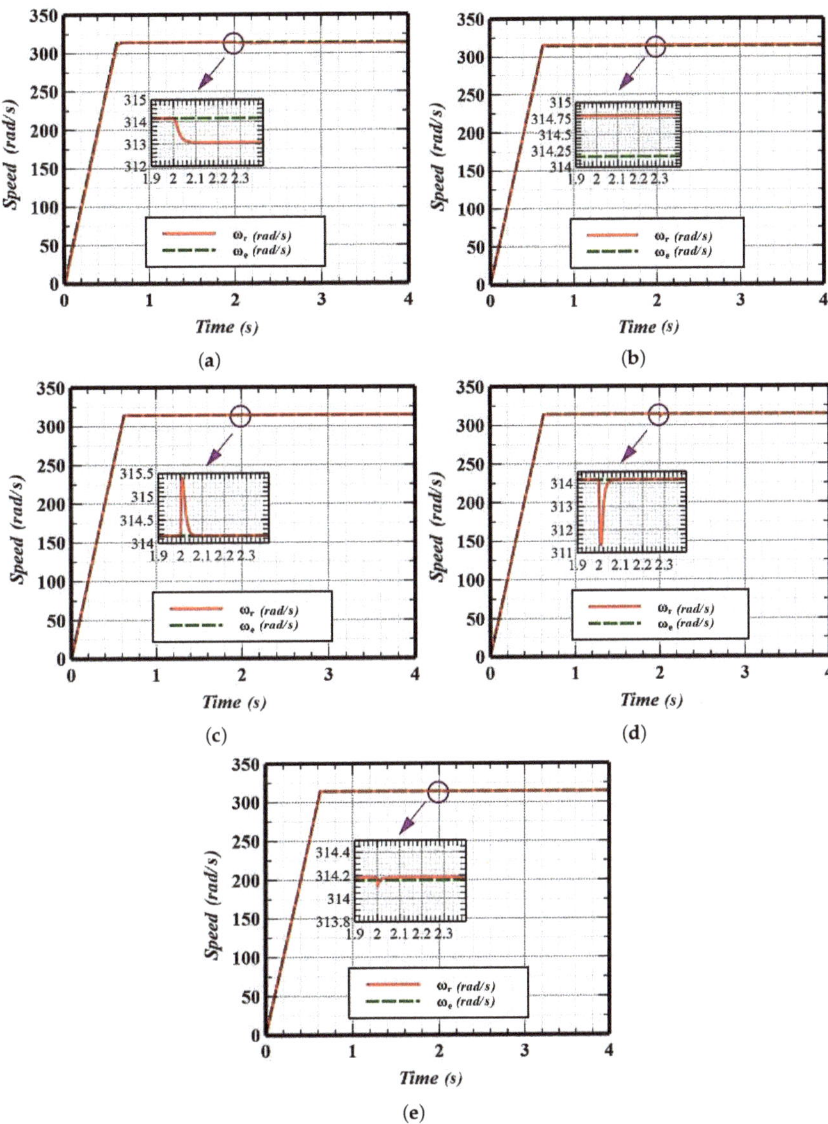

**Figure 13.** Output speed signals for DRM: (**a**) Backstepping Control, (**b**) SMC, (**c**) Back-SMC, (**d**) Back-STC, (**e**) Back-STC-EA.

**Table 4.** Numerical Evaluation—DRM.

| Control Technique | ITAE (rad/s) | IAE (rad/s) | ISE (rad/s) |
|---|---|---|---|
| Backstepping cont. | 7.217 | 4.156 | 8.417 |
| SMC | 5.065 | 2.353 | 1.457 |
| Back-SMC | 0.0424 | 0.0514 | 0.0437 |
| Back-STC | 0.1149 | 0.0888 | 0.1365 |
| Back-STC-EA | 0.1581 | 0.0790 | 0.0015 |

## 6. Conclusions and Future Work

The Induction motors are used extensively in domestic, as well as in industrial, applications. The objective of this research was to evaluate and compare the performance of different control methods when applied to a three-phase induction motors. Load disturbance rejection is the main concern in speed control applications. Firstly, the mathematical model of a three-phase induction motor was implemented in MATLAB 9.1 R2016b/Simulink software, Islamabad, Pakistan to simulate its behavior. Secondly, different control techniques were applied to study the performance of a three-phase induction motor in different scenarios, such as SSM, NOM and DRM. In SSM, Backstepping cont. The synergy of backstepping and the super twisting sliding mode control technique gave the best results when compared with other methods, owing to having the lowest values of statistical parameters. NOM, Back–STC and Back–STC–EA were the controllers with minimum ITAE. The robust qualities of the speed tracking were thoroughly investigated. Future possibilities include the hardware implementation of the designed controllers with an actual three-phase induction motor. These results could then be compared to the simulation study to further explore the topic.

**Author Contributions:** Formal analysis, S.A., Methodology, S.A., Software, S.A., Investigation, S.A., Visualization, S.A., Data curation, S.A. and A.P., Writing—original draft preparation, S.A., Validation, A.P., Writing—review and editing, A.P. and M.P., Resources, A.P., Project administration, A.P. and M.P., Funding acquisition, A.P., Conceptualization, M.P., Supervision, M.P. All authors have read and agreed to the published version of the manuscript.

**Funding:** This research was supported and funded by the Departamento de Ingeniería de Sistemas y Computación with the Universidad Católica del Norte under project 202203010029-VRIDT-UCN.

**Institutional Review Board Statement:** Not applicable.

**Informed Consent Statement:** Not applicable.

**Data Availability Statement:** Not applicable.

**Conflicts of Interest:** The authors declare no conflict of interest.

## Abbreviations

The following nomenclature was used in this manuscript:

| | |
|---|---|
| $U_{sa}$, $U_{sb}$, $U_{sc}$ | Stator Supply voltages in *abc* frame of reference (V) |
| $\omega_e$ | Base Speed (rad/s) |
| $U_{sd}^s$, $U_{sq}^s$ | Stator voltages in d-q stationary frame of reference (V) |
| $V_{sq}$, $V_{sd}$ | Stator voltages in d-q dynamic frame of reference (V) |
| $R_s$, $R_r$ | Stator & Rotor Resistances ($\Omega$) |
| $i_{sq}$, $i_{sd}$ | Stator currents in d-q dynamic frame of reference (A) |
| $\Phi_{sd}$, $\Phi_{sq}$ | Stator Flux Linkages in d-q dynamic frame of reference (Wb-t) |
| $\Phi_{sd}^s$, $\Phi_{sq}^s$ | Stator Flux Linkages in d-q Stationery frame of reference (Wb-t) |
| $i_{rd}$, $i_{rq}$ | Rotor currents in d-q dynamic frame of reference (A) |
| $\omega_r$ | Rotor's Actual Speed (rad/s) |
| $\Phi_{rd}$, $\Phi_{rq}$ | Rotor Flux Linkages in d-q dynamic frame of reference (Wb-t) |
| $F_{sd}$, $F_{sq}$ | Stator Flux Variables (Wb) |
| $F_{rd}$, $F_{rq}$ | Rotor Flux Variables (Wb) |
| $\omega_b$ | Base Frequency (Hz) |
| $L_{ls}$, $L_{lr}$ | Stator & rotor's Leakage inductances (H) |
| $L_m$ | Magnetizing Inductance (H) |
| $\Phi_{md}$, $\Phi_{mq}$ | Magnetizing flux linkage in d-q dynamic frame of reference (Wb-t) |
| $X_{ls}$, $X_{lr}$ | Stator & rotor's Leakage reactances ($\Omega$) |
| $X_m$ | Magnetizing Reactance ($\Omega$) |
| $T_L$ | Load Torque (Nm) |
| $T_e$ | Electromagnetic Torque (Nm) |
| $f$ | Electrical Frequency |
| $e_1(t)$, $e_2(t)$, $e_3(t)$, $e_4(t)$ | Error Signals for Backstepping controller |
| $k_{e1}$, $k_{e2}$, $k_{e3}$, $k_{e4}$ | Tuning parameters for Backstepping Controller |
| $V_{12}$, $V$ | Lyapunov Functions for Backstepping Controller |

| | |
|---|---|
| $F_{sd}^d,\ F_{sq}^d$ | Virtual Control Inputs for backstepping controller (Wb) |
| $V_{sd}^d,\ V_{sq}^d$ | Final Control inputs for backstepping Controller (V) |
| $F_c$ | Viscous Coefficient of Friction |
| $E(t)$ | Error signal for sliding mode controller |
| $u(t)$ | Control law for Sliding mode controller |
| $d(t)$ | Disturbance in load torque (Nm) |
| $c_1,\ c_2,\ \beta$ | Tuning parameters for sliding mode controller |
| $V(S)$ | Lyapunov Function for Sliding Mode Controller |
| $z_1,\ z_2,\ z_3,\ z_4$ | Error signals for backstepping sliding mode controller |
| $V(z_1),\ V(z_2),\ V(z_3),\ V(z_4)$ | Lyapunov Functions for backstepping sliding mode controller |
| $g_1,\ g_2,\ g_3,\ g_4$ | Tuning Parameters for backstepping sliding mode controller |
| $q_1,\ q_2,\ q_3,\ q_4$ | Tuning Parameters for backstepping sliding mode controller |
| $r_1,\ K_2$ | Tuning Parameters for backstepping Super Twisting controller |
| $U_{eq}$ | Equivalent Control for Super twisting Controller |
| $U_{swc}$ | Switching Control for Super twisting Controller |
| $S_1,\ S_2,\ S_3$ | Sliding Surfaces backstepping Super Twisting controller |
| $f(t)$ | Real time noisy signal |
| $z_0, \dot{z}_1, \dots, \dot{z}_n$ | Exact differentiators |
| $\sigma_0,\ \sigma_1, \dots, \sigma_n$ | Tuning parameters |
| $S_{app}$ | Signum Approximation |
| $k_{app}$ | Approximation constant |
| $\varnothing$ | Scaling Factor |
| $N$ | Number of Samples |
| $e(i)$ | Error for regression analysis |

## Appendix A. Derivation of Intermediate Control Inputs for the Backstepping Controller

Taking the derivative of the Lyapunov function in (20):

$$\frac{dV_{12}}{dt} = e_1 \frac{de_1}{dt} + e_2 \frac{de_2}{dt} \tag{A1}$$

where

$$\frac{de_1}{dt} = \frac{d\omega_{ref}}{dt} - k_4 F_{rd} F_{sq} + k_1 T_L \tag{A2}$$

$$\frac{de_2}{dr} = \frac{dF_{rd}^d}{dt} - \omega_b \frac{R_r}{X_{lr}} \frac{X_m}{X_{ls}} F_{sd} - \omega_b \frac{R_r}{X_{lr}} \left( \frac{X_m}{X_{lr}} - 1 \right) F_{rd} \tag{A3}$$

For the system to be stable, the Lyapunov function should be negative definite. To make the derivative of the Lyapunov function negative definite, the term $\frac{de_1}{dt}$ and $\frac{de_2}{dt}$ are made equal to $k_{e1}e_1$ and $k_{e2}e_2$, respectively:

$$\frac{d\omega_{ref}}{dt} - k_4 F_{rd} F_{sq} + k_1 T_L = k_{e1}e_1 \tag{A4}$$

$$\frac{dF_{rd}^d}{dt} - \omega_b \frac{R_r}{X_{lr}} \frac{X_m}{X_{ls}} F_{sd} - \omega_b \frac{R_r}{X_{lr}} \left( \frac{X_m}{X_{lr}} - 1 \right) F_{rd} = k_{e2}e_2 \tag{A5}$$

Rearranging these equations, we get the control inputs for subsystem 1 as,

$$F_{sd}^d = \frac{k_{e1}e_1 + \frac{d\omega_{ref}}{dt} + k_1 T_L}{k_4 F_{rd}} \tag{A6}$$

$$F_{sd}^d = \frac{k_{e2}e_2 + \frac{dF_{rd}^d}{dt} - \omega_b \frac{R_r}{X_{lr}} \left( \frac{X_m}{X_{lr}} - 1 \right) F_{rd}}{\omega_b \frac{X_m}{X_{ls}} \frac{R_r}{X_{lr}}} \tag{A7}$$

These equations constitute the intermediate control required for obtaining the desired final control input for subsystem 2. Substituting (A4) and (A5) into the derivative of the Lyapunov function, we get:

$$\frac{dV_{12}}{dt} = e_1[-k_{e1}e_1] + e_2[-k_{e2}e_2] \tag{A8}$$

$$\frac{dV_{12}}{dt} = -k_{e1}e_1^2 - k_{e2}e_2^2 \text{ (negative definite if } k_{e1} \text{ and } k_{e1} > 0) \tag{A9}$$

### Appendix B. Derivation of Final Control Inputs for the Backstepping Controller

Rearranging (25) and (26), we get

$$k_4 F_{rd} e_3 - k_{e1}e_1 = \frac{d\omega_{ref}}{dt} + k_1 T_L - k_4 F_{rd} F_{sq} = \frac{de_1}{dt} \tag{A10}$$

$$\omega_b e f e_4 - k_{e2}e_2 = \frac{dF_{rd}^d}{dt} - \omega_b \frac{R_r}{X_{lr}}\left(\frac{X_m}{X_{lr}} - 1\right) F_{rd} - \omega_b \frac{X_m}{X_{ls}} \frac{R_r}{X_{lr}} F_{sd} = \frac{de_2}{dt} \tag{A11}$$

Substituting the values of error and their derivatives in the derivative of the Lyapunov function (27), we get:

$$\frac{dV}{dt} = e_1[k_4 F_{rd} e_3 - k_{e1}e_1] + e_2\left[\omega_b \frac{X_m}{X_{ls}} \frac{R_r}{X_{lr}} e_4 - k_{e2}e_2\right] \tag{A12}$$

$$+ e_3\left[\frac{dF_{sq}^d}{dt} - \omega_b V_{sq} - \left(-\omega_e F_{sd} + \omega_b\left[\frac{R_s}{X_{ls}}\left(\frac{X_m}{X_{ls}} - 1\right)F_{sq}\right]\right)\right] \tag{A13}$$

$$+ e_4\left[\frac{dF_{sd}^d}{dt} - \omega_b V_{sd} + \left(-\omega_e F_{sq} + \omega_b\left[\frac{R_s}{X_{ls}}\left(\frac{X_m}{X_{ls}} - 1\right)F_{sd}\right]\right)\right] \tag{A14}$$

The final control inputs that satisfy the Lyapunov stability criteria are as follows:

$$V_{sq}^d = \frac{k_{e3}e_3 + \frac{dF_{sq}^d}{dt} - \left(-\omega_e F_{sd} + \omega_b\left[\frac{R_s}{X_{ls}}\left(\frac{X_m}{X_{ls}} - 1\right)F_{sq}\right]\right) + k_4 F_{rd} e_1}{\omega_b} \tag{A15}$$

$$V_{sd}^d = \frac{k_{e4}e_4 + \frac{dF_{sq}^d}{dt} - \left(\omega_e F_{sq} + \omega_b\left[\frac{R_s}{X_{ls}}\left(\frac{X_m}{X_{ls}} - 1\right)F_{sd}\right]\right) + \omega_b \frac{R_r}{X_{lr}} \frac{X_m}{X_{ls}} e_2}{\omega_b} \tag{A16}$$

Substituting these into the derivative of the Lyapunov function, we get:

$$\frac{dV}{dt} = -k_{e1}e_1^2 - k_{e2}e_2^2 - k_{e3}e_3^2 - k_{e4}e_4^2 < 0$$

(negative definite if $k_{e1}$, $k_{e2}$, $k_{e3}$, and $k_{e4} > 0$)

Hence, the condition is proved.

### Appendix C. Derivation of Final Control Inputs for Backstepping Sliding Mode Controller

**Step 1:**

$$z_1 = \omega_{rd} - \omega_r \tag{A17}$$

Taking the derivative of the error signal and substituting the value of the derivative of the rotor's speed, we get:

$$\frac{dz1}{dt} = \frac{d\omega_{rd}}{dt} - K_4 F_{rd} F_{sq} + k_1 T_L \tag{A18}$$

Considering the following Lyapunov candidate function and its derivative:

$$V(z_1) = \frac{1}{2}z_1^2 \tag{A19}$$

$$\frac{dV(z_1)}{dt} = z_1\frac{dz_1}{dt} \tag{A20}$$

$$\frac{dV(z_1)}{dt} = z_1\left(\frac{d\omega_{rd}}{dt} - K_4F_{rd}F_{sq} + k_1T_{L_1}\right) \tag{A21}$$

For the system to be stable the Lyapunov function should be negative definite. It is obtained by substituting the right-hand side of (A21) (excluding $z_1$) equal to $g_1z_1$.

$$\left(k_4F_{rd}F_{sq}\right)_d = g_1z_1 + \frac{d\omega_{rd}}{dt} + k_1T_L \tag{A22}$$

$$\frac{dV(z_1)}{dt} = -g_1z_1^2 \tag{A23}$$

Note that (A23) is negative definite if $g_1 > 0$.

**Step 2:**

Defining the error for the stator flux in the q-axis direction, we get:

$$z_2 = \left(k_4F_{rd}F_{sq}\right)_d - \left(k_4F_{rd}F_{sq}\right) \tag{A24}$$

Taking the derivative of (A24) and substituting from the Lyapunov condition (A23) yields:

$$\frac{dz_2}{dt} = \left[g_1\left(\frac{d\omega_{rd}}{dt} - k_4F_{rd}F_{sq} + k_1T_L\right) + \frac{d}{dt}\left(\frac{d\omega_{rd}}{dt}\right) + \frac{dk_1T_L}{dt}\right] \tag{A25}$$

$$- k_4\left[F_{rd}\left(\omega_bV_{sq} - \omega_eF_{sd} + \omega_b\left[\frac{R_s}{X_{ls}}\left(\frac{X_m}{X_{ls}} - 1\right)F_{sq}\right]\right)\right. \tag{A26}$$

$$\left. + \left(\omega_b\frac{R_r}{X_{lr}}\frac{X_m}{X_{ls}}F_{sd} + \omega_b\frac{R_r}{X_{lr}}\left(\frac{X_m}{X_{lr}} - 1\right)F_{rd}\right)F_{sq}\right] \tag{A27}$$

To avoid cumbersome calculations, the following arrangements are made:

$$\frac{dz_2}{dt} = \varphi_1 + \varphi_2V_{sq} \tag{A28}$$

where

$$\varphi_1 = \frac{g_1d\omega_{rd}}{dt} - g_1k_4F_{rd}F_{sq} + g_1k_1T_L + \frac{d}{dt}\left(\frac{d\omega_{rd}}{dt}\right) + \frac{dk_1T_L}{dt}$$

$$- k_4F_{rd}\left(-\omega_eF_{sd} + \omega_b\left[\frac{R_s}{X_{ls}}\left(\frac{X_m}{X_{ls}} - 1\right)F_{sq}\right]\right) \tag{A29}$$

$$- k_4\omega_b\frac{R_r}{X_{lr}}\frac{X_m}{X_{ls}}F_{sd}F_{sq} - k_4\left(\omega_b\frac{R_r}{X_{lr}}\left(\frac{X_m}{X_{lr}} - 1\right)F_{rd}\right)F_{sq}$$

$$\varphi_2 = -k_4F_{rd}\omega_b \tag{A30}$$

The Lyapunov candidate function considered in this case is:

$$V(z_2) = \frac{1}{2}z_2^2 \tag{A31}$$

Taking the derivative of the Lyapunov function, it is obtained that:

$$\frac{dV(z_2)}{dt} = z_2\frac{dz_2}{dt} \tag{A32}$$

By substituting (A28), one gets

$$\frac{dV_{z2}}{dt} = z_2\left(\varphi_1 + \varphi_2 V_{sq}\right) \tag{A33}$$

For the system to be stable, the Lyapunov function should be negative definite. It is obtained by substituting the R.H.S of (A33), (excluding $z_2$) equal to $g_2 z_2$, Then, the resulting expressions are:

$$V_{sq} = \frac{-\varphi_4 - g_2 z_2}{\varphi_2} \tag{A34}$$

and,

$$\frac{dV(z_2)}{dt} = -g_2 z_2^2, \ (\text{It is negative definite if} g_2 > 0) \tag{A35}$$

**Step 3**: Now considering subsystem 2, the error signal for the rotor flux in d-axis direction is defined as follows:

$$z_3 = F_{rd}^d - F_{rd} \tag{A36}$$

Taking the derivative and substituting the value of $\dot{F}_{rd}$, we obtain:

$$\frac{dz_3}{dt} = \frac{dF_{rd}^d}{dt} - \left[\omega_b \frac{R_r}{X_{lr}} \frac{X_m}{X_{ls}} F_{sd} + \omega_b \frac{R_r}{X_{lr}}\left(\frac{X_m}{X_{lr}} - 1\right) F_{rd}\right] \tag{A37}$$

The following Lyapunov function is being considered:

$$V(z_3) = \frac{1}{2} z_3^2 \tag{A38}$$

Taking the derivative of the Lyapunov function and substituting the value of $\dot{z}_3$, it is obtained that:

$$\frac{dV(z_3)}{dt} = z_3\left[\frac{dF_{rd}^d}{dt} - \omega_b \frac{R_r}{X_{lr}} \frac{X_m}{X_{ls}} F_{sd} - \omega_b \frac{R_r}{X_{lr}}\left(\frac{X_m}{X_{lr}} - 1\right) F_{rd}\right] \tag{A39}$$

To make it negative definite, it was selected:

$$\omega_b \frac{R_r}{X_{lr}} \frac{X_m}{X_{ls}} F_{sd} = g_3 z_3 - \omega_b \frac{R_r}{X_{lr}}\left(\frac{X_m}{X_{lr}} - 1\right) F_{rd} + \frac{dF_{rd}^d}{dt} \tag{A40}$$

Rearranging and substituting (A40) into (A39), it is obtained:

$$\frac{dV_{z3}}{dt} = -g_3 z_3^2 < 0 \ (\text{which is negative definite if } g_3 > 0) \tag{A41}$$

**Step 4**: The following error signal is selected for the stator flux in the d-axis direction:

$$z_4 = \left(\omega_b \frac{R_r}{X_{lr}} \frac{X_m}{X_{ls}} F_{sd}\right)_d - \left(\omega_b \frac{R_r}{X_{lr}} \frac{X_m}{X_{ls}} F_{sd}\right) \tag{A42}$$

Substituting the value of $\omega_b \frac{R_r}{X_{lr}} \frac{X_m}{X_{ls}} F_{sd}$ from (A41) into (A43), we get:

$$z_4 = \left(\frac{dF_{rd}^d}{dt} - \left(\omega_b \frac{R_r}{X_{lr}}\left(\frac{X_m}{X_{lr}} - 1\right) F_{rd}\right) + g_3 z_3\right) - \left(\omega_b \frac{R_r}{X_{lr}} \frac{X_m}{X_{ls}} F_{sd}\right) \tag{A43}$$

Taking the derivative of (A43) and simplifying it, we get:

$$\frac{dz_4}{dt} = \varphi_3 + \varphi_4 V_{sd} \tag{A44}$$

where

$$\varphi_3 = \omega_b \frac{R_r}{X_{lr}} \frac{X_m}{X_{ls}} \frac{dF_{sd}^d}{dt} - \omega_b \frac{R_r}{X_{lr}} \frac{X_m}{X_{ls}} \omega_e F_{sq} + \omega_b \left[ \frac{R_s}{X_{ls}} \left( \frac{X_m}{X_{lr}} F_{rd} + \left( \frac{X_m}{X_{ls}} - 1 \right) \right) F_{sd} \right] \tag{A45}$$

$$\varphi_4 = -\omega_b^2 \frac{R_r}{X_{lr}} \frac{X_m}{X_{ls}} \tag{A46}$$

A candidate Lyapunov function is selected as follows:

$$V(z_4) = \frac{1}{2} z_4^2 \tag{A47}$$

Then, taking the derivative of the Lyapunov function (A47):

$$\frac{dV(z_4)}{dt} = z_4 \frac{dz_4}{dt} \tag{A48}$$

Substituting (A44) in the derivative of the Lyapunov function, we get:

$$\frac{dV(z_4)}{dt} = z_4 [\varphi_3 + \varphi_4 V_{sd}] \tag{A49}$$

To ensure the definite negative nature of the Lyapunov function, the following value of $V_{sd}$ is selected:

$$V_{sd} = \frac{-\varphi_3 - g_4 z_4}{\varphi_4} \tag{A50}$$

Substituting (A50) in (A49), it is obtained:

$$\frac{dV(z_4)}{dt} = -g_4 z_4^2, \quad \text{which is negative definite if } g_4 > 0 \tag{A51}$$

The SMC is incorporated as follows: consider $z_2$ and $z_4$ as the sliding surfaces. Then, the derivatives of these sliding surfaces should be equal to zero, i.e., $(\dot{z}_2 = 0, \dot{z}_4 = 0)$. The following adjustments are made to achieve the final control law:

$$\frac{dz_2}{dt} = \varphi_1 + \varphi_2 V_{sq} = -q_1 \text{sign}(z_2) - q_2 z_2 = 0 \tag{A52}$$

$$V_{sd}^d = \frac{-\varphi_1 - \text{sign}(z_2) - q_2 z_2}{\varphi_2} \tag{A53}$$

$$\frac{dz_4}{dt} = \varphi_3 + \varphi_4 V_{sd} = -q_3 \text{sign}(z_4) - q_4 z_4 = 0 \tag{A54}$$

$$V_{sd}^d = \frac{-\varphi_3 - q_3 \text{sign}(z_4) - q_4 z_4}{\varphi_4} \tag{A55}$$

## References


1. Usha, S.; Tiwari, S.; Kundu, U. Robust speed control of inverter fed parallel connected induction motors for electric vehicle traction applications. *Mater. Today Proc.* **2022**, *66*, 975–981. [CrossRef]
2. Lia, L.; Yua, Q.; Jianga, Z.; Liua, Y.; Guob, M. Thermal-electromagnetic coupling simulation study of high efficiency and energy saving application of induction motor for offshore oil platform. *Energy Rep.* **2021**, *7*, 84–89. [CrossRef]
3. Rocha, E.; Junior, W.; Lucas, K.; Júnior, C.; Júnior, J.; Medeiros, R.; Nogueira, F. A fuzzy type-2 fault detection methodology to minimize false alarm rate in induction motor monitoring applications. *Appl. Soft Comput. J.* **2020**, *93*, 106373. [CrossRef]
4. Iqbal, J.; Ullah, M.; Khan, S.G.; Khelifa, B.; Ćuković, S. Nonlinear control systems—A brief overview of historical and recent advances. *Nonlinear Eng.* **2017**, *6*, 301–312. [CrossRef]
5. Hadi, N.; Ibraheem, I. Speed control of an SPMSM using a tracking differentiator-PID controller scheme with a genetic algorithm. *Int. J. Electr. Comput. Eng. (IJECE)* **2021**, *11*, 1728–1741. [CrossRef]
6. Panchade, V.; Chile, R.; Patre, B. A survey on sliding mode control strategies for induction motors. *Annu. Rev. Control* **2013**, *37*, 289–307. [CrossRef]

7. Alonge, F.; Cirrincione, M.; Pucci, M.; Sferlazza, A. Input–output feedback linearizing control of linear induction motor taking into consideration the end-effects. Part II: Simulation and experimental results. *Control Eng. Pract.* **2015**, *36*, 142–150. [CrossRef]

8. Aktas, M.; Awaili, K.; Ehsani, M.; Arisoy, A. Direct torque control versus indirect field-oriented control of induction motors for electric vehicle applications. *Eng. Sci. Technol. Int. J.* **2020**, *23*, 1134–1143. [CrossRef]

9. Roubache, T.; Chaouch, S.; Nait Said, M.S. Backstepping Fault Tolerant Control for Induction Motor. In Proceedings of the 2014 International Symposium on Power Electronics, Electrical Drives, Automation and Motion, Ischia, Italy, 18–20 June 2014. [CrossRef]

10. Mossa, M.A.; Echeikh, H. A novel fault tolerant control approach based on backstepping controller for a five phase induction motor drive: Experimental investigation. *ISA Trans.* **2021**, *112*, 373–385. [CrossRef]

11. Doria-Cerezo, A.; Olm, J.; Repecho, V.; Biel, D. Complex-valued sliding mode control of an induction motor. *IFAC-PapersOnLine* **2020**, *53*, 5473–5478. [CrossRef]

12. Fateh, M.; Abdellatif, R. Comparative study of integral and classical backstepping controllers in IFOC of induction motor fed by voltage source inverter. *Int. J. Hydrogen Energy* **2017**, *42*, 17953–17964. [CrossRef]

13. Mai, T.; Tran, H. An adaptive robust backstepping improved control scheme for mobile manipulators robot. *ISA Trans.* **2023**. [CrossRef] [PubMed]

14. Benlaloui, I.; Drid, S.; Alaoui, L.C.; Benoudjit, D. Sensorless Speed Backstepping Control of Induction Motor Based on Sliding Mode Observer: Experimental Results. In Proceedings of the 15th International Conference on Sciences and Techniques of Automatic Control & Computer Engineering, 2014. [CrossRef]

15. Yang, Z.; Yue, Q. Induction Motor Speed Control Based on Model Reference. *Procedia Eng.* **2012**, *29*, 2376–2381. [CrossRef]

16. Toumi, I.; Meghni, B.; Hachana, O.; Azar, A.; Boulmaiz, A.; Humaidi, A.; Ibraheem, I.; Kamal, N.; Zhu, Q.; Fusco, G.; Bahgaat, N. Robust Variable-Step Perturb-and-Observe Sliding Mode Controller for Grid-Connected Wind-Energy-Conversion Systems. *Entropy* **2022**, *24*, 731. [CrossRef]

17. Masumpoor, S.; Yaghobi, H.; Ahmadieh Khanesar, M. Adaptive sliding-mode type-2 neuro-fuzzy control of an induction motor. *Expert Syst. Appl.* **2015**, *42*, 6635–6647. [CrossRef]

18. Abderazak, S.; Farid, N. Comparative study between Sliding mode controller and Fuzzy Sliding mode controller in a speed control for doubly fed induction motor. In Proceedings of the 2016 4th International Conference on Control Engineering & Information Technology, Hammamet, Tunisia, 16–18 December 2016. [CrossRef]

19. Suchithra, R.; Ezhilsabareesh, K.; Samad, A. Optimization based higher order sliding mode controller for efficiency improvement of a wave energy converter. *Energy* **2019**, *187*, 116111. [CrossRef]

20. Morfin, O.; Valenzuela, F.; Ramirez Betancour, R.; Castaneda, C.; Ruiz-Cruz, R.; Valderrabano-Gonzalez, A. Real-Time SOSM Super-Twisting Combined With Block Control for Regulating Induction Motor Velocity. *IEEE Access* **2018**, *6*, 25898–25907. [CrossRef]

21. Kleindienst, M.; Reichhartinger, M.; Horn, M.; Usai, E. An application of computer aided parameter tuning of a super-twisting sliding mode controller. In Proceedings of the 2014 13th International Workshop on Variable Structure Systems (VSS), Nantes, France, 29 June 2014–2 July 2014. [CrossRef]

22. Rivera Dominguez, J.; Mora-Soto, C.; Ortega, S.; Raygoza, J.; De La Mora, A. Super-twisting control of induction motors with core loss. In Proceedings of the 2010 11th International Workshop on Variable Structure Systems (VSS), Mexico City, Mexico, 26–28 June 2010. [CrossRef]

23. Nettari, Y.; Labbadi, M.; Kurt, S. Adaptive Backstepping Integral Sliding Mode Control Combined with Super Twisting Algorithm For Nonlinear UAV Quadrotor System. *IFAC Pap. OnLine* **2022**, *55*, 264–269. [CrossRef]

24. Prieto, P.; Cazarez-Castro, N.; Aguilar, L.; Cardenas-Maciel, S. Chattering existence and attenuation in fuzzy-based sliding mode control. *Eng. Appl. Artif. Intell.* **2017**, *61*, 152–160. [CrossRef]

25. Zhang, S.; Wang, Q.; Yang, G.; Zhang, M. Anti-disturbance backstepping control for air-breathing hypersonic vehicles based on extended state observer. *ISA Trans.* **2019**, *92*, 84–93. [CrossRef]

26. Bellure, A.; Aspalli, M.S. Dynamic d-q Model of Induction Motor Using Simulink. *Int. J. Eng. Trends Technol. (IJETT)* **2015**, *24*, 5.

27. Ozpineci, B.; Tolbert, L.M. Simulink Implementation of Induction Machine Model-A Modular Approach. In Proceedings of the IEEE International Electric Machines and Drives Conference, Madison, WI, USA, 1–4 June 2003. IEMDC.2003.1210317. [CrossRef]

28. Singh, B.; Malar, S. Implementation of fractional pi controller for optimal speed control of induction motor fed with quasi z-source converter. *Microprocess. Microsyst.* **2020**, 103323. [CrossRef]

29. Ali, K.; Mehmood, A.; Iqbal, J. Fault-tolerant scheme for robotic manipulator—Nonlinear robust back-stepping control with friction compensation. *PLoS ONE* **2021**, *16*, e0256491. [CrossRef] [PubMed]

30. Awan, Z.; Ali, K.; Iqbal, J.; Mehmood, A. Adaptive Backstepping Based Sensor and Actuator Fault Tolerant Control of a Manipulator. *J. Electr. Eng. Technol.* **2019**, *14*, 2497–2504. [CrossRef]

31. Irfan, S.; Mehmood, A.; Razzaq, M.; Iqbal, J. Advanced sliding mode control techniques for Inverted Pendulum: Modelling and simulation. *Eng. Sci. Technol. Int. J.* **2018**, *21*, 753–759. [CrossRef]

32. Mechali, O.; Iqbal, J.; Xie, X.; Xu, L.; Senouci, A. Robust Finite-Time Trajectory Tracking Control of Quadrotor Aircraft via Terminal Sliding Mode-Based Active Antidisturbance Approach: A PIL Experiment. *Int. J. Aerosp. Eng.* **2021**, *2021*, 5522379. [CrossRef]

33. Chen, C.; Yu, H. Backstepping sliding mode control of induction motor based on disturbance observer. *IET Electr. Power Appl.* **2020**, *14*, 2537–2546. [CrossRef]
34. Levant, A. Higher-order sliding modes, differentiation and output-feedback control. *Int. J. Control* **2003**, *76*, 924–941. [CrossRef]
35. Svečko, R.; Gleich, D.; Sarjaš, A. The Effective Chattering Suppression Technique with Adaptive Super-Twisted Sliding Mode Controller Based on the Quasi-Barrier Function; An Experimentation Setup. *Appl. Sci.* **2020**, *10*, 595. [CrossRef]
36. Ul Islam, R.; Iqbal, J.; Khan, Q. Design and Comparison of Two Control Strategies for Multi-DOF Articulated Robotic Arm Manipulator. *Control Eng. Appl. Informatics* **2014**, *16*, 28–39.

 *electronics*

*Article*

# Integral Windup Resetting Enhancement for Sliding Mode Control of Chemical Processes with Longtime Delay

Alvaro Javier Prado [1,*], Marco Herrera [2], Xavier Dominguez [3], Jose Torres [2] and Oscar Camacho [4]

[1]  Departamento de Ingeniería de Sistemas y Computación, Universidad Católica del Norte, Antofagasta 1270709, Chile
[2]  Departamento de Automatización y Control Industrial, Escuela Politénica Nacional, Quito 170525, Ecuador
[3]  Product Development, Plexigrid S.L., 33510 Gijón, Spain
[4]  Colegio de Ciencias e Ingenierías "El Politécnico", Universidad San Francisco de Quito USFQ, Quito 170157, Ecuador
*   Correspondence: alvaro.prado@ucn.cl

**Citation:** Prado, A.J.; Herrera, M.; Dominguez, X.; Torres, J.; Camacho, O. Integral Windup Resetting Enhancement for Sliding Mode Control of Chemical Processes with Longtime Delay. *Electronics* **2022**, *11*, 4220. https://doi.org/10.3390/electronics11244220

Academic Editors: Jamshed Iqbal, Ali Arshad Uppal and Muhammad Rizwan Azam

Received: 31 October 2022
Accepted: 12 December 2022
Published: 18 December 2022

**Publisher's Note:** MDPI stays neutral with regard to jurisdictional claims in published maps and institutional affiliations.



**Abstract:** The effects of the windup phenomenon impact the performance of integral controllers commonly found in industrial processes. In particular, windup issues are critical for controlling variable and longtime delayed systems, as they may not be timely corrected by the tracking error accumulation and saturation of the actuators. This work introduces two anti-windup control algorithms for a sliding mode control (SMC) framework to promptly reset the integral control action in the discontinuous mode without inhibiting the robustness of the overall control system against disturbances. The proposed algorithms are intended to anticipate and steer the tracking error toward the origin region of the sliding surface based on an anti-saturation logistic function and a robust compensation action fed by system output variations. Experimental results show the effectiveness of the proposed algorithms when they are applied to two chemical processes, i.e., (i) a Variable Height Mixing Tank (VHMT) and (ii) Continuous Stirred Tank Reactor (CSTR) with a variable longtime delay. The control performance of the proposed anti-windup approaches has been assessed under different reference and disturbance changes, exhibiting that the tracking control performance in the presence of disturbances is enhanced up to 24.35% in terms of the Integral Square Error (ISE) and up to 88.7% regarding the Integral Time Square Error (ITSE). Finally, the results of the proposed methodology demonstrated that the excess of cumulative energy by the actuator saturation could reduce the process resources and also extend the actuator's lifetime span.

**Keywords:** anti-windup compensator; input saturation; sliding mode control; longtime delay; chemical processes; ISE; ITSE

## 1. Introduction

One of the major challenges when designing controllers for industrial processes is the time delay, which comprehends an inherent characteristic in several dynamics systems that impacts the information synchronization and control performance. This is the case of food, chemical, biological, and agricultural processes, which usually include systems that exhibit variable and longtime delays due to the accumulation of low transient responses [1–3]. The time delay, so-called dead-time, can be found either in the process input or output, where lag dynamics can directly impair the nominal performance of the control loop [4]. When this time issue appears in the process input, it indicates that the system dynamics are associated with delays in the actuation channels. Therefore, the control action can hardly be applied on time, and the control efficiency may be deteriorated against disturbances [5]. On the other hand, the delay in the process output can be associated with the measurement process and sensors/instrumentation systems, under which the controller could acquire outdated information. In both cases, delays typically impose strict limitations on feasibility and performance [6]. For instance, high-torque mixing machines take considerable time to recover their inertia [7], exhibiting a time delay in the process output. This delay is usually propagated to other processes or instrumentation and communication systems, in which controllers could be fed by non-updated system information [8]. In this scenario, the presence of delays typically demands conditions on the computational activity and control performance [9].

The sliding mode control (SMC) strategy arises as a suitable control technique to deal with delayed systems by its simplicity [10–12]. The SMC strategy is appealing since the design procedure usually considers a reduced order model concerning the original one, simplifying the decoupling of delayed dynamics in linear and nonlinear processes [13–15]. The SMC strategy also can maintain control robustness against uncertainties and modeling mismatch errors when the system states remain within a sliding mode surface, thus excluding the need for exact modeling [16,17]. However, robust control performance can be compromised in processes with time-varying parameters, such as systems with variable dead time [18,19]. In practice, hybrid control strategies combining SMC and PID (i.e., Proportional, Integral, and Derivative) control are usually used in industrial processes to complement a proper tracking performance with stability. Nevertheless, such integration within a sliding surface usually presents an unbounded growth of the control action due to the need for error correction, thus leading to the saturation of the actuators, a phenomenon known as windup. The windup occurrence is precisely related to the lack of a quick reaction from a controller against the continuous accumulation of the tracking error [20]. As a consequence of the windup phenomenon, the control performance is degraded, generating a large overshoot, long settling time, and system instability [21,22].

To overcome the effects of the windup phenomenon on the error accumulation, several anti-windup and anti-saturation techniques have been proposed in the literature [23–25] which may be categorized into three perspectives [26]: (i) conditional integration, (ii) tracking back calculation and (iii) bounded integration of the tracking error. In the conditional integration approach [27], stand-by conditions are activated in order to suspend the integral control action as the control input increases up to a certain maximum control limit or up to the control input is close to being saturated. When there is no presence of saturation in the actuators or the output variables, the integral term of the original controller is idled. Under the back calculation scheme, the control does not instantaneously reset the integral action but dynamically with a given time constant before the system output saturates. Under this approach, the difference between saturated and unsaturated input signals is used to produce feedback signals that inhibit the integrator output. Particularly, using transient responses to lessen the integral action is preferable rather than frequently re-turning gains of controllers because they change performance but do not reduce the tracking error accumulation [28]. On the other hand, the bounded integration approach does not allow for a wide range of compensation but shortens the error correction [29]. The integrator value is bounded here with high-gain dead zones to assure a linear operation behavior [30]. For instance, the authors in [31] present a comparison between classic and current anti-windup SMC approaches with event-triggering strategies to reduce the number of control updates. The authors in [32] exposed an anti-windup approach with an auxiliary feedback loop that feeds an integration block with the saturation error; however, it was not accounted for the system output dynamics. In [33], an integral sliding mode approach is used along with an ancillary controller to reject disturbances and track nominal trajectories (disturbance-free). Similarly, in [34], a static windup correction method was evaluated to decelerate an integrator when approaching the saturation region in an SMC.

According to the current state of the art, none of the early works consider rebooting the integral control action included in an SMC control framework to avoid the windup effects. Thus, the main contribution of this work lies in: (a) developing, testing, and validating a resetting strategy of the integral control action in the discontinuous component of the proposed SMC and (b) limiting the amplitude and repeatability of the windup based on an anti-saturation logistic function and a robust compensation action fed by the system output variations. Moreover, the proposed approach also copes with the windup phenomenon in variable time-delay processes, as found in thermal stream mixing systems of the industrial field. This work comprises an extended version of the work in [20], incorporating here the validity of the anti-windup algorithms over a wide variety of control approaches and a series of experiments on a Variable Height Mixing Tank (VHMT) and on a Continuous Stirred Tank Reactor (CSTR). In addition, the control performance is evaluated against process disturbances by using the Integral Square Error (ISE), and the Integral Time-Square Error (ITSE) as assessment metrics [35,36].

The remainder of this paper is structured as follows: Section 2 presents some theoretical concepts of SMC that allow for raising the problem formulation. Then, Section 3 exposes the proposed windup resetting algorithms to correct the integral control action and avoid the saturation of the actuators. The process models in which the anti-windup algorithms were tested are presented in Section 4. Results of the proposed anti-windup strategies tested on the VHMT and CSTR are detailed in Sections 5 and 6, respectively. Finally, conclusions of this work are drawn in Section 7.

## 2. Problem Formulation

The SMC is a robust control technique derived from a Variable Structure Control (VSC) scheme. It is appealing due to its capability to handle time-varying systems, modeling mismatch, or system parameter variations [37]. Insights from this control strategy lie in preparing a sliding surface $\sigma(t)$, which defines the required dynamics of the closed-loop system. The current state trajectory drives along a given surface until the final state is reached, generally designed to be the system reference.

After the sliding surface is selected, the control output is computed as the sum of a pair of control actions, i.e., (a) a continuous control action $U_C(t)$, which stands for the sliding mode, and (b) a discontinuous or switching control action $U_D(t)$:

$$U(t) = U_C(t) + U_D(t) \tag{1}$$

where the continuous part $U_C(t)$ is responsible for maintaining the trajectory of the system dynamics within the sliding mode surface until reaching the desired system state. This part of the control action is usually obtained by the Filippov's equivalent [38] and under the null conditions of the error and its derivative (i.e., $\sigma(t) = 0$, and $\dot{\sigma}(t) = 0$).

The discontinuous control action, on the other hand, represents the reachability mode of the controller. This mode aims to rapidly achieve the system state beginning from the initial condition to $\sigma(t)$ and constantly pointing out to the sliding mode surface. Furthermore, the discontinuous part of the SMC can be formulated as a non-linear switching function as follows:

$$U_D(t) = K_D \mathrm{sgn}(\sigma(t)) \tag{2}$$

The chattering phenomenon is the main concern in SMC because of the high control activity produced by the control output variations (2) along the sliding surface $\sigma(t)$ [15]. To reduce such effects, several approaches have been proposed (see [6] and references therein). One of the most widely used strategies lies in smoothing the switching Equation (2) using soft logistic activation functions, such as the sigmoid function [23]. Therefore, the discontinuous control part $U_D$ from (2) using the smoothing sigmoid function can be rewritten as follows:

$$U_D(t) = K_D \frac{\sigma(t)}{|\sigma(t)| + \delta} \tag{3}$$

where $K_D$ and $\delta$ are parameters obtained from a previous setting to fine-tune the discontinuous region of $\sigma$ [6]. In particular, the parameter $\delta$ stands for the width of the sliding mode surface where the chattering effect may occur. Note that a wide variety of softening functions could be used, such as ReLU, max, hyperbolic, geometrical functions, or others [23]; nevertheless, the proposed softening function has been selected here since it fits the control design without compromising computational burden.

### 2.1. Integral Control Law for the SMC

A First Order Plus Dead-Time (FOPDT) process model is proposed to represent the nonlinear delayed system dynamics and design the SMC control strategy. The time-delayed process model is simplified with Taylor's approximation since high-order structures could introduce complex formulations [30]. Therefore, the process model is assumed to be as follows:

$$G(s) = \frac{Y(s)}{U(s)} \frac{K}{(\tau s + 1)(1 + t_0 s)} \tag{4}$$

where $G(s)$ is the process model composed of the model output $Y(s)$ and model input $U(s)$; $K$ is the process gain; $\tau$ denotes the process time constant; and $t_0$ is the process dead-time. Using $G(s)$ from (4), the control law [30] becomes:

$$U(t) = \left(\frac{\tau t_0}{K}\right)\left[\frac{Y(t)}{\tau t_0} + \lambda_0 e(t)\right] + K_D \frac{\sigma(t)}{|\sigma(t)| + \delta} \tag{5}$$

where $e(t)$ describes the trajectory tracking error; $t_0$ denotes the transportation lag; $Y(t)$ is the process variable. The sliding mode surface can be written as follows:

$$\sigma(t) = \text{sgn}(K)\left(-\frac{dY(t)}{dt} + \lambda_1 e(t) + \lambda_0 \int e(t)dt\right) \tag{6}$$

where $\lambda_1 = \frac{t_0 + \tau}{\tau t_0}$ and $\lambda_0 < \frac{\lambda_1^2}{4}$ are the adjustable parameters of the sliding mode control surface [30]; $e(t)$ is the tracking error between the reference $Y^{\text{ref}}(t)$ and $Y(t)$. Note that $\text{sgn}(K)$ in (6) is only a sign function of the static gain $K$ to indicate the controller action (direct or reverse); therefore, it does not affect the sliding mode region or change the control switching over $U(t)$. In addition, note that an integral term in the sliding mode surface is included in the control formulation. Thus, as the tracking error accumulation increases without constraints, the sliding mode surface could also grow unbounded and the control output could become saturated. Hence, an integral windup resetting algorithm is required.

### 2.2. Stability Analysis

In order to analyze stability of the sliding surface, the Lyapunov criteria has been used in this work [30], where if the projection of the system trajectories on the sliding surface is stable, then the system is stable.

**Theorem 1.** *If there exists a candidate Lyapunov function $V = \frac{1}{2}\sigma^2(t)$, which is positive definite function and its derivative is negative everywhere except for the discontinuity surface, then the following inequality must satisfy the Lyapunov stability condition:*

$$\frac{dV}{dt} = \sigma(t)\frac{\sigma(t)}{dt} < 0 \tag{7}$$

*Considering $\frac{dY^{\text{ref}}}{dt} = 0$ and $e(t) = Y^{\text{ref}} - Y$, the derivative of the sliding surface (6) is given by:*

$$\frac{d\sigma(t)}{dt} = \text{sgn}(K)\left[-\frac{d^2Y(t)}{d^2t} - \lambda_1 \frac{dY(t)}{dt} + \lambda_0 e(t)\right] \tag{8}$$

From (4), the derivative of higher order is solved and then substituted in (8), resulting in the following:

$$\frac{d^2Y(t)}{dt} = -\lambda_1 \frac{dY(t)}{dt} + \lambda_0 e(t) + \frac{KK_D}{\tau t_0}\frac{\sigma(t)}{|\sigma(t)| + \delta} \tag{9}$$

Now, (9) is replaced in (8):

$$\frac{d\sigma(t)}{dt} = -\text{sgn}(K)\left[\frac{KK_D}{\tau t_0}\frac{\sigma(t)}{|\sigma(t)| + \delta}\right] \tag{10}$$

It is worth of mention here that after replacing (9) in (8), the proceeding term from the integral control $\lambda_0\, e(t)$ in (6) is dropped from (10) regardless any action of anti-windup reset. Thus, this stability analysis is also valid for the proposed SMC controllers that incorporate windup resetting algorithms. The reachability condition is then given as follows:

$$\sigma(t)\frac{d\sigma(t)}{dt} = -\mathrm{sgn}(K)\left[\frac{KK_D}{\tau t_0}\,\frac{\sigma^2(t)}{|\sigma(t)|+\delta}\right] < 0 \tag{11}$$

where the parameters $\tau > 0, t_0 > 0, \delta > 0$, and $\mathrm{sgn}(K)K > 0$. Notice that to ensure stability of the closed-loop control system, it is necessary to satisfy $K_D > 0$.

## 3. Anti-Windup Algorithms

This Section presents details of the proposed windup resetting algorithms. First, the anti-windup techniques consider that the integral control action can be expressed as an explicit function of the sliding mode surface $\sigma$ within the discontinuous control part of the SMC. Then, the evolution of the discontinuous part in (5) as a function of $\sigma$ in (6) is presented in Figure 1. As $U_D(t)$ is associated with the transient response of the system dynamics, the control tendency can be characterized by an inactive operating region where its slope is close to zero. The discontinuous part of the SMC is constant and does not represent changes in the integral term of $\sigma(t)$. Indeed, when the integral term begins to increase cumulatively to apply error corrections, the discontinuous component of the control law also traces the same tendency (see Figure 1b). At last, the discontinuous component becomes activated, and the control aggressiveness remains unchanged to compensate for disturbances (see Figure 1a).

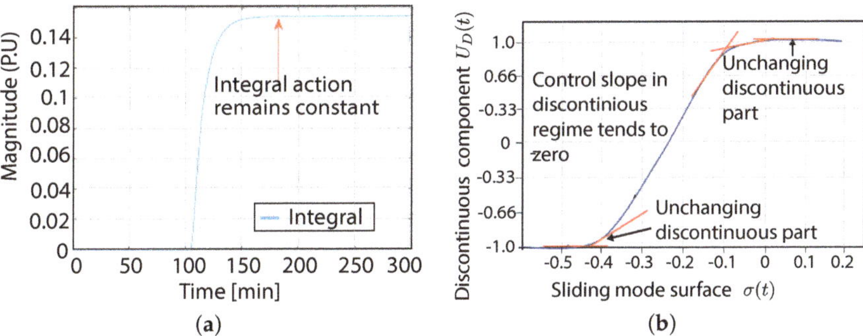


**Figure 1.** Effect of integral control saturation on the discontinuous part of SMC. (**a**) shows that, as the slope is close to zero, the error integral in the sliding surface does not grow unbounded, enabling the controller to counteract the windup effect. (**b**) shows as the output slope $U_D(t)$ tends to be null as a particular value of $\sigma(t)$ is reached. (**a**) Evolution of the integral control action in discontinuous mode. (**b**) Discontinuous control output with respect to the sliding mode surface $\sigma$.

Two windup resetting algorithms are proposed to overcome the performance deterioration under the cumulative tracking error produced by the windup phenomenon. The first approach is the SMC with Windup Instantaneous Reset (i.e., SMC-WIR) and the other is SMC with Windup Conditional Reset (i.e., SMC-WCR). The SMC-WIR and SMC-WCR approaches work on resetting the error accumulation generated by the integrator in the sliding surface of SMC controllers; however, they could be applied to other control structures as well. As the integral term is reset, the discontinuous part of the controller works here on the highest point of the sliding surface (i.e., zero slope or analogously derivative zero). Then, the algorithm is addressed to shorten the settling time and improve control performance. The proposed SMC-WIR and SMC-WCR windup resetting algorithms are as follows.

### 3.1. SMC with Windup Instantaneous Reset

It is worth noting that, in a PID controller, while the proportional and derivative control actions return to their natural zero values, the integral action may remain unchanged and no longer return to the origin. Beyond the accumulation of the integral control action, it may also result in an additional delay on the system actuation. For the algorithm design, the integral term should remain within the same correction values to allow the process output to reach the reference steady-state. Still, the discontinuous part of the controller must be able to steer the integral action toward its initial condition to keep correcting new tracking errors and not accumulate unnecessary error integration. In particular, since the discontinuous part of the SMC is associated with the temporal response of the process, this part is manipulated to reset the integral term of the controller. To do so, an SMC strategy with windup instantaneous reset (SMC-WIR) of the integral action in discontinuous control mode is proposed. The SMC-WIR technique works on the process variable to automatically steer the system state to its original operating point. Considering that the process output is supposed to be measured, the transient response can be incorporated into the analysis to give an idea of how rapidly the tracking error varies due to the windup phenomenon. In this scenario, the system output slope is computed as its first derivative at the current time instant. It is worth mentioning that as the system approaches the reference, the output is kept constant and the derivative control action tends to be zero. Therefore, it indicates that once the system output slope is nearby zero, the integrator can be instantaneously restarted and the P, I terms can return to the original condition.

A diagram of the SMC with the proposed WIR approach is shown in Figure 2. Similarly, the proposed WIR strategy is exhibited in Algorithm 1. The SMC with the WIR approach works as follows. At the initial state (see lines 1–2 of Algorithm 1), the control parameters of the SMC controller under test are first tuned and set for required specifications of control performance, including a suitable transitory system response with zero integral control action from a pre-defined sliding surface $\sigma(t)$. In code line 3, it is computed the saturation overflow $sat_1$ between the control output and saturation as the windup resetting conditional. Then, it is also monitored if $e(t)$ is within an error margin $\pm\epsilon_1$ and also if its derivative has increased within a range $\pm\epsilon_2$ to reboot the integral term. Furthermore, it is verified if it only kept working with the SMC in continuous control mode, as presented in pseudo-code lines 4–8 of Algorithm 1. The core of the algorithm is in lines 9–18. Briefly, once the tracking error increase has been detected, the process output derivative is systematically obtained, and it is verified if a value close to zero (see lines 9–12) has been reached. Consequently, if the process output in steady state did not achieve null values (code lines 13–15), the integral term is re-initialized to zero values after verifying if the clamped energy has decayed and, thus, the error is expected to decrease (see line 16). Finally, since the integral term is rebooted to the original conditions, it is necessary to hold the discontinuous control value of the test controller (SMC in this case) to be consistent with the new instantaneous checking of the output derivative, as described in lines 19–20.

### 3.2. SMC with Windup Conditional Reset

The proposed Windup Conditional Reset (WCR) method is intended to periodically restart the integral control action over longer frames, reducing in this way prompt changes of the sliding surface and monotonous corrections of the integral term in SMC. The WCR will work only under programmed conditions. In particular, rebooting is automatically triggered once the reset action is operation-free and the error keeps increasing boundless. The control framework of SMC-WCR can be seen in Figure 3, while the WCR strategy is detailed in Algorithm 2.

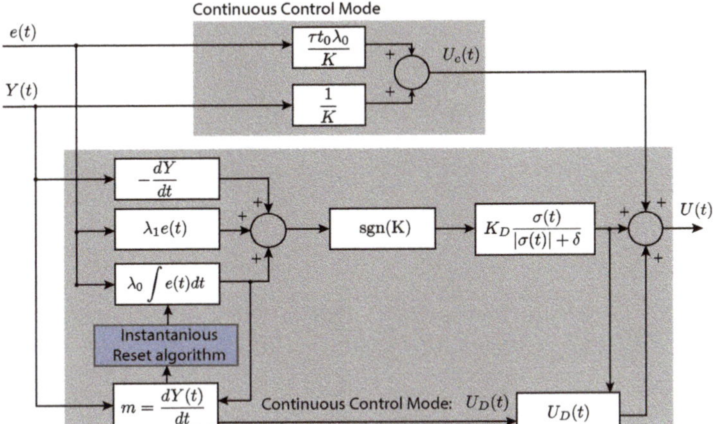



**Figure 2.** General scheme of the proposed SMC-WIR method.

---

**Algorithm 1** Windup Instantaneous Reset—WIR. Anti-windup algorithm to instantaneously reboot the integral control action in an SMC

1: Set initial control and model parameters, i.e., $K_D$, $\delta$, $\lambda_0$, $\lambda_1$.
2: Set the integrator of the SMC by suppressing previous cumulative values of the sliding mode surface $\sigma(t)$.
3: Compute the saturation overflow $sat_1$ between control output and pre-set saturation.
4: **if** $|e(t)| < \epsilon_1$ and $|de(t)/dt| < \epsilon_2$ and $sat_1 < \epsilon_3$ **then**
5:     Maintain monitoring the tracking error $e(t)$.
    (Continuous control mode.)
6: **else**
7:     Maintain operating in continuous mode of the SMC.
    (Perform Window Instantaneous Reset.)
8: **end if**
9: Compute the output derivative $m_1 = \frac{dY}{dt}$.
10: **if** The process output derivative $m_1$ differs from zero **then**
11:     Maintain verifying the process output derivative.
12: **else**
13:     **if** The process output in steady-state does not tend to zero **then**
14:         Monitor the output slope without restarting the integral control action.
15:     **else**
16:         Reinitialize the integrator to zero in Equation (6). until the clamped energy fully decays.
17:     **end if**
18: **end if**
19: Maintain the last discontinuous control value $U_D$.
20: Go back at the beginning of the WIR algorithm (code line 3).

---

The operation of Algorithm 2 for the WCR approach is as follows. Analogous to the previous resetting algorithm, the initial conditions of the test controller require to be set; however, it is assumed here an additional variable to account for the number of reboots settled by the operator. The number of reset counts $n_r$ is configured to assess the persistence of entering in windup mode (see code line 3 of Algorithm 2). Pseudo-code line 4 is in charge of computing the saturation excess $sat_2$ to begin with the windup resetting condition. Lines 5–7 are devoted to verify if the tracking error has increased and if its derivative is within an error speed bounding range $\epsilon_2$ to operate the re-initialization of the integral control action in the discontinuous part of the test controller. Unlike the previous anti-windup algorithm, now it is monitored if the number of reboots $n_r$ has not exceeded the counting of resets of the integral control term to prevent the tracking error from being repeatedly reset (see the core of this algorithm in lines 10–23). Then, the reset of the integral control action in the discontinuous control part is conditioned to the number of resets, avoiding in this way the recurrent and unnecessary initialization of the integral control action. Finally, the algorithm is repeated after holding the previous discontinuous control value, as shown in lines 24–25.

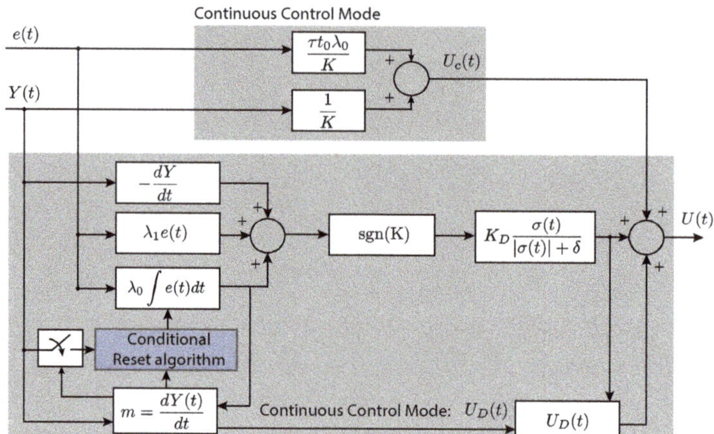

**Figure 3.** General scheme of the proposed SMC-WCR method.

---

**Algorithm 2** Windup Conditional Reset—WCR. Anti-windup algorithm to reboot the integrator in a controller by conditionally evaluating the process output derivative

---

1: Set initial parameters an variables of the controller. In this case, $K_D$, $\delta$, $\lambda_0$, $\lambda_1$.
2: Configure the integrator in $\sigma(t)$ by initially adjusting zero values.
3: Let $n_r$ be the reboot counts prescribed by the designer.
4: Compute the saturation overflow $sat_2$ between the control output and predefined saturation value.
5: **if** $|e(t)| < \epsilon_1$ and $|de(t)/dt| < \epsilon_2$ and $sat_2 < \epsilon_3$ **then**
6:     Maintain monitoring for the tracking tracking error
    (continuous control mode).
7: **else**
8:     Start the Window Conditional Reset
    (discontinuous control mode).
9: **end if**
10: Compute the process output derivative with $m_2 = \frac{dY}{dt}$
11: **if** The output derivative $m_2$ does not tend to zero **then**
12:     Keep searching for output zero-slope condition on $m_2$.
13: **else**
14:     **if** The process output in steady-state does not tend to zero **then**
15:         Monitor the process output derivative without restarting the integral control action.
16:     **else**
17:         **if** The reset counts do not overpass the number of reboots $n_r$ **then**
18:             Monitor the process output derivative $m_2$ avoiding to tend zero values.
19:         **else**
20:             Reinitialize the integrator to zero in Equation (6) until the clamped energy fully decays.
21:         **end if**
22:     **end if**
23: **end if**
24: Maintain the last discontinuous control value $U_D$.
25: Go back at the beginning of the WCR algorithm (code line 4).

---

**Remark 1.** *When switching among the anti-windup algorithms, there are changes in the integral term slope of the WCR and WIR, as shown in Figure 4. Note that when the algorithms run for times greater than 300 min, the slope $m_1$ of the WIR is faster than the slope $m_2$ for the WCR algorithm, despite the fact that WIR algorithm updates the integration faster than in the WCR; therefore, the computational activity could increase.*

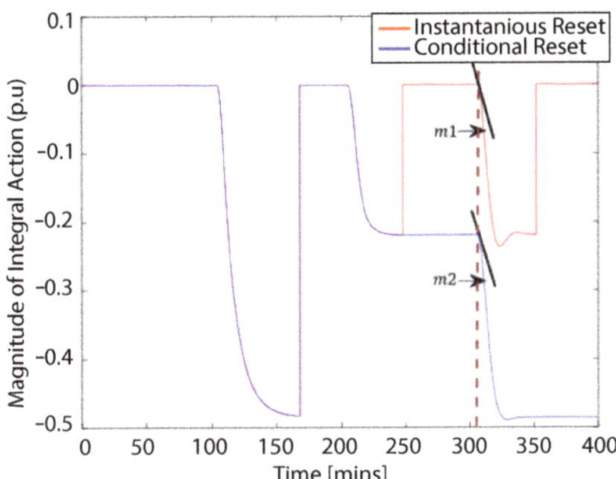


**Figure 4.** Evolution of the integrator slope on the sliding surface in instantaneous and conditional reset.

## 4. Modeling of the Mixing Process

This Section presents two process models where the proposed anti-windup algorithms were implemented.

### 4.1. Mixing Tank Process with Long Time-Delay

Consider the mixing tank shown in Figure 5, the operation mode consists of the concurrently entering of hot $W_1(t)$ and cold product flow $W_2(t)$ as inputs to the process with temperatures $T_1(t)$ and $T_2(t)$, respectively. The output $T_4(t)$ is the mixture temperature measured at a point 125 ft downstream from the mixing tank. The Fail-Closed (FC) actuator is in charge of regulating the cold stream to maintain the desired temperature $T_3$ within the mixing tank.

The control objective consists on maintaining the required mixing temperature $T_3(t)$ (despite disturbances of hot flow $W_1(t)$) through the control output $u(t)$ of the SMC acting on the FC valve position. In general, the product mixing model considers the following characteristics:

- The mixing tank and the entire guidance pipe are ideally isolated.
- The product within the repository is completely homogenized.
- The TT is calibrated in an operating range [100, 200] °F.
- The TT is assumed to provide variations of the process temperature output.
- The height and volumetric properties of the product inside the mixing tank remain constant during the whole test due to its internal control system.

The mathematical formulation of the process dynamics is based on the principle of energy balance, as follows:

$$W_1(t)C_{p1}(t)T_1(t) + W_2(t)C_{p2}(t)T_2(t) - (W_1(t) + W_2(t)) \times$$
$$C_{p3}(t)T_3(t) = V\rho v_3 \frac{dT_3(t)}{dt} \tag{12}$$

where the output temperature $T_4(t)$, after transporting the product with initial output temperature $T_3(t)$ from the mixing tank towards the location of the transmitter, is given by:

$$T_4(t) = T_3(t - t_0) \tag{13}$$

where the time-delay $t_0$ depends on the transport pipe length $L$, the cross-section of the pipe $A$, and the hot $W_1$ and cold $W_2$ product flow. Then, this time delay can be quantified:

$$t_0 = \frac{LA\rho}{W_1(t) + W_2(t)} \tag{14}$$

where $\rho$ is the density of the mixed product. On the other hand, the output of the temperature transmitter $TO(t)$ is represented according to the following system output dynamics:

$$\frac{dTO(t)}{dt} = \frac{1}{\tau_T}\left[\frac{T_4(t) - 100}{100} - TO(t)\right] \tag{15}$$

where $\tau_T$ is the constant time of the temperature sensor. The servo-valve positioning is driven by the controller according to the following equation:

$$\frac{V_p(t)}{dt} = \frac{1}{\tau_{V_p}}\left[m_p(t) - V_p\right] \tag{16}$$

where $m_p(t)$ is the process input to be controlled. The temperature output after the servo-valve with respect to cold product flow also can be represented by:

$$W_2(t) = \frac{500}{60}C_{VL}V_p(t)\sqrt{G_f \Delta P_v} \tag{17}$$

where $C_{VL}$ the valve flow coefficient; $G_f$ the specified acceleration gravity; and $\Delta P_v$ the pressure drop across the servo-valve. A complete description of the model parameters in stationary state and the system variables in (12)–(17) can be found in [6,20].

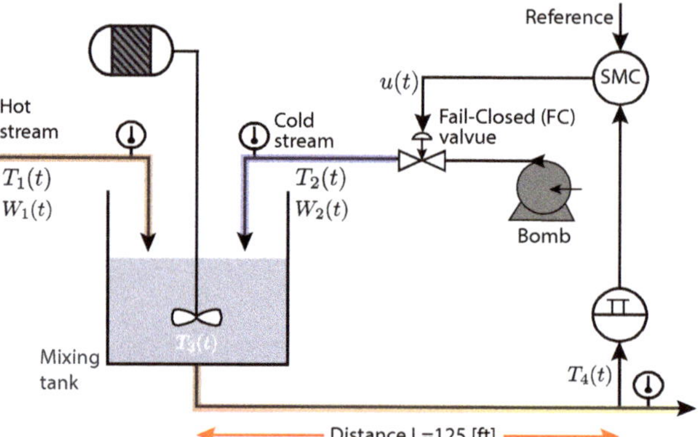


**Figure 5.** Mixing Tank Process. An SMC controller receives temperature measurements $T_4(t)$ of the mixing tank through the Temperature Transmitter (TT). The control output $u(t)$ is applied to a Fail-Closed (FC) typed servo-valve so that regulates the output temperature of the product with cold flow input $T_2(t)$. Hot flow disturbs the system temperature, whereas the temperature output and process delay depend on the distance between the mixing tank and TT.

To test the proposed anti-windup algorithms, the FOPDT approximation [20] of the mixing tank described in (12)–(17) was considered. To obtain the reduced order model, the reaction curve procedure is used [39]. Then, by inspection, the necessary gain was found, as well as the time constant and time delay parameters of the FOPDT model as detailed in the transfer function $G(s)$ of the process model given by:

$$G(s) = \frac{Y(s)}{U(s)} = -\frac{0.81\,e^{-3.14s}}{2.29s + 1} \tag{18}$$

where $\frac{Y(s)}{U(s)}$ is the output/input relation for the FOPDT model of the thermal process under test. The model output is associated with the output temperature $T_4(t)$, whereas the model input stands for the control variable $m(t)$, which is related to the valve position. The control parameters, on the other hand, were adjusted using experimentation trials and the reaction curve of the system according to the Nelder–Mead method for processes that can be approximated to a delayed first-order system [6]. By doing so, these were the results: $\lambda_0 = 0.14$, $\lambda_1 = 0.50$, $K_D = 0.75$, and $\delta = 0.75$.

*4.2. Continuous Stirred Tank Reactor Process with Long Time-Delay*

The Continuous Stirred Tank Reactor (CSTR) is used in chemical reaction processes, whose dynamics present an inverse response. The CSTR comprises a stirred tank in which the reaction material and reacted material flow are continuously mixed [40] (see Figure 6). The reactor is covered by a jacket in which a coolant inlet allows to flow of a cooling liquid to maintain the temperature constant and guarantees that the heat produced for the reaction is low. Therefore, the effluent stream contains the same composition as the contents.

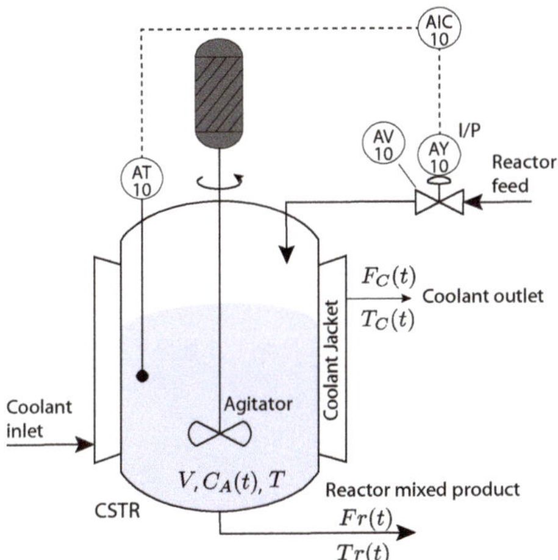


**Figure 6.** Continuous Stirred Tank Reactor—CSTR.

For analysis of the system dynamics, the following assumptions were considered:

- Heat and density capacities of the reactants are constant;
- The heat loss in a coolant jacket is considered negligible;
- The reaction heat and volume remain constant;
- The reaction and reacted material are uniformly mixed.

The mathematical model of the CSTR is obtained based on exothermic reactions according to [41]. A reaction process in a CSTR can be described as follows:

$$A \xrightarrow{K_1} B \xrightarrow{K_2} C \tag{19}$$

$$2A \xrightarrow{K_3} D \tag{20}$$

where $A \xrightarrow{K_1} B$ stands for an exothermic reaction. From mass balance on reactants $A$ and $B$, the system can be described as follows:

$$\frac{dC_A(t)}{dt} = \frac{F_r(t)}{V}[C_{Ai} - C_A(t)] - k_1\, C_A(t) - k_3\, C_A^2(t) \tag{21}$$

$$\frac{dC_B(t)}{dt} = -\frac{F_r(t)}{V}C_B(t) + k_1\, C_A(t) = k_2\, C_B(t) \tag{22}$$

Reactor temperature dynamics associated with the energy balance in the jacket:

$$\frac{T_c}{dt} = K_1(T(t) - T_c(t)) - \frac{F_c(t)}{V}(T_c(t) - T_{ci}(t))$$

For practical purposes, it should be considered that the control range of the concentration B is from 0 to 1.5714 mol $L^{-1}$; the range of variation of the flow is from 0 to 634.1719 L $min^{-1}$. In addition, the transmitter signal $y$, process input $u$, and flow through the reactor $F_r$ are presented in percentage. The model of the temperature transmitter for the reactor is given by: The sensor–transmitter element takes the form:

$$\frac{dTO(t)}{dt} = \frac{1}{\tau_T}\left(\frac{T(t) - 80}{20} - TO(t)\right)$$

Similar to the previous model, the CSTR process is characterized by an approximated FOPDT model, and its parameters were obtained with the procedure in [6]. The model in transfer function is given by:

$$H(s) = \frac{Y_R(s)}{U_R(s)} = \frac{1.66}{12.34s + 1}e^{-3.16s} \tag{23}$$

where the output $Y_R$ is associated with the reactor temperature and $U_R$ represents the control valve position. Variables of the CSTR model are detailed in the Nomenclature Section. The steady-state conditions for each variable of the CSTR process can be found in Table 1. For more details regarding the CSTR model, the reader is referred to [40].

**Table 1.** Initial values of the continuous stirred reactor tank.

| Model Parameters | Values | Units |
|---|---|---|
| $k_1$ | 5/6 | $min^{-1}$ |
| $k_2$ | 5/3 | $min^{-1}$ |
| $k_3$ | 1/6 | L. $mol^{-1}$. $min^{-1}$ |
| $C_{Ai}$ | 10 | mol $L^{-1}$ |
| $V$ | 700 | L |
| $C_{Ao}$ | 2.9175 | mol $L^{-1}$ |
| $C_{Bo}$ | 1.1 | mol $L^{-1}$ |
| $u_{0\%}$ | 60 | % |

## 5. Results for the VHMT Process

This Section presents the results of assessing tracking and robust control performance on the VHMT process.

### 5.1. System Response against Disturbances

This test was performed using six trials to achieve robust performance. The first test was carried out using three control versions, i.e., the SMC per se and the other two with the proposed SMC-WIR and SMC-WCR control strategies. The three remaining trials were carried out with the classical PID controller, including the proposed windup resetting approaches. In all six scenarios, the operating temperature of the product is set to $T_3(t) = 15\,°F$ with a stepped decrease variation of hot flow mass $W_1(t)$ acting as disturbance and changing from 250 lb/min to 110 lb/min.

Results of robustness tests are shown in Figure 7, while the system response against hot flow disturbances is presented in Figure 7a. By supervision, the process output is compensated for disturbances through the three versions of the SMC controllers along the whole testing time; however, the settling time and peak response slightly increase in all cases as hot flow temperature decreases while product temperature consequently decreases. The SMC-WIR and SMC-WCR controllers allow the system output to consistently stay on track (at $T_4(t) = 150\,°F$) despite decreasing the hot flow temperature and system dead time. Indeed, the SMC-WCR can provide a system output with a better transient response than the SMC and SMC-WIR controllers. This effect could be possible because of two reasons. First, the reset frequency of the integral action in SMC-WCR is reduced with respect to the other controllers. Second, the integral action for the discontinuous part of the controller is higher than the ones from the other proposed approaches. Indeed, both proposed algorithms work on the sliding mode surface of the SMC controller, but when new disturbances occur, the control output of the SMC-WIR framework becomes faster than the others at expense of larger control input overshoot, as shown in Figure 7b. This increase in the transient of the control action in the SMC-WIR is a result of the integral action updating faster than the ones of the other two controllers.

Both approaches (SMC-WIR and SMC-WCR) enhance the control performance by limiting the magnitude of the integral term of the sliding surface, as shown in Figure 7c. In particular, for the case of SMC-WIR, an increase in the reaction speed can be observed, although it presents a more oscillatory response. Meanwhile, in the SMC-WCR, the evolution of the integrator speed is restricted to avoid consecutively rebooting the integral control action. Then, the discontinuous part of the controller works under the operating limits and the tracking error could not increase unbounded despite the temperature disturbances, counteracting in this way the windup effect while enhancing control performance.

A phase diagram of the error and its derivative is depicted in Figure 8. There it can be observed that the inclusion of the anti-windup algorithm rapidly conducts to error trajectory towards the origin. In particular, the error trajectory of the PID and PID-WIR presented similar response and took common settling behavior, unlike those results obtained with SMC-WCR. The difference between these two responses lies on the ability to counteract disturbances within the control framework, i.e., the common scheme of a PID controller does not allow to directly compensate for disturbances on the discontinuous part, as the case in SMC.

The overall control performance during a 500 min simulation time for each trial is quantified with ISE and ITSE metrics [36] (see Table 2). This Table indicates that the performance of the SMC alone was improved along the trials with the proposed algorithms by mitigating the impact of the windup effect in a 24.35% and 17.05% of the error when using the SMC-WIR and SMC-WCR, respectively. It is worth noticing that this result is hard to reach with common algorithms such as PID, where the ISE is reduced from 2.39 to 2.25 for the best cases (i.e., PID-WIR with respect to PID) [18]. Similarly, based on ITSE when comparing against the SMC, the SMC-WCR has reduced the time spent to overcome the anti-windup effect by 88.7%, while the SMC-WIR approaches by 29.51%.

*5.2. System Response against Reference Changes*

This test consisted in changing the temperature reference in order to assess the control performance when tracking a given trajectory. Similarly, the same six control configurations of the previous experiments have also been considered here. When doing so, besides including variations in the system dynamics with disturbances of hot product flow, the product temperature in the mixing tank was increased by 3 °F for the six scenarios. As shown in Figure 9, the temperature output of the process tracks the reference change set at 270 min in the controllers under test, except for the PID in which an oscillatory response during a long time does not allow to stabilize the process dynamics. For the SMC with both proposed anti-windup approaches, the tracking performance is similar. However, lower peak response and shorter settling time are achieved with the SMC-WCR. Although the reference change does not saturate the servo-valve actuation of the cold product flow, the control action is increased and the temperature output takes longer to reach the steady state. For both proposed anti-windup strategies, the settling time has been reduced since the integral control action has been increased in correspondence with the peak response of the temperature. At this point, it is worth mentioning that the presented anti-windup strategies allow the test controllers, regardless of their control structure, to reduce the impact of the windup effect. Thus, the proposed algorithms comprise an effective approach for integral controllers that seek to shrink the impact of the windup phenomenon.

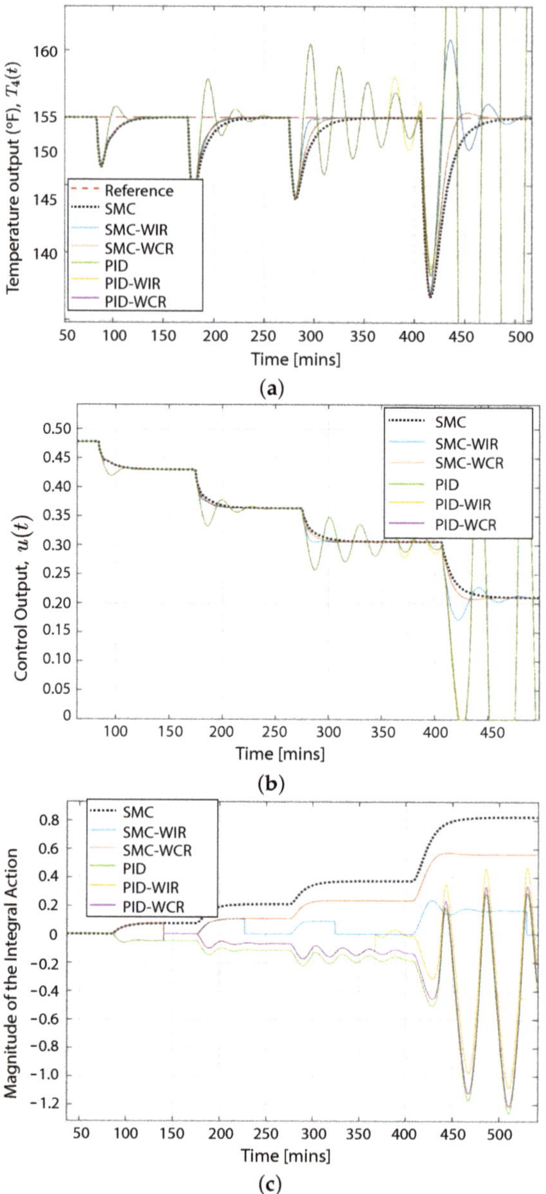

**Figure 7.** Robustness test. Process response using the proposed anti-windup algorithms for different control strategies against systematic disturbances acting on a mixing tank process (see (**a**)). It is noticeable how the control action for the SMC-WIR and SMC-WCR is increased as much as the hot flow disturbance impacts the output stream temperature (see (**b**)). By inspection, the integral action of the SMC-WIR is the one that increases the least, generating a suitable condition to avoid the tracking error accumulation (see (**c**)). (**a**) Process temperature output. (**b**) Control output against disturbances. (**c**) Evolution of integral control action.

On the other hand, regarding the oscillatory behavior in the system output using the PID strategy with both the WIR and WCR anti-windup algorithms, the trajectory error is longer than those of the SMC-WIR and SMC-WCR as shown in Figure 10. Hence, it is evidenced in the closed-up of Figure 10 that the controller takes shorter time to reach the origin in the SMC-WCR with respect to the SMC since the proposed controller occasionally resets the integral control action and reduces the tracking error only when the anti-windup action is activated.

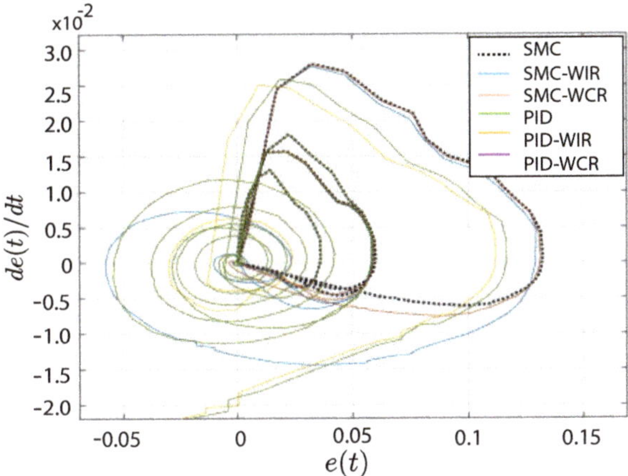

**Figure 8.** Phase diagram of the error derivative for the two test controllers and the two proposed windup correction strategies on the mixing tank process.

**Table 2.** Outcome of robustness tests on the VHMT process by comparing performance in terms of ISE and ITSE.

| | ISE | ITSE | Variation of ISE (Percentage) | Variation of ITSE (Percentage) |
|---|---|---|---|---|
| SMC | 0.357 | 95.454 | - | - |
| SMC-WIR | 0.270 | 67.343 | 24.35% (of reduction) | 29.51% (of reduction) |
| SMC-WCR | 0.302 | 79.301 | 17.05% (of reduction) | 88.7% (of reduction) |
| PID | 2.394 | 933.536 | - | - |
| PID-WIR | 2.259 | 933.235 | 532.77% (of increase) | 497.18% (of increase) |
| PID-WCR | 2.569 | 935.781 | 619.19% (of increase) | 548.06% (of increase) |

As in the previous assessment, to quantitatively evaluate the control performance of the test controllers with the proposed anti-windup algorithms under changes of reference, the ISE and ITSE metrics were also considered, as exhibited in Table 3. In general, the control performance of the two test controllers (i.e., SMC-WIR and SMC-WCR) derived in an enhanced performance as expected. Although such improvement is not quite noticeable for the PID controller since the reduction in the ISE is low when compared to the ISE obtained in the SMC with all anti-windup algorithms. Moreover, even though the SMC-WCR has reached lower ISE (5.36%) with respect to the one of the SMC-WIR (25.62%), the SMC-WCR presents a higher reduction in the ITSE (21.31%) than the SMC-WIR (13.83%). This suggests that the integral action allows the controller to mitigate the anti-windup effect faster than the SMC-WIR performs. As the robustness of the PID controller has not improved against disturbances unlike the SMC, the PID controller with any of the proposed algorithms shows a reduction in the tracking error during the trials. Actually, it shows a 0.05% increase in the ISE for the PID-WCR.

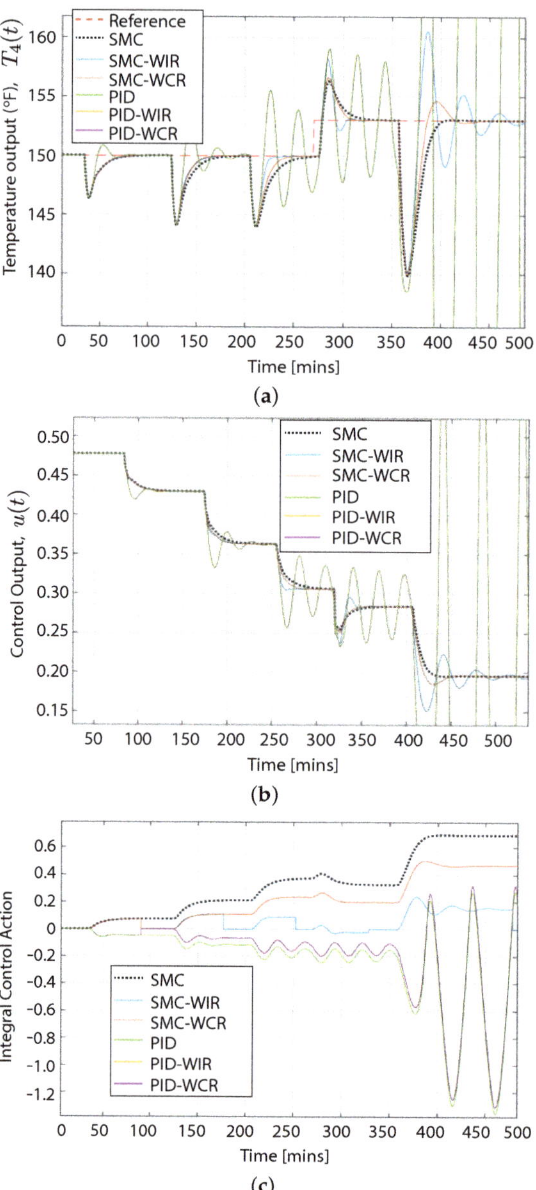

**Figure 9.** Tracking performance test. Process response using the anti-windup algorithms for different control strategies against reference changes of product temperature on a mixing tank process. (**a**) Process temperature output. (**b**) Control output against disturbances. (**c**) Evolution of integral control action.

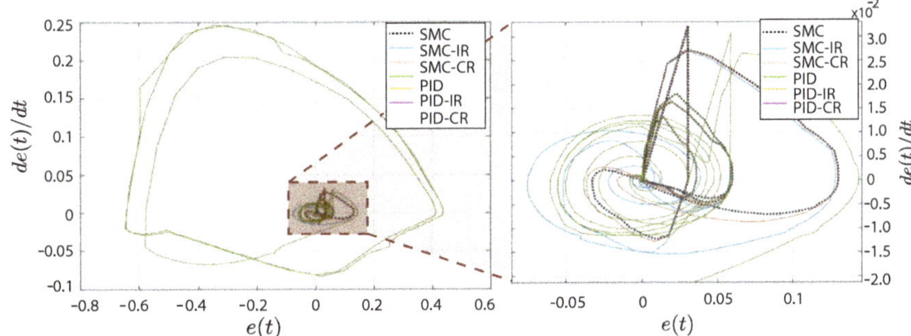

**Figure 10.** Closed-up of the phase diagram for the derivative of the tracking error comparing the two test controllers with the two proposed anti-windup strategies on the mixing tank process.

**Table 3.** Outcome of trajectory tracking tests from different SMC approaches on the VHMT process by comparing control performance in terms of ISE and ITSE.

|  | ISE | ITSE | Variation of ISE (Percentage) | Variation of ITSE (Percentage) |
|---|---|---|---|---|
| SMC | 0.56 | 161.34 | - | - |
| SMC-WIR | 0.41 | 139.02 | 27.78% (of reduction) | 13.83% (of reduction) |
| SMC-WCR | 0.53 | 143.08 | 5.36% (of reduction) | 21.31% (of reduction) |
| PID | 23.66 | 1166.08 | - | - |
| PID-WIR | 23.64 | 1155.66 | 0.05% (of reduction) | 0.89% (of reduction) |
| PID-WCR | 23.69 | 1160.73 | 0.05% (of increase) | 0.45% (of reduction) |

## 6. Results for the CSTR Process

This Section presents the results of assessing tracking and robust control performance on the CSTR process.

### 6.1. System Response against Disturbances

Similar to the previous trials, this test was performed with six trials. Each one of the trials was carried out with the proposed anti-windup algorithms along with the two test controllers. In all six scenarios, the temperature of the flow through the reactor was set to $T(t) = 88\ °C$ with a coolant temperature increase of 10% acting as an external disturbance.

Figure 11 shows the results obtained for the robustness test. The process output is uniformly compensated for the external disturbances for the three SMC controllers, as shown in Figure 11a. Indeed, The SMC-WIR and SMC-WCR controllers lead the system output to keep the reference temperature of $T(t) = 88\ °C$; nevertheless, by inspection, the transient response of the CSTR process when using SMC-WIR is the lowest when compared to the other test controllers. Similarly, the settling time is reduced with the proposed anti-windup techniques, reaching a minimum of 45.7 min and 47.9 min for PID-WCR and SMC-WCR, respectively. Regarding the control output, the control action becomes faster with the two proposed approaches (see Figure 11b) since the integral control action is not cumulatively increasing, as shown in Figure 11c. Unlike the previous test process, the integral control action does not grow unbounded despite the external disturbances because the integral term has been rebooted promptly.

A comparative phase diagram of the tracking error is depicted in Figure 12. This figure presents the error trajectory on which the process variable approaches the temperature reference. Note here that any of the two proposed controllers with the anti-windup algorithms approaches zero due to the integral control action is not cumulative by the continuous and instantaneous reboot action. Nevertheless, the PID-WCR and PID-WIR approach the origin faster than the other proposed SMC controllers.

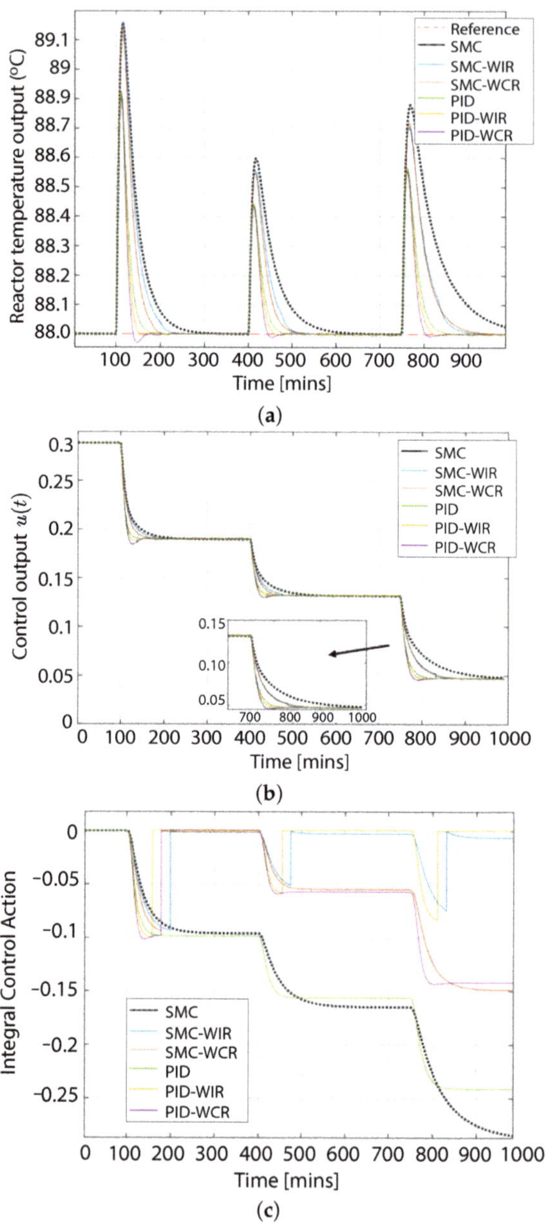

**Figure 11.** Robustness test for the CSTR process. (**a**) shows the regulation response against coolant temperature disturbances. The control actions for the SMC-WIR and SMC-WCR are increased as much as the disturbance impacts reactor temperature (see (**b**)). By inspection, the integral control actions of SMC-WCR and SMC-WIR remain fixed as the tracking error is accumulated over time (see (**c**)). (**a**) Process temperature output. (**b**) Control output against disturbances. (**c**) Evolution of integral control action.

The overall control performance during a 1000 min simulation time for each trial is quantified with ISE and ITSE metrics [36] (see Table 4). This table shows a performance enhancement with any of the proposed anti-windup algorithms reducing the ISE at 82.57% and 85.22% when using the SMC-WIR and SMC-WCR, respectively. Similarly, the SMC-WIR has reduced the time spent to overcome the anti-windup effect by 26.9%, while the SMC-WIR approach by 33.34%.

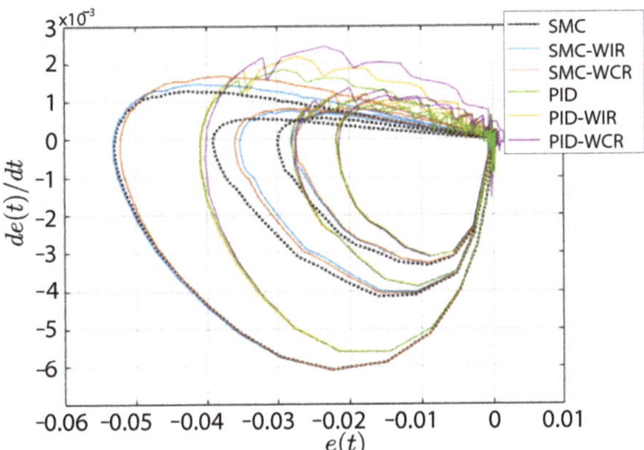


**Figure 12.** A comparative phase diagram of the error derivative for the proposed windup correction strategies on the CSTR process.

**Table 4.** Outcome of robustness tests on the CSTR process by comparing performance in terms of ISE and ITSE.

| | ISE | ITSE | Variation of ISE (Percentage) | Variation of ITSE (Percentage) |
|---|---|---|---|---|
| SMC | 0.786 | 139.111 | - | - |
| SMC-WIR | 0.137 | 101.557 | 82.57% (of reduction) | 26.99% (of reduction) |
| SMC-WCR | 0.124 | 92.736 | 84.22% (of reduction) | 3.34% (of reduction) |
| PID | 0.867 | 95.188 | - | - |
| PID-WIR | 0.815 | 89.528 | 10.31% (of increase) | 35.64% (of reduction) |
| PID-WCR | 0.603 | 85.901 | 23.28% (of reduction) | 38.25% (of reduction) |

### 6.2. System Response against Reference Changes

This test consisted on changing the temperature reference to evaluate control performance. By doing so, the reference temperature experienced a step variation from the steady-state value of $T(t) = 88\ °C$ to $90\ °C$. As shown in Figure 13, the process output is capable of tracking the new temperature reference; however, different transient responses can be seen for the proposed controllers. For example, the PID controls with any of the anti-windup algorithms approach reduced peak responses when compared to the standard PID alone. Similarly, the amplitude of the temperature output of the reactor is higher with the SMC controller than those of the SMC-WIR and SMC-WCR (see Figure 13a). Regarding the settling time, both proposed WIR and WCR strategies lead the PID and SMC controllers to achieve faster process responses; however, the controllers with instantaneous reset perform faster than those of conditional reset since they operate as soon as the integral error begins to grow unbounded (see Figure 13b). The resetting action is depicted in Figure 13b, where integral control action is rapidly steered to zero with the PID-WIR, PID-WCR, SMC-WIR, and SMC-WCR. Note that this is not the case for the PID and SMC as they remain with zero values after the reference change.

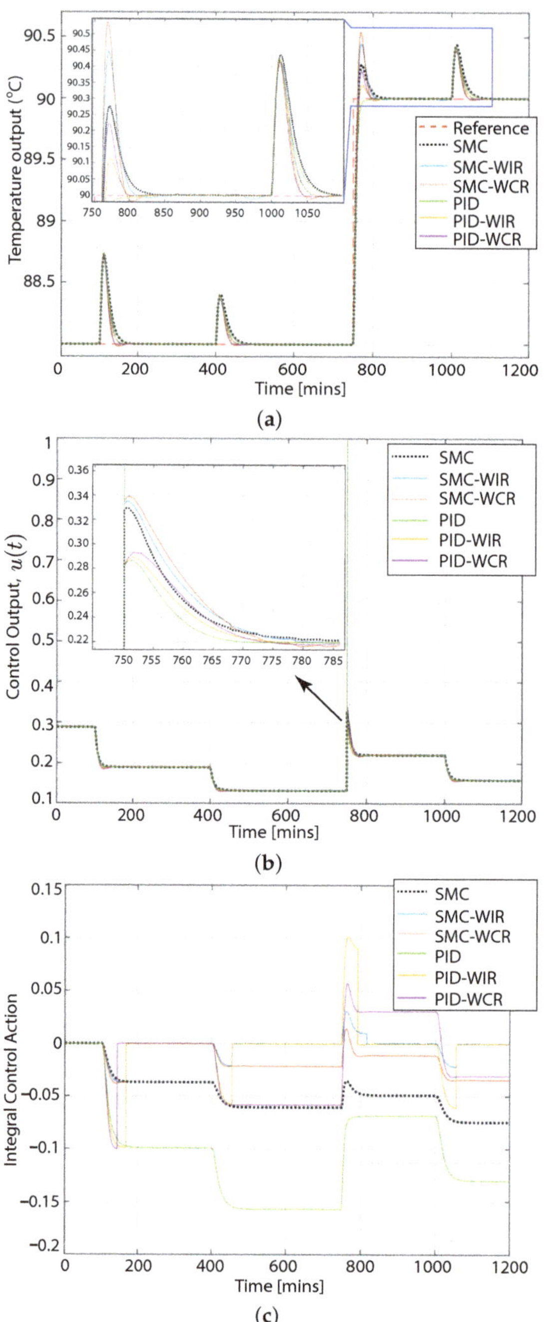

**Figure 13.** Tracking performance test. Responses of the CSTR process using the anti-windup algorithms for different control strategies against reference changes. (**a**) Process temperature output. (**b**) Control output against disturbances. (**c**) Evolution of integral control action.

Control performance obtained with ISE and ITSE metrics is briefly included in Table 5. In general, the control performance of the proposed SMC-WIR and SMC-WCR approaches achieved reasonable performance in terms of ISE and ITSE. In particular, the SMC-WIR demonstrated a greater reduction of ISE with respect to its SMC counterparts, i.e., 81.11% with respect to SMC and 73.33 with respect to SMC-WCR. Similar results are obtained with the PID controller. Regarding the ITSE metric, some similar results were obtained for the PID and SMC using both anti-windup algorithms, achieving a reduction of 35.8% average.

**Table 5.** Outcome of trajectory tracking tests from different SMC approaches on the CSTR process by comparing control performance in terms of ISE and ITSE.

| | ISE | ITSE | Variation of ISE (Percentage) | Variation of ITSE (Percentage) |
|---|---|---|---|---|
| SMC | 0.9 | 128.22 | - | - |
| SMC-WIR | 0.17 | 78.18 | 81.11% (of reduction) | 39.03% (of reduction) |
| SMC-WCR | 0.24 | 83.21 | 73.33% (of reduction) | 35.10% (of reduction) |
| PID | 0.33 | 99.76 | - | - |
| PID-WIR | 0.21 | 87.91 | 76.67% (of reduction) | 31.44% (of reduction) |
| PID-WCR | 0.38 | 89.26 | 57.78% (of reduction) | 30.39% (of reduction) |

## 7. Conclusions

This work proposed, implemented, and validated two anti-windup algorithms for sliding mode control (SMC) and for typical controllers (i.e., PIDs) found in industrial processes with changing time-delay natures. The results of the experiments demonstrated that the SMC, along with the proposed Windup Instantaneous Reset (SMC-WIR) and Windup Conditional Reset (SMC-WCR) algorithms, can avoid saturation in the actuators with reasonable overshoot and suitable settling time despite disturbances that increase the tracking error. Moreover, robust performance was achieved without loss of system stability. The outcome of SMC, with WCR at the integral control action, presented a smoother system response with respect to SMC-WIR. The proposed anti-windup strategies allowed the SMC to improve the performance up to 15.28% in terms of ISE before saturating actuators in the presence of recurrent disturbances during the experimentation. The SMC-WCR also allowed for experiencing an 84% reduction in the settling time in the ITSE metric, thus implying a faster system response while tracking a changing reference. To summarize, the results evidenced that the proposed algorithms can be implemented without changing the original scheme of test controllers and can reduce the anti-windup effect in SMC control strategies for industrial processes with a long time delay.

**Author Contributions:** Conceptualization, A.J.P., J.T., and O.C.; Methodology, A.J.P., J.T., and O.C.; Software, J.T. and O.C.; Validation, A.J.P., J.T., and O.C.; Formal analysis, A.J.P. and J.T.; Investigation, A.J.P. and J.T.; Resources, A.J.P., X.D., and O.C.; Data curation, A.J.P. and O.C.; Writing—original draft preparation, A.J.P., J.T., and O.C.; Writing—review and editing, A.J.P. and O.C.; Visualization, A.J.P. and J.T.; Supervision, A.J.P., M.H., and O.C.; Project administration, A.J.P. and O.C.; Funding acquisition, A.J.P. All authors have read and agreed to the published version of the manuscript.

**Funding:** This research was supported and funded by the Departamento de Ingeniería de Sistemas y Computación with the Universidad Católica del Norte under project 202203010029-VRIDT-UCN. Oscar Camacho thanks the Colegio de Ciencias e Ingenierías, Universidad San Francisco de Quito USFQ, for the supporting given.

**Data Availability Statement:** Not applicable

## Nomenclature

The following nomenclature is used in this work:

| | |
|---|---|
| $U_D$ | Discontinuous control action |
| $U_C$ | Continuous control action. |
| $\sigma$ | Sliding mode surface. |
| $\text{sgn}(K)$ | Sign of the control gain $K$. |
| $K_D$ | Tuning gain for discontinuous control. |
| $\delta$ | Tuning parameter for discontinuous control. |
| $Y, Y^{\text{ref}}$ | Output and reference of the process output, respectively |
| $U$ | Process control input. |
| $G(s)$ | Process model in transfer function of variable $s$. |
| $\tau$ | Time constant of the process (min). |
| $t_0$ | Time delay of the process (min). |
| $\lambda_{1,2}$ | Adjustable parameters of the sliding surface. |
| $\pm\epsilon$ | Allowable tracking error boundary. |
| $m$ | Derivative of the process output. |
| $sat$ | Anti-windup saturation function. |
| $W_{1,2}$ | Hot and cold product flow, respectively, (lb/min). |
| $T_{1,2}$ | Hot and cold stream temperature, respectively, (°C). |
| $T_{4,3}$ | Temperature of the process output with and without delay (°C). |
| $C_p$ | Stream heat capacity at constant pressure, Btu/lb$-$°F. |
| $C_v$ | Stream heat capacity at constant volume (Btu/lb$-$°F). |
| $\rho$ | Density of the mixing tank content in lb/ft$^3$. |
| $A_3$ | Cross section of the mixing tank (ft$^2$). |
| $h_3$ | Product level in the mixing tank (ft). |
| $C_{VL3}$ | Manual valve flow coefficient (gpm/ft$^{1/2}$). |
| $C_{VL}$ | Control valve flow coefficient in (gpm/psi$^{1/2}$). |
| $V_{(p)}$ | Valve position from 0 (valve closed) to 1 (valve open). |
| $TO$ | Output signal of the temperature transmitter (0 to 1 p.u.). |
| $m_p$ | Mass of the process to control (0 to 1 p.u.). |
| $G_f$ | Specific gravity, dimensionless. |
| $\Delta P_v$ | Pressure loss through the control valve (psi). |
| $\tau_{V_p}$ | Control valve time constant (min). |
| $A$ | Cross section of the transportation pipe (ft$^2$). |
| $L$ | Length of the transportation pipe (ft). |
| $F_r$ | Flow through the reactor (m$^3$/min) |
| $V$ | Reactor volume that will remain constant during operation (m$^3$) |
| $C_{A,B}$ | Concentration of A or B in the reactor (mol L$^{-1}$) |
| $C_{Ao,Bo}$ | Concentration of $A_o$ or $B_o$ in steady-state (mol L$^{-1}$) |
| $F_c$ | Coolant flow (m$^3$/s) |
| $TO(t)$ | Normalized output signal of the transmitter (0 to 1 p.u) |
| $T_c$ | Coolant temperature (°C). |
| $k_i \ (i = 1, 2, 3)$ | Reaction rate constants for three reactions. |

## References


1. Mejía, C.; Salazar, E.; Camacho, O. A comparative experimental evaluation of various Smith predictor approaches for a thermal process with large dead time. *Alex. Eng. J.* **2022**, *61*, 9377–9394. [CrossRef]
2. Yan, H.; Zhou, X.; Zhang, H.; Yang, F.; Wu, Z.-G. A Novel Sliding Mode Estimation for Microgrid Control With Communication Time Delays. *IEEE Trans. Smart Grid* **2019**, *10*, 1509–1520. [CrossRef]
3. Liu, Z.X.; Wang, Z.Y.; Wang, Y.; Ji, Z.-C. Optimal Zonotopic Kalman Filter-based State Estimation and Fault-diagnosis Algorithm for Linear Discrete-time System with Time Delay. *Int. J. Control Autom. Syst.* **2022**, *20*, 1757–1771. [CrossRef]
4. Qi, X.; Liu, W.; Lu, J. Observer-based Finite-time Adaptive Prescribed Performance Control for Nonlinear Systems with Input Delay. *Int. J. Control Autom. Syst.* **2022**, *20*, 1428–1438. [CrossRef]
5. Camacho, O.; Leiva, H. Impulsive semilinear heat equation with delay in control and in state. *Asian J. Control* **2020**, *22*, 1075–1089. [CrossRef]
6. Espín, J.; Castrillon, F.; Leiva, H.; Camacho, O. A modified Smith predictor based-sliding mode control approach for integrating processes with dead time. *Alex. Eng. J.* **2022**, *61*, 10119–10137. [CrossRef]

7. Mancisidor, I.; Pena-Sevillano, A.; Dombovari, Z.; Barcena, R.; Munoa, J. Delayed feedback control for chatter suppression in turning machines. *Mechatronics* **2019**, *63*, 102276. [CrossRef]

8. Gao, F.; Chen, W. Disturbance Rejection in Singular Time-delay Systems with External Disturbances. *Int. J. Control Autom. Syst.* **2022**, *20*, 1841–1848. [CrossRef]

9. Cajamarca, B.; PatiñO, K.; Camacho, O.; Chavez, D.; Leica, P.; Pozo, M. A comparative analysis of sliding mode controllers based on internal model for a nonminimum phase buck and boost converter. In Proceedings of the 2019 International Conference on Information Systems and Computer Science (INCISCOS), Quito, Ecuador, 20–22 November 2019; pp. 189–195. [CrossRef]

10. Salinas, L.R.; Santiago, D.; Slawiñski, E.; Mut, V.; Chavez, D.; Leica, P.; Camacho, O. P+ d plus sliding mode control for bilateral teleoperation of a mobile robot. *Int. J. Control Autom. Syst.* **2018**, *16*, 1927–1937. [CrossRef]

11. Báez, E.; Bravo, Y.; Leica, P.; Chávez, D.; Camacho, O. Dynamical sliding mode control for nonlinear systems with variable delay. In Proceedings of the 2017 IEEE 3rd Colombian Conference on Automatic Control (CCAC), Cartagena, Colombia, 18–20 October 2017; pp. 1–6. [CrossRef]

12. Owais, K.; Mahmood, P.; Ejaz, A.; Jamshed, I. On the derivation of novel model and sophisticated control of flexible joint manipulator. *Rev. Roum. Sci. Tech.-Ser. Electrotech. Energ.* **2017**, *62*, 103–108.

13. Prado, A.J.; Cheein, F.A.; Torres-Torriti, M. Probabilistic approaches for self-tuning path tracking controllers using prior knowledge of the terrain. In Proceedings of the 2016 IEEE/RSJ International Conference on Intelligent Robots and Systems (IROS), Daejeon, Republic of Korea, 9–14 October 2016; pp. 3095–3100. [CrossRef]

14. Wu, X.; Lin, Z. On Immediate, Delayed and Anticipatory Activation of Anti-Windup Mechanism: Static Anti-Windup Case. *IEEE Trans. Autom. Control* **2012**, *57*, 771–777. [CrossRef]

15. Peng, Y.; Vrancic, D.; Hanus, R. Anti-windup, bumpless, and conditioned transfer techniques for PID controllers. *IEEE Control Syst. Mag.* **1996**, *16*, 48–57. [CrossRef]

16. Chen, L.; Fang, J. Adaptive Continuous Sliding Mode Control for Fractional-order Systems with Uncertainties and Unknown Control Gains. *Int. J. Control Autom. Syst.* **2022**, *20*, 1509–1520. [CrossRef]

17. Iqbal, J. Modern Control Laws for an Articulated Robotic Arm: Modeling and Simulation. *Eng. Technol. Appl. Sci. Res.* **2019**, *9*, 4057–4061. [CrossRef]

18. Herrera, M.; Morales, L.; Rosales, A.; Garcia, Y.; Camacho, O. Processes with variable dead time: Comparison of hybrid control schemes based on internal model. In Proceedings of the 2017 IEEE Second Ecuador Technical Chapters Meeting (ETCM), Salinas, Ecuador, 16–20 October 2017; pp. 1–6. [CrossRef]

19. Prado, J.; Yandun, F.; Torres Torriti, M.; Auat Cheein, F. Overcoming the Loss of Performance in Unmanned Ground Vehicles Due to the Terrain Variability. *IEEE Access* **2018**, *6*, 17391–17406. [CrossRef]

20. Torres, J.; Chavez, D.; Aboukheir, H.; Herrera, M.; Prado, J.; Camacho, O. Anti-Windup Algorithms for Sliding Mode Control in Processes with Variable Dead-Time. In Proceedings of the 2020 IEEE ANDESCON, Quito, Ecuador, 13–16 October 2020; pp. 1–6. [CrossRef]

21. Camacho, O.; Díaz, R.; Obando, C. Hardware in the loop simulation for sliding mode control schemes for deadtime systems. In Proceedings of the International Conference on Information Systems and Computer Science (INCISCOS), Quito, Ecuador, 20–22 November 2019; pp. 183–188. [CrossRef]

22. Aichi, B.; Bourahla, M.; Kendouci, K. Real-Time Hybrid Control of Induction Motor Using Sliding Mode and PI Anti-Windup. In Proceedings of the 2018 International Conference on Electrical Sciences and Technologies in Maghreb (CISTEM), Algiers, Algeria, 28–31 October 2018; pp. 1–6. [CrossRef]

23. Guo, X.-G.; Ahn, C.K. Adaptive Fault-Tolerant Pseudo-PID Sliding-Mode Control for High-Speed Train with Integral Quadratic Constraints and Actuator Saturation. *IEEE Trans. Intell. Transp. Syst.* **2021**, *22*, 7421–7431. [CrossRef]

24. Zhang, X.; Su, C. Stability Analysis and Antiwindup Design of Switched Linear Systems with Actuator Saturation. *Int. J. Control Autom. Syst.* **2018**, *16*, 1247–1253. [CrossRef]

25. Zhao, D.; Wang, Y.; Xu, L.; Wu, H. Adaptive Robust Control for a Class of Uncertain Neutral Systems with Time Delays and Nonlinear Uncertainties. *Int. J. Control Autom. Syst.* **2021**, *19*, 1215–1227. [CrossRef]

26. Prusty, S.B.; Seshagiri, S.; Pati, U.C.; Mahapatra, K.K. Sliding mode control of coupled tank systems using conditional integrators. *IEEE/CAA J. Autom. Sin.* **2020**, *7*, 118–125. [CrossRef]

27. Sun, T.; Liu, C.; Wang, X. Anti-windup neural network-sliding mode control for dynamic positioning vessels. In Proceedings of the 2021 8th International Conference on Information, Cybernetics, and Computational Social Systems (ICCSS), Beijing, China, 10–12 December 2021; pp. 356–361. [CrossRef]

28. Scholl, T.H.; Groll, L. Stability Criteria for Time-Delay Systems from an Insightful Perspective on the Characteristic Equation. *IEEE Trans. Autom. Control* 2022, *ahead of print*. [CrossRef]

29. Sajjadi-Kia, S.; Jabbari, F. Modified Anti-windup compensators for stable plants: Dynamic Anti-windup case. In Proceedings of the 48h IEEE Conference on Decision and Control (CDC) held jointly with 2009 28th Chinese Control Conference, Shanghai, China, 15–18 December 2009; pp. 2795–2800. [CrossRef]

30. Camacho, O.; Smith, C. Sliding mode control: An approach to regulate nonlinear chemical processes. *ISA Trans.* **2000**, *39*, 205–218. [CrossRef]

31. Nath, K.; Bera, M.K. Adaptive Integral Sliding Mode Control: An Event-Triggered Approach. *IFAC-PapersOnLine* **2022**, *55*, 604–609. [CrossRef]

32. Nath, U.M.; Dey, C.; Mudi, R.K. Designing of anti-windup feature for internal model controller with real-time performance evaluation on temperature control loop. In Proceedings of the 2019 2nd International Conference on Intelligent Computing, Instrumentation and Control Technologies (ICICICT), Kannur, India, 5–6 July 2019; pp. 787–790. [CrossRef]
33. Bandyopadhyay, B.; Deepak, F.; Park, Y.J. A Robust Algorithm Against Actuator Saturation using Integral Sliding Mode and Composite Nonlinear Feedback. *IFAC Proc. Vol.* **2008**, *41*, 14174–14179. [CrossRef]
34. Septanto, H.; Syaichu-Rohman, A.; Mahayana, D. Static anti-windup compensator design of linear Sliding Mode Control for input saturated systems. In Proceedings of the 2011 International Conference on Electrical Engineering and Informatics, Bandung, Indonesia, 17–19 July 2011; pp. 1–4. [CrossRef]
35. Guanghui, Z.; Huihe, S. Anti-windup design for the controllers of integrating processes with long delay. *J. Syst. Eng. Electron.* **2007**, *18*, 297–303. [CrossRef]
36. Proaño, P.; Capito, L.; Rosales, A.; Camacho, O. Sliding Mode Control: Implementation Like PID for Trajectory-Tracking for Mobile Robots. In Proceedings of the 2015 Asia-Pacific Conference on Computer Aided System Engineering, Quito, Ecuador, 14–16 July 2015; pp. 220–225. [CrossRef]
37. Coronel, W.; Camacho, O. A Dynamic Sliding Mode Controller using a Rotating Type Moving Sliding Surface for Chemical Processes with Variable Delay. In Proceedings of the 2021 IEEE CHILEAN Conference on Electrical, Electronics Engineering, Information and Communication Technologies (CHILECON), Valparaiso, Chile, 6–9 December 2021; pp. 1–6. [CrossRef]
38. Utkin, V.; Poznyak, A.; Orlov, Y.V.; Polyakov, A. *Road Map for Sliding Mode Control Design*; Springer: Cham, Switzerland, 2020. [CrossRef]
39. Smith, C.; Corripio, A.B. *Principles and Practices of Automatic Process Control*, 3rd ed.; Wiley: Hoboken, NJ, USA, 2005, ISBN: 978-0-471-43190-9.
40. Dogu, T.; Dogu, G. *Fundamentals of Chemical Reactor Engineering: A Multi-Scale Approach*; John Wiley & Sons: Hoboken, NJ, USA, 2021.
41. Victor, M.A.; Pedro, B.; Orlando, A. Robustness Considerations on PID Tuning for Regulatory Control of Inverse Response Processes. *IFAC Proc. Vol.* **2012**, *45*, 193–198. [CrossRef]

 *electronics*

*Article*

# Dynamic Fractional-Order Nonsingular Terminal Super-Twisting Sliding Mode Control for a Low-Cost Humanoid Manipulator

Rong Hu [1,2], Xiaolei Xu [1,2], Yi Zhang [1,2,*] and Hua Deng [1,2,*]

1 School of Mechanical and Electrical Engineering, Central South University, Changsha 410017, China
2 The State Key Laboratory of High Performance and Complex Manufacturing, Changsha 410083, China
* Correspondence: zhangyicsu@csu.edu.cn (Y.Z.); hdeng@csu.edu.cn (H.D.)

**Abstract:** Prosthetic humanoid manipulators manufacturing requires light overall weight, small size, compact structure, and low cost to realize wearing purpose. These requirements constrain hardware configuration conditions and aggravate the nonlinearity and coupling effects of manipulators. A dynamic fractional-order nonsingular terminal super-twisting sliding mode (DFONTSM-STA) control is proposed to realize multi-joints coordination for a low-cost humanoid manipulator. This method combines a dynamic fractional-order nonsingular terminal sliding mode (DFONTSM) manifold with the super-twisting reaching law, which can enhance the entire control performance by dynamically changing the position of the sliding mode manifold. By hiding the sign function in a higher-order term, chattering can be effectively suppressed. The stability of the low-cost humanoid manipulator system has been proven based on the Lyapunov stability theory. Experimental results show that the terminal trajectory tracking accuracy of DFONTSM-STA control was promoted by 53.3% and 23.7% respectively compared with FONTSM control and FONTSM-STA control. Thus, the DFONTSM-STA controller is superior in error convergence speed, chattering suppression, and accurate position tracking performance.

**Keywords:** chattering suppression; fractional-order nonsingular terminal sliding mode controllers; higher-order sliding mode controllers; humanoid manipulator

**Citation:** Hu, R.; Xu, X.; Zhang, Y.; Deng, H. Dynamic Fractional-Order Nonsingular Terminal Super-Twisting Sliding Mode Control for a Low-Cost Humanoid Manipulator. *Electronics* **2022**, *11*, 3693. https://doi.org/10.3390/electronics11223693

Academic Editor: Jamshed Iqbal

Received: 14 October 2022
Accepted: 8 November 2022
Published: 11 November 2022

**Publisher's Note:** MDPI stays neutral with regard to jurisdictional claims in published maps and institutional affiliations.



## 1. Introduction

Humanoid manipulators have the characteristics of light weight, small size, strong nonlinearity, coupling, friction, and clearance [1–3]. Among them, friction and clearance are disturbances that significantly influence the control performances of humanoid manipulators. As sliding mode control has the excellent characteristic of efficiently disturbances settlement [4–7], it is widely used for humanoid manipulators' position-tracking control [8–10]. A low-cost humanoid manipulator further constrains its hardware configuration conditions. Complex and high-precision sensors or motors cannot be used when designing a humanoid manipulator because they are generally large in volume. The sensors and motors with small volumes and high precision are generally quite expensive. Teleoperation systems are also not included for cost-saving purposes [11]. All the above features aggravate the nonlinearity, coupling, and complexity of the low-cost humanoid manipulator's dynamics. Benefiting from their variable structure characteristics, sliding mode controllers possess excellent robustness and anti-disturbance ability [12,13]. However, the alternation between the reaching phase and the sliding phase of a sliding mode controller always accompanies by a phenomenon called chattering [14]. The production of the chattering is related to the sliding mode boundary layer thickness [15]. The system states trajectory always has a certain speed when reaching the manifold, and the inertia makes the system states move across the manifold. After that, the controller input generates a reverse signal to pull the system states back to the sliding mode manifold. However, it must take time. This time difference makes the system states trajectory deviate from the sliding mode manifold for a certain distance and intensifies the thickness of the boundary layer. In this way, the system states trajectory repeatedly passes through the sliding mode manifold, which is reflected in the macroscopic view as a chattering phenomenon [16,17]. Chattering may cause energy consumption and mechanical system damage problems. Therefore, the chattering problem must be first solved when designing sliding mode controllers, especially for a low-cost humanoid manipulator.

The sign function term in a sliding mode reaching law is not only the key to the rapid convergence of sliding functions but also one of the main causes of chattering [18]. Replacement of the sign function with the saturation function is one way to suppress chattering [19]. Hiding the sign function term in a higher-order term is another effective way to eliminate chattering [20–24]. As the system states enter the quasi-sliding mode, the super-twisting sliding mode (STSM) gains the characteristics of the sign function term and ensures that the sliding function and its derivative converge to zero within a limited time [25–27]. The STSM control which belongs to the second-order sliding mode control only requires information on the sliding function [28]. It contains a power rate term that can reduce the time of the system states arriving in quasi-sliding mode and suppress the chattering more effectively compared with the proportional term [29]. Moreno and Osorio used the Lyapunov stability theory to analyze the stability of the super-twisting algorithm (STA) [30]. The proof process has been used in many STSM controllers. Kali et al. proposed an STSM control for manipulators with uncertainties. [31]. Tayebi-Haghighi et al. designed a high-order STSM controller to address multiple control challenges of robots with verification. [20].

Apart from introducing super-twisting algorithms, adding a fractional-order operator to the sliding mode controller is also an effective way to suppress chattering and improve motion control performance [32]. Applying a fractional-order (FO) operator to a sliding mode controller can enhance the flexibility of the controller, and improve its control performances [32–36]. Among all the fractional-order controllers, fractional-order nonsingular terminal sliding mode (FONTSM) controllers are widely used in manipulators to obtain precise position-tracking performance [37–43]. Tran et al. designed an adaptive fuzzy FONTSM control strategy for a two-degree-of-freedom manipulator and used simulation results to illustrate its control performances [40]. Nojavanzadeh et al. proposed an adaptive FONTSM controller to control robots with uncertainty and external interference and used simulation to verify its effectiveness [41]. Su et al. proposed an adaptive FONTSM controller for a cable-driven manipulator with external interferences, which was verified by simulations [42]. Wang et al. formulated an adaptive sliding mode controller that integrates the advantages of PID and FOTNSM manifold [43]. However, most of the FONTSM controllers were verified by simulations. Only a few papers showed the experimental verification of the FONTSM control for manipulators [44–46]. In the previous work of the authors, a dynamic FONTSM controller was proposed for a class of second-order nonlinear systems with simulation verification [47].

To summarize, super-twisting sliding mode control can effectively accelerate the speed of system states reaching the quasi-sliding mode and suppress the chattering. FONTSM control has superiority in motion control accuracy. The fractional-order nonsingular super-twisting sliding mode (FONTSM-STA) control combines the advantages of the FONTSM manifold and the STA, which not only meets the demand for high tracking accuracy but also suppresses the chattering [45]. However, simply combining these two methods has little effect on the joint error converge speed which is a vital factor for humanoid manipulators to mimic arm movements.

Aiming at the above-mentioned problems, a dynamic fractional-order nonsingular terminal super-twisting sliding mode (DFONTSM-STA) control scheme is proposed for a low-cost humanoid manipulator in this study. The main contributions of this paper are summarized as follows:

(1) A DFONTSM-STA controller is formulated for the control of the low-cost humanoid manipulator by combining the dynamic fractional-order nonsingular terminal sliding mode (DFONTSM) manifold with the super-twisting reaching law, which can effectively improve the control accuracy, quickly force the tracking error of each joint to converge and significantly suppress the chattering. The stability and convergence of the low-cost humanoid manipulator control system are proven based on the Lyapunov stability theory.

(2) Experiments illustrate the superiority and feasibility of the proposed DFONTSM-STA control for the low-cost humanoid manipulator. Compared with FONTSM and FONTSM-STA control, the DFONTSM-STA controller has superior control performance. Its terminal position tracking accuracy is increased by 53.3% and 23.7% respectively. Its chattering of joints one to four is decreased by 54.1%, 51.1%, 46.2%, and 55.1% compared with FONTSM control. Its error convergence speed is accelerated significantly. Joints one and two are converged at the beginning, and joints three and four are accelerated by 43.5% and 33.6%, 72.7%, and 54.6% respectively compared with the FONTSM and FONTSM-STA control.

## 2. Model Description and Problem Analysis

The low-cost humanoid manipulator is limited by cost, size, and weight, which uses hall sensors that generate three pulses in one turn to read the angle position information. The position is detected by sensors whose resolution converted to the joint is $0.3°$. Figure 1 shows the three dimensional (3D) model of the low-cost humanoid manipulator and the Denavit-Hartenberg (D-H) coordinate of each joint. Table 1 shows its D-H parameters. Coordinate $x_0y_0z_0$ represents the base coordinate; $x_1y_1z_1$ and $x_2y_2z_2$ represent the coordinate of shoulder joints one and two; $x_3y_3z_3$ and $x_4y_4z_4$ represent the coordinate of elbow joints three and four; $x_ty_tz_t$ represents the terminal coordinate. The dynamic equation of the four degrees-of-freedoms (DOFs) low-cost humanoid manipulator with friction model is described as follows:

$$M(q)\ddot{q} + C(q,\dot{q}) + G(q) + f(q,\dot{q}) = \tau_r \tag{1}$$

$$f(q,\dot{q}) = f_c sign(\dot{q}) + f_v \dot{q} \tag{2}$$

where $q, \dot{q}$, and $\ddot{q} \in R^4$ respectively represent the position, velocity, and acceleration vectors of every joint; $M(q) \in R^{4\times4}$ is the non-singular inertia matrix; $C(q,\dot{q}) \in R^{4\times4}$ is the centrifugal and Coriolis matrix; $G(q) \in R^4$ represents the gravitational vector; $f(q,\dot{q}) \in R^4$ is the vector of the viscous friction torque at joints; $\tau \in R^4$ denotes the torque input vectors; $f_c = diag(f_{c1}, \quad f_{c2}, \quad f_{c3}, \quad f_{c4})$ denotes every joint coulomb friction torque; $f_v = diag(f_{v1}, \quad f_{v2}, \quad f_{v3}, \quad f_{v4})$ denotes every joint viscous friction coefficient.

**Table 1.** The parameter list of D-H.

| Link $i$ | $\alpha_{i-1}$ | $a_{i-1}$ | $\theta_i$ | $d_i$ |
|---|---|---|---|---|
| 1 | 0 | 0 | $\theta_1\ (0°)$ | 0 |
| 2 | 90° | 0 | $\theta_2\ (-90°)$ | 0 |
| 3 | 90° | 0 | $\theta_3\ (-90°)$ | $d_3 = -313$ mm |
| 4 | 90° | 0 | $\theta_4\ (-180°)$ | 0 |
| Terminal | 0° | 0 | $\theta_5\ (0°)$ | $d_5 = -441$ mm |

The low-cost humanoid manipulator contains four joints to mimic shoulder and elbow movements. To suppress chattering and improve both the error convergence speed and the tracking accuracy, it is necessary to design a new sliding mode control method. As the STSM control hides the sign function in higher order, it possesses a better effect on chattering suppression. Apart from the chattering suppression requirement, the low-cost humanoid manipulator also needs precise trajectory tracking accuracy for mimicking human arm movements. The DFONTSM manifold can enhance the entire control performance by dynamically changing the position of the sliding mode manifold [47]. The steady-state accuracy can be improved because the fractional-order operator has the characteristics of heredity and memory, which can describe detailed information and increase the flexibility of the control law. Therefore, to solve the trajectory tracking performances of the low-cost humanoid manipulator, this paper proposes a DFONTSM-STA control method that combines a DFONTSM manifold with a super-twisting reaching law.

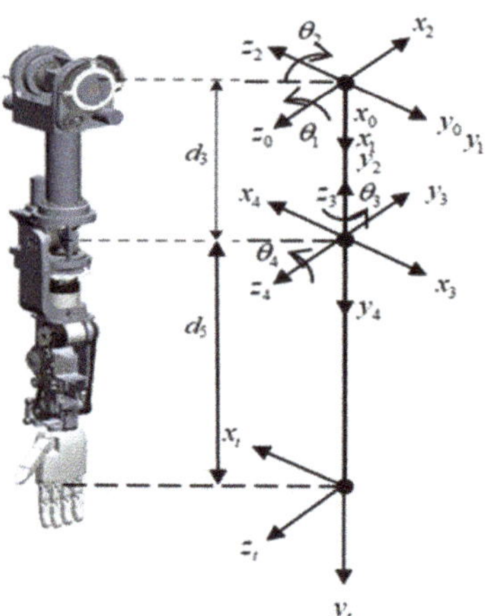

**Figure 1.** Three dimensional (3D) model and Denavit-Hartenberg (D-H) coordinate of the low-cost humanoid manipulator.

### 3. DFONTSM-STA Control for the Low-Cost Humanoid Manipulator

This section will briefly present some definitions and lemmas that are needed in the stability proof of the proposed DFONTSM-STA controller.

**Definition 1.** *The general representation of the FO derivative-integral operator is expressed by Equation (3), [48]*

$$_{t_0}D_t^\lambda = \begin{cases} \frac{d^\lambda}{dt^\lambda} & \lambda > 0 \\ 1 & \lambda = 0 \\ I^\lambda = \int_0^t (dt)^{-\lambda} & \lambda < 0 \end{cases} \tag{3}$$

*where $\lambda$ is the FO and $t_0$ is the initial time. The operator $_{t_0}D_t^\lambda$ is the symbol for the FO, integral and constant operator.*

**Definition 2.** *The $\lambda$ th-order Riemann–Liouville (RL) fractional and integral are presented as Equations (4) and (5) [48].*

$$D^\lambda f(t) = \frac{d^\lambda f(t)}{dt^\lambda} = \frac{1}{\Gamma(m-\lambda)} \bullet \frac{d^m}{dt^m} \bullet \int_{t_0}^t \frac{f(\tau)}{(t-\tau)^{\lambda-m+1}} d\tau \tag{4}$$

$$_{t_0}I_t^\lambda f(t) = \frac{1}{\Gamma(\alpha)} \bullet \int_{t_0}^t \frac{f(t)}{(t-\tau)^{1-\lambda}} d\tau \tag{5}$$

*where $m - 1 < \lambda < m$, $m \in N$.*

**Property 1.** *[48] if $\alpha \in C$, $\beta \in C$, $\Re e(\alpha) > 0$, $\Re e(\beta) > 0$ and $f(a,b) \in Lp(a,b)$ ($1 \leq p \leq \infty$), then $I_{t-}^\alpha I_{t-}^\beta f(x) = I_{t-}^{\alpha+\beta} f(x)$, $I_{t+}^\alpha I_{t+}^\beta f(x) = I_{t+}^{\alpha+\beta} f(x)$; $D_{t-}^\alpha I_{t-}^\alpha f(x) = f(x)$, $D_{t+}^\alpha I_{t+}^\alpha f(x) = f(x)$.*

**Lemma 1.** *[44]. Supposing parameters $a_1, a_2, \cdots, a_n$; $0 < p < 2$ are positive, then the following inequality about $a_1, a_2, \cdots, a_n$ and $p$ can be obtained:*

$$\left(a_1^2 + a_2^2 + \cdots + a_n^2\right)^p \leq \left(a_1^p + a_2^p + \cdots + a_n^p\right)^2$$

**Lemma 2.** *[38,44]. For a Lyapunov function $V(x)$, assuming that $V_0$ is its initial value and $\alpha \in C$, $\beta \in C$, then*

$$\dot{V}(x) + \alpha V(x) + \beta V^\gamma(x) \leq 0, \alpha, \beta > 0, 0 < \gamma < 1 \tag{6}$$

*The corresponding settling time can be calculated as*

$$T \leq \alpha^{-1}(1-\gamma)^{-1} \ln\left(1 + \alpha\beta^{-1} V_0^{1-\gamma}\right) \tag{7}$$

The FONTSM manifold parameter transforms into the function $k^*$ that calculates the error by exponential function is designed as Equation (9). Then the DFONTSM manifold (8) is formulated for the system described in the model (1) [47].

$$s = \dot{e} + k^* D^{\lambda-1}[sig(e)^\alpha] \tag{8}$$

$$k^* = \frac{k_0}{\delta_0 + \rho_0 \cdot \delta_0 \cdot \exp^{-\alpha_\kappa |e|}} \tag{9}$$

where $0 < \lambda_i < 1$, $i = 1\sim4$, $D^\lambda$ is the FO operator; $\delta_0 = diag(\delta_{0i})$ is the diagonal matrix and $\delta_{0i} \in (0,1)$; $\rho_0 = diag(\rho_{0i})$ and $\alpha_\kappa = diag(\alpha_{\kappa i})$ are tuning matrices and $\rho_{0i}, \alpha_{\kappa i} \in (0, +\infty)$; $e = [q_{d1} - q_1, \quad \cdots \quad q_{d4} - q_4] \in R_4$ denotes the tracking error vector between the target joint rotation angle and the actual joint rotation angle. $k_0 = diag(k_{0i})$ is a positive diagonal matrix. $i$ can be chosen from one to four.

$$sig(e)^\alpha = \left[|e_1|^{\alpha_1} sgn(e_1), \quad \cdots, \quad |e_4|^{\alpha_4} sgn(e_4)\right]^T$$

where $\alpha$ is the exponential parameter, which satisfies the relation: $1 > \alpha > 0$. A super-twisting reaching law from [31] is adopted.

$$\begin{aligned}\dot{s} &= -\rho_1|s|^{1/2} sgn(s) + \omega \\ \dot{\omega} &= -\rho_3 sgn(s)\end{aligned} \tag{10}$$

where $k_1 = diag(k_{1i})$, $k_2 = diag(k_{2i})$ are constant matrices and $0 < b = b_1 = \cdots = b_4 < 1$, $i = 1\sim4$; With the above reaching law (10), the error vector $(e, \dot{e})$ can be forced to zero. Derivating the DFONTSM manifold (8), we have:

$$\dot{s} = \ddot{e} + k^* \cdot D^\lambda[sig(e)^\alpha] + \dot{k}^* \cdot D^{\lambda-1}[sig(e)^\alpha] \tag{11}$$

where:

$$\dot{k}^* = -\rho_0 \cdot \delta_0 \cdot k_0 \cdot \alpha_\kappa \cdot \dot{e} \cdot \left(\delta_0 + \rho_0 \cdot \delta_0 \cdot \exp^{-\alpha_\kappa \cdot |e|}\right)^{-2} \cdot \exp^{-\alpha_\kappa \cdot |e|} \cdot sign(e) \tag{12}$$

Combining (11) with (10), the control law (14) can be obtained by Equation (13).

$$\ddot{e} + k^* \cdot D^\lambda[sig(e)^\alpha] + \dot{k}^* \cdot D^{\lambda-1}[sig(e)^\alpha] = -\rho_1|s|^{1/2} sgn(s) + \omega \tag{13}$$

$$\tau_r = M(q)\left(\ddot{q}_d + k^* \cdot D^\lambda[sig(e)^\alpha] + \dot{k}^* \cdot D^{\lambda-1}[sig(e)^\alpha] + \rho_1|s|^{1/2} sgn(s) + \rho_3 \int_0^t sgn(s)dt\right) + C(q,\dot{q}) + G(q) \tag{14}$$

## 4. Stability Proof of DFONTSM-STA Control

The states $\ddot{q}$ can be obtained from the dynamic model (1):

$$\ddot{q} = M(q)^{-1}\left(\tau_r - C(q,\dot{q}) - G(q) - f(q,\dot{q})\right) \tag{15}$$

where $\|f(q,\dot{q})\| \le \|f_{\max}(x,t)\| \le \overline{F}$. The constant diagonal matrix of $\overline{F}$ represents the friction interference upper boundary. Substituting (14) and (15) into (13), yields:

$$\left(\ddot{q}_d - M(q)^{-1}\cdot\left(M(q)\left(\ddot{q}_d + k^*\cdot D^\lambda[sig(e)^\alpha] + \dot{k}^*\cdot D^{\lambda-1}[sig(e)^\alpha] + \rho_1|s|^{1/2}sgn(s) + \int_0^t \rho_3 sgn(s)dt\right) + C(q,\dot{q}) + G(q)\right.$$
$$\left. -C(q,\dot{q}) - G(q) - f(q,\dot{q})\right) + k^*\cdot D^\lambda[sig(e)^\alpha] + \dot{k}^*\cdot D^{\lambda-1}[sig(e)^\alpha] = \dot{s}$$

By simplifying the above equation, error dynamics (16) can be determined:

$$M(q)^{-1}\cdot f(q,\dot{q}) - \rho_1|s|^{1/2}sgn(s) - \int_0^t \rho_3 sgn(s)dt = \dot{s} \tag{16}$$

Rewrite error dynamics (16) as the form as follows:

$$\begin{aligned}\dot{s} &= -\rho_1|s|^{1/2}sgn(s) + \omega \\ \dot{\omega} &= -\rho_3 sgn(s) + \dot{\varepsilon}\end{aligned} \tag{17}$$

where $\varepsilon$ is formulated from (16) and (17) as $\varepsilon = M(q)^{-1}\bullet f(q,\dot{q}) \le M(q)^{-1}\bullet \overline{F}$, satisfying the inequality equation $|\dot{\varepsilon}| \le \delta_\varepsilon$, where $\delta_\varepsilon > 0$. For each joint, the error dynamics can be written as:

$$\begin{aligned}\dot{s}_i &= -\rho_{1i}|s_i|^{1/2}sgn(s_i) + \omega_i \\ \dot{\omega}_i &= -\rho_{3i}sgn(s_i) + \dot{\varepsilon}_i\end{aligned} \tag{18}$$

where $\varepsilon_i = M(q_i)^{-1}\cdot f(q_i,\dot{q}_i)$, and $\dot{\varepsilon}_i$ satisfying $|\dot{\varepsilon}_i| \le \delta_{\varepsilon i}$, where $\delta_{\varepsilon i} > 0$. The following Lyapunov function is chosen:

$$V = 2\rho_3|s| + \frac{1}{2}\omega^2 + \frac{1}{2}\left(\rho_1|s|^{1/2}sgn(s) - \omega\right)^2 \tag{19}$$

Then the Lyapunov function $V_i$ of each joint of the low-cost humanoid manipulator can be determined as:

$$V_i = 2\rho_{3i}|s_i| + \frac{1}{2}\omega_i^2 + \frac{1}{2}\left(\rho_{1i}|s_i|^{1/2}sgn(s_i) - \omega_i\right)^2 \tag{20}$$

where $i$ represents joint $i$, and $V_i$ satisfies $V_i \ge 0$.

Supposing $\eta_i = \left[|s_i|^{1/2}sgn(s_i)\quad \omega_i\right]^T$ and $P_i = \begin{bmatrix} 2\rho_{3i} + \frac{\rho_{1i}^2}{2} & -\frac{\rho_{1i}}{2} \\ -\frac{\rho_{1i}}{2} & 1 \end{bmatrix}$, then (20) can be rewritten as:

$$V_i = \eta_i^T P_i \eta_i \tag{21}$$

$V_i$ satisfies the following inequality equation:

$$\lambda_{\min}\{P_i\}\|\eta_i\|^2 \le V_i \le \lambda_{\max}\{P_i\}\|\eta_i\|^2 \tag{22}$$

where $\lambda_{\min}\{P_i\}$ denotes minimum eigenvalues of $P_i$ and $\lambda_{\max}\{P_i\}$ denotes the maximum eigenvalue. $\|\eta_i\|$ is the Euclidean norm. By rearranging Equation (22), we obtain:

$$\frac{V_i^{\frac{1}{2}}}{\lambda_{\max}^{\frac{1}{2}}\{P_i\}} \le \|\eta_i\| \le \frac{V_i^{\frac{1}{2}}}{\lambda_{\min}^{\frac{1}{2}}\{P_i\}} \tag{23}$$

Derivating the Lyapunov function, $\dot{V}_i$ can be expressed as:

$$\dot{V}_i = 2\rho_{3i}\dot{s}_i sgn(s_i) + \omega_i\dot{\omega}_i + \left(\rho_{1i}|s_i|^{1/2}sgn(s_i) - \omega_i\right)\left(\frac{\rho_{1i}}{2}|s_i|^{-1/2}\dot{s}_i - \dot{\omega}_i\right)$$

Substituting error dynamics (18) into the above equation, yields:

$$\dot{V}_i = -2\rho_{1i}\rho_{3i}|s_i|^{1/2} + 2\rho_{3i}\omega_i sgn(s_i) - \rho_{3i}\omega_i sgn(s_i) + \omega_i\dot{\varepsilon}_i + \left(\rho_{1i}|s_i|^{1/2}sgn(s_i) - \omega_i\right)$$
$$\cdot\left(\frac{\rho_{1i}}{2}|s_i|^{-1/2}\left(-\rho_{1i}|s_i|^{1/2}sgn(s_i) + \omega_i\right) - \left(-\rho_{3i}sgn(s_i) + \dot{\varepsilon}_i\right)\right)$$

$$\dot{V}_i = \frac{1}{|s_i|^{1/2}}\left(-\tfrac{1}{2}\rho_{1i}^3|s_i| + \rho_{1i}^2|s_i|^{1/2}sgn(s_i) - \tfrac{\rho_{1i}}{2}\omega_i^2 - \rho_{1i}\rho_{3i}|s_i|\right) + \dot{\varepsilon}_i\left(-\rho_{1i}|s_i|^{1/2}sgn(s_i) + 2\omega_i\right)$$

Simplify the above equation into matrix multiple forms:

$$\dot{V}_i = -\frac{1}{|s_i|^{1/2}}\begin{bmatrix} |s_i|^{1/2}sgn(s_i) & \omega_i \end{bmatrix}\begin{bmatrix} \tfrac{1}{2}\rho_{1i}^3 + \rho_{1i}\rho_{3i} & -\tfrac{\rho_{1i}^2}{2} \\ -\tfrac{\rho_{1i}^2}{2} & \tfrac{\rho_{1i}}{2} \end{bmatrix}\begin{bmatrix} |s_i|^{1/2}sgn(s_i) \\ \omega_i \end{bmatrix}$$
$$+\dot{\varepsilon}_i\begin{bmatrix} -\rho_{1i} & 2 \end{bmatrix}\begin{bmatrix} |s_i|^{1/2}sgn(s_i) \\ \omega_i \end{bmatrix} \tag{24}$$

Supposing: $A_i = \begin{bmatrix} \tfrac{1}{2}\rho_{1i}^3 + \rho_{1i}\rho_{3i} & -\tfrac{\rho_{1i}^2}{2} \\ -\tfrac{\rho_{1i}^2}{2} & \tfrac{\rho_{1i}}{2} \end{bmatrix}$, $B_i = \begin{bmatrix} 0 \\ 1 \end{bmatrix}$, $C_i = \begin{bmatrix} 1 \\ 0 \end{bmatrix}$, then Equation (24) can be rewritten as:

$$\dot{V}_i = -\frac{1}{|s_i|^{1/2}}\eta_i^T A_i \eta_i + \frac{2\dot{\varepsilon}_i}{|s_i|^{1/2}}|s_i|^{1/2}B_i^T P_i \eta_i \tag{25}$$

If $\|\dot{\varepsilon}\| \leq \|\delta\|$, $|\dot{\varepsilon}_i| \leq \delta_i$, then $2\dot{\varepsilon}_i|s_i|^{1/2}B_i^T P_i \eta_i$ satisfies:

$$2\dot{\varepsilon}_i|s_i|^{1/2}B_i^T P_i \eta_i \begin{aligned} &\leq \dot{\varepsilon}_i^2|s_i| + \eta_i^T P_i B_i B_i^T P_i^T \eta_i \\ &\leq \delta_{\varepsilon i}^2\eta_i^T C_i C_i^T \eta_i + \eta_i^T P_i B_i B_i^T P_i^T \eta_i \end{aligned} \tag{26}$$

By substituting inequality (26) into rearranged $\dot{V}_i$ (25), we obtain the following inequality (27):

$$\dot{V}_i \leq -\frac{1}{|s_i|^{1/2}}\eta_i^T\left(A_i - \delta_{\varepsilon i}^2 C_i C_i^T - P_i B_i B_i^T P_i^T\right)\eta_i = -\frac{1}{|s_i|^{1/2}}\eta_i^T Q_i \eta_i \tag{27}$$

where:

$$Q_i = \begin{bmatrix} \tfrac{1}{2}\rho_{1i}^3 + \rho_{1i}\rho_{3i} - \delta_{\varepsilon i}^2 - \tfrac{\rho_{1i}^2}{4} & -\tfrac{\rho_{1i}^2}{2} + \tfrac{\rho_{1i}}{2} \\ -\tfrac{\rho_{1i}^2}{2} + \tfrac{\rho_{1i}}{2} & \tfrac{\rho_{1i}}{2} - 1 \end{bmatrix} \tag{28}$$

If $\rho_{1i}$ and $\rho_{3i}$ satisfy the following conditions:

$$\rho_{1i} > 2$$
$$\rho_{3i} > \frac{\rho_{1i}^3 + 4\delta_{\varepsilon i}^2(\rho_{1i}-2)}{4\left(\rho_{1i}^2 - 2\rho_{1i}\right)}$$

then the error dynamics for each joint (18) meet the demands of the Lyapunov stability and the matrix $Q_i$ is symmetrical and positive. Obviously $|s_i|^{1/2} \leq \|\eta_i\|$, substituting Equation (23) into Equation (27), $\dot{V}_i$ satisfies:

$$\dot{V}_i \leq -\frac{1}{\|\eta_i\|}\lambda_{\min}\{Q_i\}\|\eta_i\|^2 \leq -\frac{\lambda_{\min}\{Q_i\}}{\lambda_{\max}^{\frac{1}{2}}\{P_i\}}V_i^{\frac{1}{2}}$$

where term $\lambda_{\min}\{Q_i\}$ denotes the minimum eigenvalues of the matrix $Q$, $\lambda_{\max}\{Q_i\}$ denotes the maximum eigenvalues. Thus, the converge time $T_{ci}$ for each joint of the low-cost humanoid manipulator satisfies the following inequality which can be obtained through Lemmas 1 and 2:

$$T_{ci} \leq 2\lambda_{max}^{1/2}\{P_i\}\lambda_{min}^{-1}\{Q_i\}V_i^{1/2}(0)$$

Therefore, the control law (14) can ensure the error dynamics of the low-cost humanoid manipulator to be stable.

## 5. Simulation and Discussion

To reveal the superior performances of the proposed DFONTSM-STA control, we perform simulation verification on the low-cost humanoid manipulator. The coulomb and viscous friction parameters are selected as $f_{c1} = f_{c2} = f_{c4} = 1.25N$, $f_{c3} = 0.125N$, $f_{v1} = f_{v2} = f_{v4} = 1$, $f_{v3} = 0.1$ [44]. The model information of the low-cost humanoid manipulator is listed in Tables 2–4.

**Table 2.** The center mass position of the connecting rod.

| Center Mass Position (m) | Joint One | Joint Two | Joint Three | Joint Four |
|---|---|---|---|---|
| X | 0 | $-0.04 \times 10^{-3}$ | 0 | $-0.94 \times 10^{-3}$ |
| Y | 0 | $125.16 \times 10^{-3}$ | $-7 \times 10^{-3}$ | $166.42 \times 10^{-3}$ |
| Z | $11.57 \times 10^{-3}$ | $7.71 \times 10^{-3}$ | $-0.26 \times 10^{-3}$ | $6.97 \times 10^{-3}$ |

**Table 3.** Mass of connecting rod of the humanoid manipulator.

| Mass (kg) | Rod One | Rod Two | Rod Three | Rod Four |
|---|---|---|---|---|
| M | $728.22 \times 10^{-3}$ | $7016.49 \times 10^{-3}$ | $382.18 \times 10^{-3}$ | $3341.64 \times 10^{-3}$ |

**Table 4.** Connecting rod inertia of humanoid manipulator.

| Moment of Inertia $(kg \cdot m^2)$ | Rod One | Rod Two | Rod Three | Rod Four |
|---|---|---|---|---|
| Ixx | $1,709,909.71 \times 10^{-9}$ | $26,245,6649.98 \times 10^{-9}$ | $625,787.92 \times 10^{-9}$ | $158,795,139.03 \times 10^{-9}$ |
| Iyy | $875,916.10 \times 10^{-9}$ | $20,863,623.56 \times 10^{-9}$ | $544,829.79 \times 10^{-9}$ | $2,996,242.06 \times 10^{-9}$ |
| Izz | $1,375,449.61 \times 10^{-9}$ | $249,837,989.90 \times 10^{-9}$ | $270,182.90 \times 10^{-9}$ | $157,591,605.74 \times 10^{-9}$ |
| Ixy | $24.45 \times 10^{-9}$ | $-46,664.65 \times 10^{-9}$ | $13.85 \times 10^{-9}$ | $-1,391,395.68 \times 10^{-9}$ |
| Iyz | $-47.57 \times 10^{-9}$ | $17,384,004.90 \times 10^{-9}$ | $279.02 \times 10^{-9}$ | $707,363.35 \times 10^{-9}$ |
| Ixz | $16.53 \times 10^{-9}$ | $185,311.42 \times 10^{-9}$ | $-2.14 \times 10^{-9}$ | $107,519.45 \times 10^{-9}$ |

### 5.1. Comparison of Simulation Results

To better compare and analyze the performances of the FONTSM control (29) [44], FONTSM-STA control (30) [45], and DFONTSM-STA control (14), we select control parameters listed in Tables 5 and 6. The simulation performances are described in Figures 2–6.

$$\tau_r = M(q)\left(\ddot{q}_d + k \cdot D^\lambda[sig(e)^\alpha] + k_2 sig(s)^b + k_1 s\right) + C(q,\dot{q}) + G(q) \tag{29}$$

$$\tau_r = M(q)\left(\ddot{q}_d + k \cdot D^\lambda[sig(e)^\alpha] + \rho_1|s|^{1/2}sgn(s) + \rho_3\int_0^t sgn(s)dt\right) + C(q,\dot{q}) + G(q) \tag{30}$$

**Table 5.** Parameters list of dynamic fractional-order nonsingular terminal super-twisting sliding mode (DFONTSM-STA) control.

|  | $\rho_1$ | $\rho_2$ | $\alpha_\kappa$ | $\delta_0$ | $k_0$ | $\alpha$ | $\lambda$ | $\rho_0$ |
|---|---|---|---|---|---|---|---|---|
| Joint one | 40 | 4 | 1 | 0.1 | 2 | 0.8 | 0.9 | 2 |
| Joint two | 40 | 4 | 1 | 0.1 | 2 | 0.8 | 0.9 | 2 |
| Joint three | 70 | 4 | 1 | 0.1 | 2 | 0.8 | 0.9 | 0.8 |
| Joint four | 70 | 4 | 1 | 0.1 | 2 | 0.8 | 0.9 | 0.8 |

**Table 6.** Parameters list of fractional-order nonsingular terminal super-twisting sliding mode (FONTSM-STA) control and fractional-order nonsingular terminal sliding mode (FONTSM) control.

|  | $k_1$ | $\rho_1$ | $k_2$ | $\rho_2$ | $k_0$ | $\alpha$ | $\lambda$ |
|---|---|---|---|---|---|---|---|
| Joint one | 60 | 40 | 6 | 4 | 1 | 0.8 | 0.9 |
| Joint two | 60 | 40 | 6 | 4 | 1 | 0.8 | 0.9 |
| Joint three | 80 | 70 | 6 | 4 | 4 | 0.8 | 0.9 |
| Joint four | 80 | 70 | 6 | 4 | 4 | 0.8 | 0.9 |

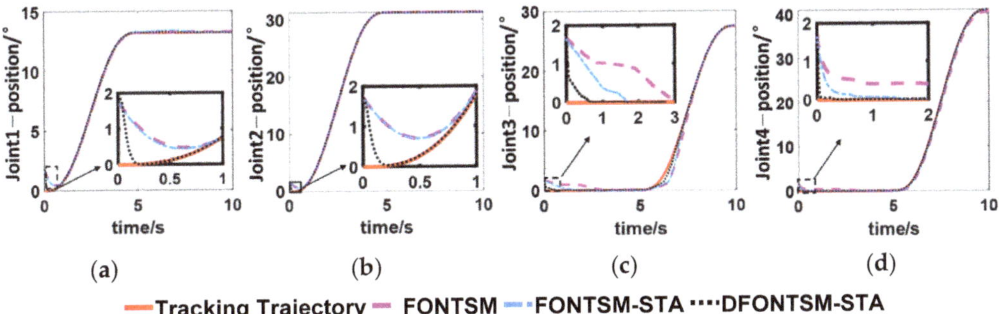

**Figure 2.** Position tracking trajectory of each joint. (**a**): joint one; (**b**): joint two; (**c**): joint three; (**d**): joint four.

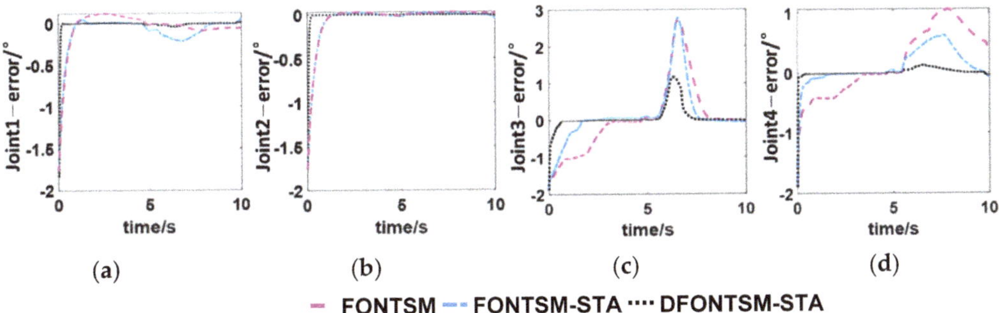

**Figure 3.** The position tracking error of each joint. (**a**): joint one; (**b**): joint two; (**c**): joint three; (**d**): joint four.

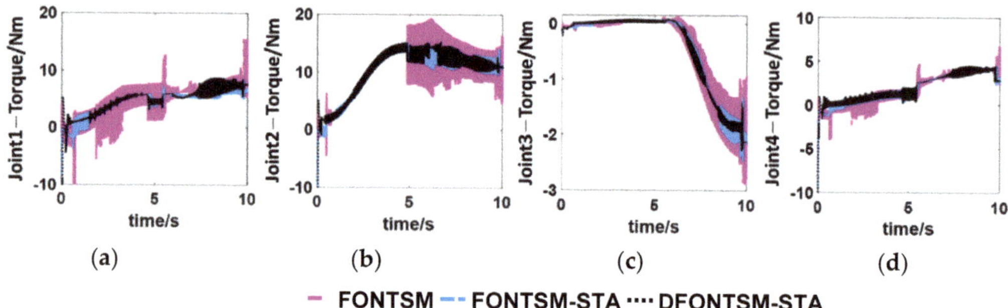



**Figure 4.** Control input torque of each joint (**a**): joint one; (**b**): joint two; (**c**): joint three; (**d**): joint four.

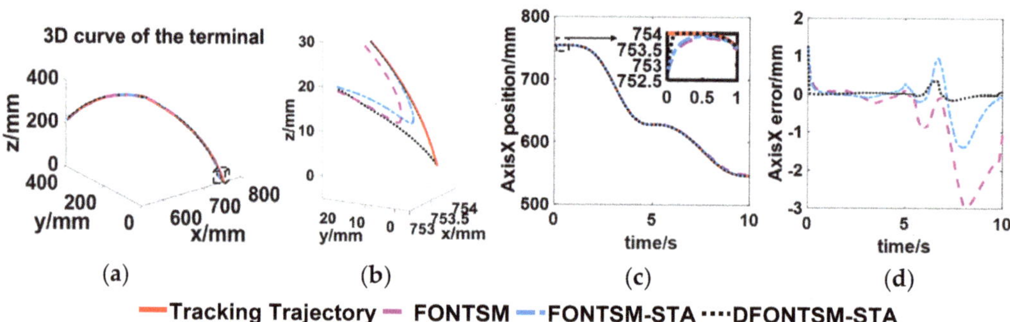



**Figure 5.** Terminal motion trajectory and its tracking error (**a**): Terminal trajectory tracking curve under base coordinate; (**b**): Terminal trajectory tracking error; (**c**): axis X position; (**d**): axis X error.

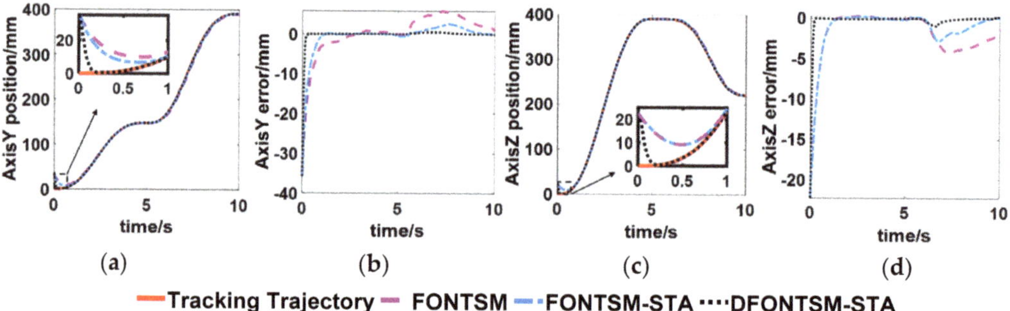



**Figure 6.** Terminal motion trajectory and its tracking error on Y axis and Z axis of the base coordination. (**a**): axis Y position; (**b**): axis Y error; (**c**): axis Z position; (**d**): axis Z error.

### 5.2. Simulation Analysis

The convergence time is calculated when the tracking error accuracy is within $0.5°$ (shoulder joints) or $1°$ (elbow joints), which is shown in Table 7. Table 8 shows the maximum amplitude of control torque variations in a small time interval of each joint. The variations of torque in a small interval reflect the degree of chattering suppression, we defined an indicator shown as equation (31) to facilitate analysis and to directly evaluate the controller's chattering suppression performance.

$$|\tau|_{\max} = \left|\tau(t) - \tau\left(t - t_p\right)\right| \tag{31}$$

where $t_p$ is the time interval between the two calculated torques and $t_p = 20$ ms is defined in this simulation.

**Table 7.** Convergence time for each joint.

| $\lvert e \rvert \leq 0.5°$ $\lvert e \rvert \leq 1°$ | DFONTSM-STA $(t_{conv})$s | FONTSM-STA $(t_{conv})$s | FONTSM $(t_{conv})$s |
|---|---|---|---|
| Joint one | 0.101 | 0.409 | 0.443 |
| Joint two | 0.090 | 0.464 | 0.464 |
| Joint three | 0.036 | 0.515 | 1.309 |
| Joint four | 0.021 | 0.067 | 0.098 |

**Table 8.** The maximum amplitude of the low-cost humanoid manipulator control torque variations.

| $\lvert \tau \rvert_{max}$ $t_p = 20$ ms | DONTSM-STA | FONTSM-STA | FONTSM |
|---|---|---|---|
| Joint one | 4.2875 | 3.7355 | 12.7760 |
| Joint two | 3.5760 | 4.3541 | 7.4774 |
| Joint three | 0.7294 | 0.6397 | 0.9683 |
| Joint four | 2.0093 | 2.1479 | 4.2029 |

As shown in Table 7, the tracking error convergence speed of the DFONTSM-STA control is faster than that of FONTSM and FONTSM-STA control. Compared with FONTSM control, the error convergence time by DFONTSM-STA control from joints one to four is reduced by 77.2%, 80.6%, 97.2%, and 78.6% respectively. Compared with the FONTSM-STA control, the convergence time is reduced by 75.3%, 80.6%, 93.0%, and 68.7% respectively. It can be concluded that under the control of DFONTSM-STA, each joint of the low-cost humanoid manipulator takes the least time to converge.

The DFONTSM-STA control and the FONTSM-STA control also have better chattering suppression effects compared with the FONTSM control. When $t_p = 20$ ms, the maximum chattering variations of joints one to four under FONTSM-STA control are reduced by 70.8%, 41.8%, 33.9%, and 48.9% compared with the FONTSM control. While under DFONTSM-STA control, the maximum chattering variations for joints one to four are reduced by 66.4%, 52.1%, 24.7%, and 52.2% compared with the FONTSM control. From Figure 4, it can be seen that the control inputs under the FONTSM control have the most severe high-frequency chattering among the three control methods. If the unmodeled dynamics of the low-cost humanoid manipulator are excited by high-frequency chattering, the moving trajectory may diverge, which will destroy the machine structure and cause security compromises.

Compared with the FONTSM-STA and FONTSM methods, DFONTSM-STA adds dynamic characteristics to the sliding manifold, which can make the sliding manifold change dynamically according to the actual error. Therefore, among these three methods, the DFONTSM-STA controller has the strongest ability to maintain the error at zero, and the best joint error convergence performance.

From Table 9, the data of terminal position root mean square error indicates that compared with the FONTSM control, the terminal root mean square tracking error ($S_{rmste}$) under DFONTSM-STA control are decreased by 90.9%, 57.4%, and 57.1% in the x-direction, y-direction, and z-direction of the low-cost humanoid manipulator. Furthermore, under DFONTSM-STA control, the $S_{rmste}$ in X, Y, and Z directions are further reduced by 78.2%, 43.7%, and 51.3% with the comparison of the FONTSM-STA control. The $S_{rmste}$ of the distance between the terminal and the target under DFONTSM-STA control is limited within 3 mm and decreased by 46.7% and 58.0% with the comparison of FONTSM-STA and FONTSM control respectively. Thus, adding dynamic characteristics can not only improve the joint error convergence but also greatly improve the terminal error convergence.

**Table 9.** Terminal position root mean square error.

| $S_{rmste}$ (mm) | DFONTSM-STA | FONTSM-STA | FONTSM |
|---|---|---|---|
| Axis x | 0.1126 | 0.5165 | 1.2403 |
| Axis y | 2.3329 | 4.1434 | 5.4771 |
| Axis z | 1.5838 | 3.2514 | 3.6918 |
| Distance between terminal and target | 2.8220 | 5.2921 | 6.7206 |

To sum up, the DFONTSM-STA control has the biggest possibility to realize the terminal trajectory tracking with smooth and accurate movements.

## 6. Experimental Verification

### 6.1. Experiment Setting

Figure 7 displays the experimental platform. The core components of the low-cost humanoid manipulator system are as follows: the PC; the hardware controller; EPOS4 drivers; Maxon DC motors. The PC calculates each joint's control input. The hardware controller has data transmission functions. It uses EtherCAT as the communication channel to exchange position and torque data with the PC; EPOS4 drivers support the CANopen protocol. Its communication speed can reach to 1 Mbps. The hall sensor whose resolution converted to joints is 0.3° for obtaining angular position information was inserted in the Maxon DC motor. By debugging the above hardware, a real-time system has been constructed to control the low-cost humanoid manipulator. The parameters of the three controllers in experiments are listed in Tables 10 and 11.

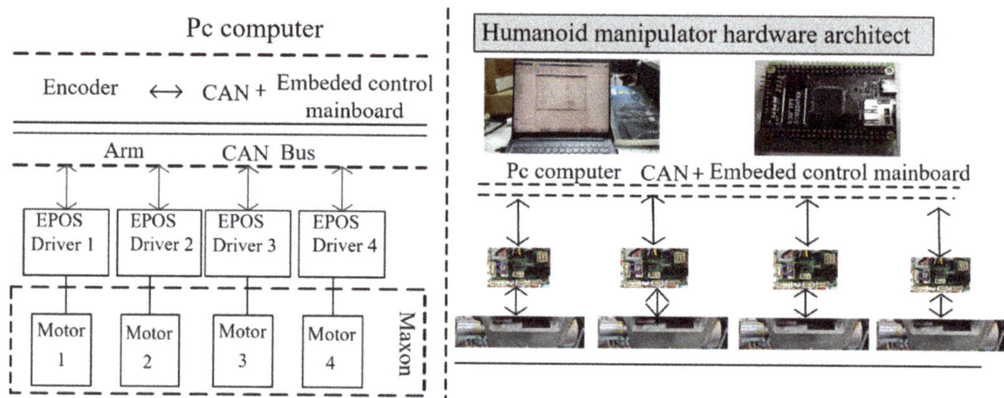


**Figure 7.** The architecture of the humanoid manipulator system.

**Table 10.** Parameters of DFONTSM-STA control.

| | $\rho_1$ | $\rho_2$ | $\alpha_\kappa$ | $\delta_0$ | $k_0$ | $\alpha$ | $\lambda$ | $\rho_0$ |
|---|---|---|---|---|---|---|---|---|
| Joint one | 6 | 6 | 1 | 0.05 | 1 | 0.8 | 0.9 | 1 |
| Joint two | 6 | 6 | 1 | 0.05 | 1 | 0.8 | 0.9 | 1 |
| Joint three | 3 | 3 | 1 | 0.05 | 1 | 0.8 | 0.9 | 2 |
| Joint four | 3 | 3 | 1 | 0.05 | 1 | 0.8 | 0.9 | 2 |

**Table 11.** Parameters of FONTSM-STA and FONTSM control.

|  | $k_1$ | $\rho_1$ | $k_2$ | $\rho_2$ | $k_0$ | $\alpha$ | $\lambda$ |
|---|---|---|---|---|---|---|---|
| Joint one | 9.5 | 6 | 6 | 6 | 15 | 0.8 | 0.9 |
| Joint two | 9.5 | 6 | 6 | 6 | 15 | 0.8 | 0.9 |
| Joint three | 6.5 | 3 | 3 | 3 | 15 | 0.8 | 0.9 |
| Joint four | 6.5 | 3 | 3 | 3 | 15 | 0.8 | 0.9 |

The experiments adapt the trajectories in the simulation as the target of the low-cost humanoid manipulator. The control performances of DFONTSM-STA control, FONTSM-STA control, and FONTSM control are compared. The simulation movements and physical schematic of the low-cost humanoid manipulator are shown in Figure 8.

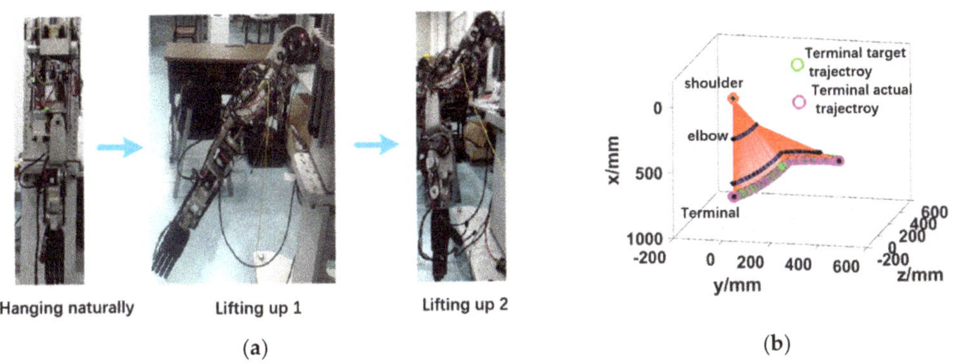


**Figure 8.** Simulation and physical schematic of humanoid manipulator movement. (**a**): Physical schematic (**b**): The simulation movement.

### 6.2. Experiment Results

Figures 9–18 present the experimental performances of the low-cost humanoid manipulator.

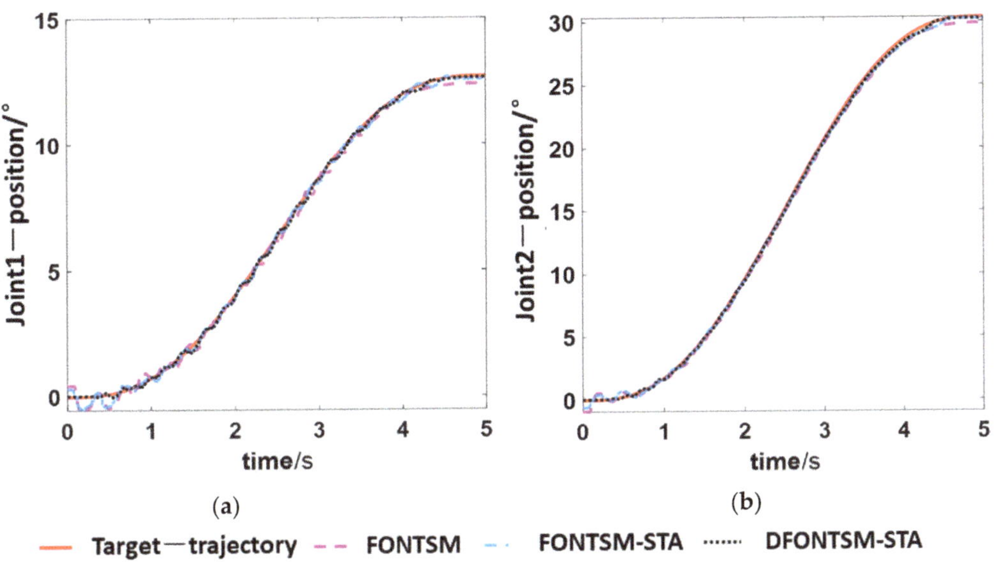


**Figure 9.** Position tracking trajectories of the shoulder joints. (**a**): joint one; (**b**): joint two.

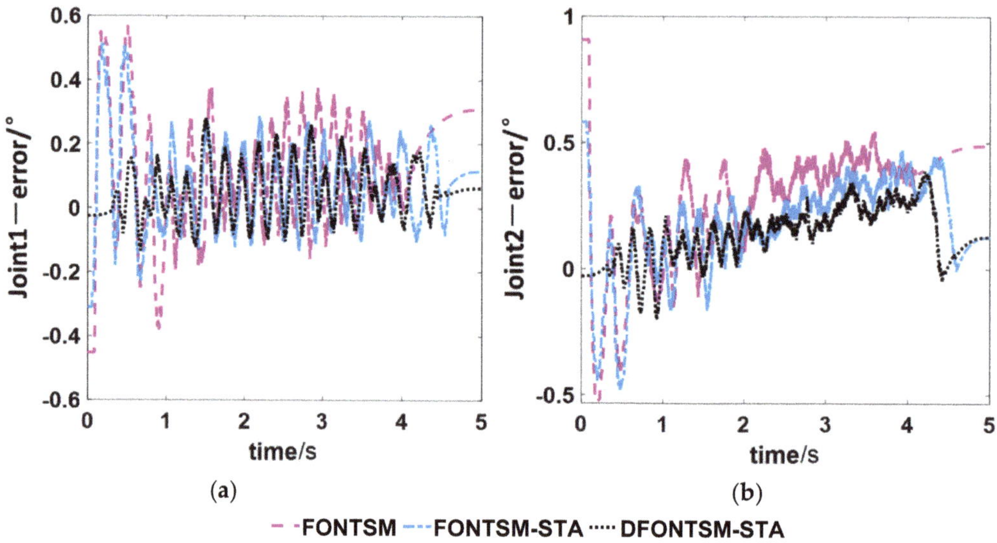



**Figure 10.** Position tracking errors of the shoulder joints. (**a**): joint one; (**b**): joint two.

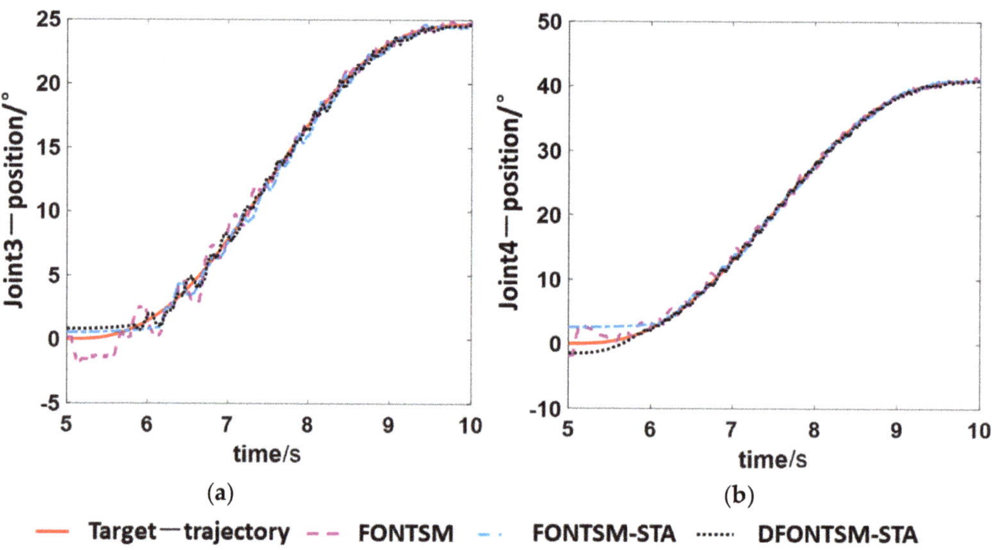



**Figure 11.** Position tracking trajectories of the elbow joints. (**a**): joint three; (**b**): joint four.

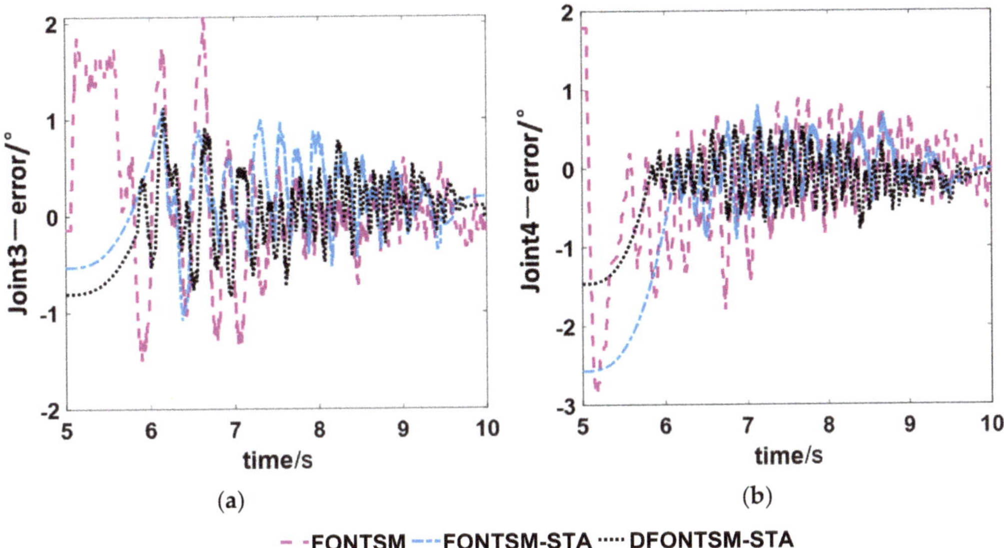



**Figure 12.** The position tracking error of the elbow joints. (**a**): joint three; (**b**): joint four.

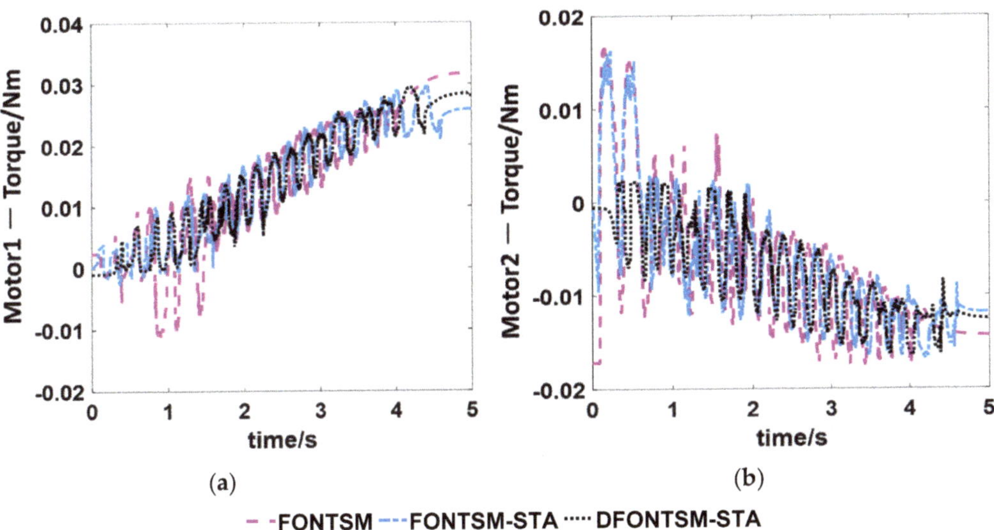



**Figure 13.** Input torques of the shoulder joints. (**a**): joint one; (**b**): joint two.

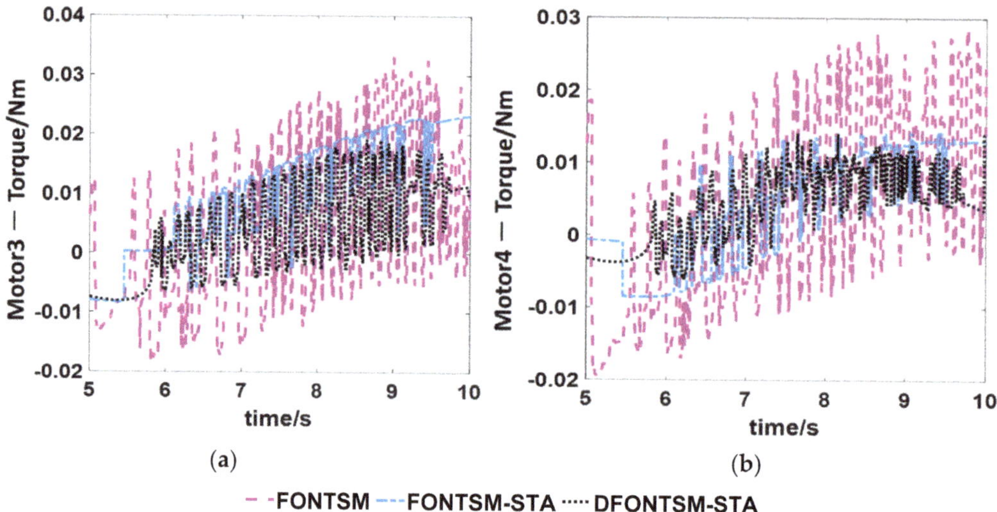



**Figure 14.** Input torques of elbow joints. (**a**): joint three; (**b**): joint four.

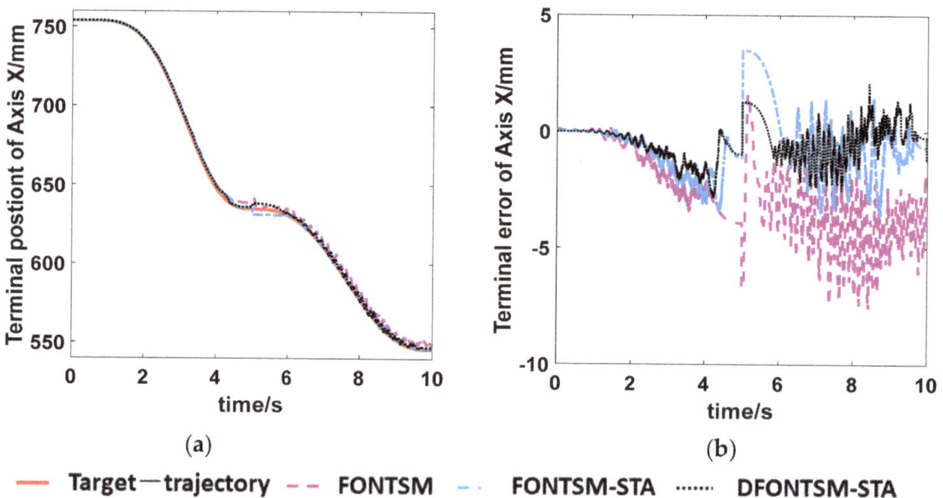



**Figure 15.** Terminal trajectory and its tracking error. (**a**): terminal trajectory on the X axis; (**b**): terminal tracking error on the X axis.

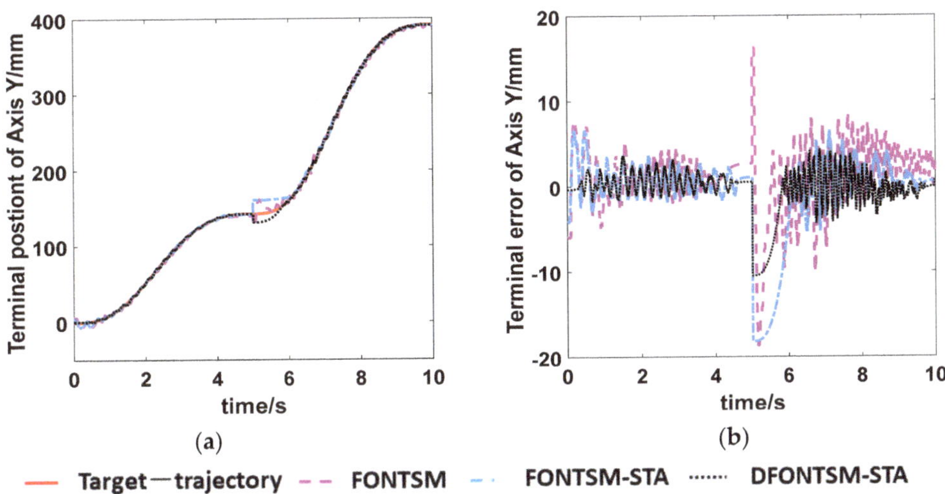



**Figure 16.** Terminal trajectory and its tracking error. (**a**): terminal trajectory on the Y axis; (**b**): terminal tracking error on the Y axis.

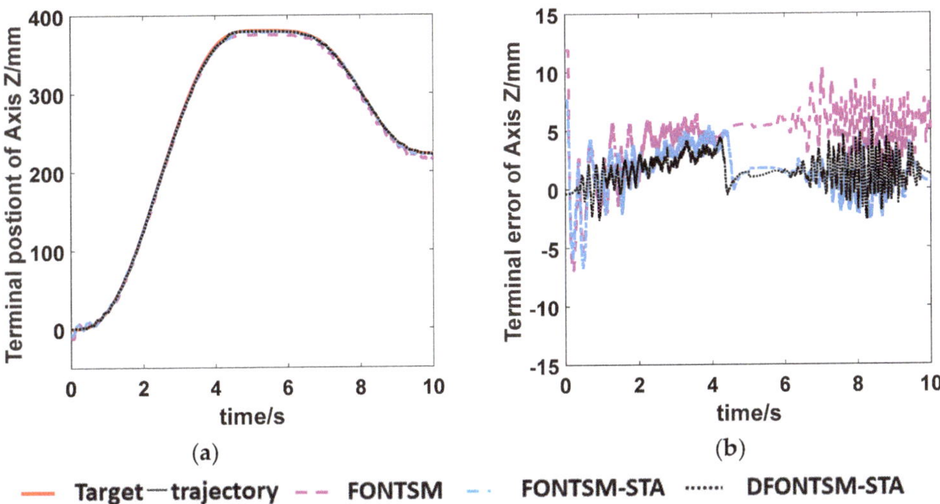



**Figure 17.** Terminal trajectory and its tracking error. (**a**): terminal trajectory on the Z axis; (**b**): terminal tracking error on the Z axis.

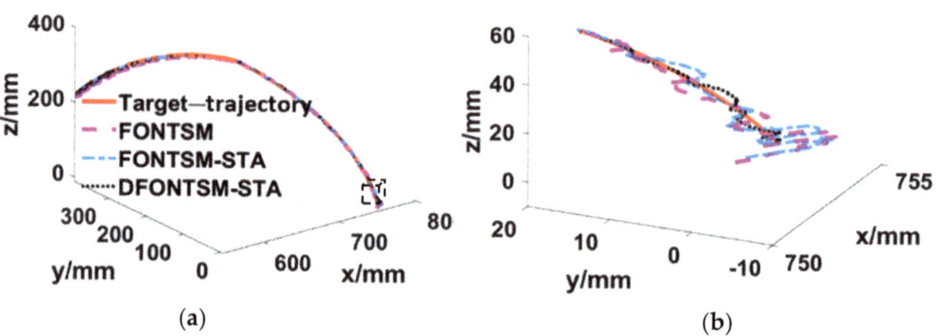



**Figure 18.** Terminal motion under base coordination system. (**a**): Terminal trajectory tracking curve; (**b**): initial part of the terminal trajectory tracking curve.

*6.3. Experiment Analysis*

The convergence time is calculated when the absolute value of the tracking error is within the 1° (elbow joints) or 0.5° (shoulder joints), which is shown in Table 12.

**Table 12.** Convergence time for each joint.

| $|e| \leq 0.5°$ or $|e| \leq 1°$ | DFONTSM-STA<br>($t_{conv}$)s | FONTSM-STA<br>($t_{conv}$)s | FONTSM<br>($t_{conv}$)s |
|---|---|---|---|
| Joint one | 0 | 0.474 | 0.548 |
| Joint two | 0 | 0.075 | 0.229 |
| Joint three | 6.188 | 6.395 | 7.101 |
| Joint four | 5.562 | 5.934 | 7.057 |

Figures 10 and 12 are the angular tracking error of the three methods, which show the robustness of the FOSM control. Through experimental data, DFONTSM-STA control is superior in accurate and smooth position tracking performance. Compared with the FONTSM control and FONTSM-STA control, the error convergence time of joint one under DFONTSM-STA control is reduced by 0.548 s and 0.474 s respectively, and the error convergence time of joint two is reduced by 0.229 s and 0.075 s respectively. The error convergence speed of joint three under DFONTSM-STA control is speeded up by 43.5% and 33.6% respectively, and the error convergence speed of joint four is speeded up by 72.7% and 54.6% respectively. The DFONTSM-STA control can significantly improve the error convergence by dynamically changing the spatial position of the sliding mode manifold.

Figures 13 and 14 show the control torques required to drive the low-cost humanoid manipulator. Table 13 shows the maximum variation of the control torque chattering, which can reveal the chattering suppression effects of the three controllers in real-time position control. From Table 13, it can be concluded that the chattering of joints one to four under the FONTSM-STA control is suppressed by 47.8%, 9.6%, 52.2%, and 34.5% respectively compared with the FONTSM control, and the chattering of joints one to four under the DFONTSM-STA control is suppressed by 54.1%, 51.1%, 46.2%, and 55.1% respectively, with the comparison of the FONTSM control. Thus, the chattering suppression performance of DFONTSM-STA control is ahead of the other two control methods.

**Table 13.** The maximum variation of control torque in a small interval.

| $\lvert\tau\rvert_{\max}$ ($t_p$ = 20 ms) | DFONTSM-STA | FONTSM-STA | FONTSM |
|---|---|---|---|
| Joint one | 0.0095 | 0.0108 | 0.0207 |
| Joint two | 0.0112 | 0.0207 | 0.0229 |
| Joint three | 0.0204 | 0.0181 | 0.0379 |
| Joint four | 0.0151 | 0.022 | 0.0336 |

Figures 15–18 show the terminal motion performances of the low-cost humanoid manipulator. According to Figure 18, the DFONTSM-STA control can make the terminal faster and more accurate to tracking the target trajectory. The terminal tracking process of DFONTSM-STA control is smoother than FONTSM-STA control and FONTSM control, which can be illustrated by the terminal tracking trajectory that is decomposed to the X, Y, and Z direction of the base coordination. Compared with the FONTSM control, the motion accuracy of joints one to four of the DFONTSM-STA control embodied by the root mean square of joint tracking error is improved by 55.4% 51.1% 38.5%, and 32.8% respectively. Compared with the FONTSM-STA control, the motion accuracy of joints one to four is improved by 32.1%, 35.4%, 2.9%, and 42.6% respectively.

Table 14 shows the terminal root mean square error of the low-cost humanoid manipulator ($S_{rmste}$). The $S_{rmste}$ of FOTNSM-STA control and FONTSM control between the terminal and the target trajectory surpasses 5 mm, which indicates the tracking performances of the anthropomorphic trajectory are not quite well. Compared with FONTSM-STA control, the $S_{rmste}$ on the X, Y, and Z coordinate axes of the base coordination under DFONTSM-TSM control is reduced by 35.5%, 45.4%, and 20.9%, respectively. Compared with FONTSM-STA control, the $S_{rmste}$ on the X, Y, and Z coordinate axes of the base coordination under DFONTSM-TSM control is reduced by 69.4%, 36.8%, and 61.2%, respectively. The $S_{rmste}$ of DFOTNSM-TSM control between the terminal and the target trajectory is limited to 3 mm and reduced by 23.7% and 53.3% compared with FONTSM-STA and FONTSM control.

**Table 14.** Terminal root mean square error of the humanoid manipulator.

| $S_{rmste}$ (mm) | DFONTSM-STA | FONTSM-STA | FONTSM |
|---|---|---|---|
| Axis x | 0.983 | 1.524 | 3.212 |
| Axis y | 2.584 | 4.730 | 4.090 |
| Axis z | 1.984 | 2.509 | 5.111 |
| Distance between the terminal and the target | 3.403 | 5.567 | 7.292 |

To sum up, the experiments confirm that The DFONTSM-STA control, compared with FONTSM-STA control and FONTSM control, can further improve the accuracy and smoothness of the terminal tracking performances and basically realize anthropomorphic motion of the low-cost humanoid manipulator.

## 7. Conclusions

A DFONTSM-STA control was proposed for a low-cost humanoid manipulator by combining the dynamic fractional-order nonsingular terminal sliding mode manifold with the super-twisting reaching law. Experiments were conducted to control the low-cost humanoid manipulator and showed that the proposed control method can effectively improve error convergence speed, tracking accuracy, and chattering suppression ability. In error convergence speed, the errors of joints one and two were converged at the beginning while joint three was speeded up by 43.5% and 33.6%, and joint four was speeded up by 72.7% and 54.6% compared with FONTSM and FONTSM-STA controllers. In chattering suppressing, suppression performances of joints, one to four were enhanced by 54.1%, 51.1%, 46.2%, and 55.1%, respectively, compared with FONTSM control. In trajectory tracking performance, compared with FONTSM control and FONTSM-STA control, the tracking accuracy of DFONTSM-STA control was promoted by 53.3% and 23.7% respectively. Simulation and experimental results illustrate the superiority of the proposed DFONTSM-STA control for the low-cost humanoid manipulator.

The proposed DFONTSM-STA controller is applicable to humanoid manipulators that require fast error convergence and accurate trajectory tracking movements. The dynamic model of the low-cost humanoid manipulator is complex, and there exists the problem of big calculations in solving control inputs. Therefore, the adaptive fuzzy method will be studied to estimate model uncertainties and disturbances in future research work.

**Author Contributions:** Conceptualization, H.D.; Data curation, R.H., X.X. and Y.Z.; Funding acquisition, Y.Z. and H.D.; Methodology, R.H.; Project administration, X.X., Y.Z. and H.D.; Software, R.H. and X.X.; Supervision, H.D.; Validation, R.H., X.X. and Y.Z.; Writing—original draft, R.H.; Writing—review & editing, R.H. and H.D. All authors have read and agreed to the published version of the manuscript.

**Funding:** This research was funded in part by the National Key Research and Development Program of China under grant number 2018YFB1307203, in part by National Natural Science Foundation of China under Grant 52275297, and in part by the Project of State Key Laboratory of High Performance Complex Manufacturing, Central South University under Grant ZZYJKT2021-17.

**Data Availability Statement:** Not applicable.

**Conflicts of Interest:** The authors declare no conflict of interest.

### Abbreviations

The following abbreviations are used in this manuscript:

| Acronyms | Definition |
| --- | --- |
| D-H | Denavit-Hartenberg |
| PC | Personal computer |
| 3D | Three dimensional |
| RL | Riemann-Liouville |
| EtherCAT | Ethernet control automation technology |
| CANopen | Controller area network open |
| FO | Fractional-order |
| FOSM | Fractional-order sliding mode |
| FONTSM | Fractional-order nonsingular terminal sliding mode |
| DFONTSM | Dynamic fractional-order nonsingular terminal sliding mode |
| STA | Super-twisting algorithm |
| STSM | Super-twisting sliding mode |
| FONTSM-STA | Fractional- order nonsingular terminal super-twisting sliding mode |
| DFONTSM-STA | Dynamic fractional- order nonsingular terminal super-twisting sliding mode |

## References

1. Kaneko, K.; Kaminaga, H.; Sakaguchi, T.; Kajita, S.; Morisawa, M.; Kumagai, I.; Kanehiro, F. Humanoid robot HRP-5P: An electrically actuated humanoid robot with high-power and wide-range joints. *IEEE Robot. Autom. Lett.* **2019**, *4*, 1431–1438. [CrossRef]
2. Zhang, B.; Xiong, R.; Wu, J. Kinematics and trajectory planning of a novel humanoid manipulator for table tennis. In Proceedings of the 2011 International Conference on Electrical and Control Engineering, Yichang, China, 16–18 September 2011; pp. 3047–3051. [CrossRef]
3. Khusainov, R.; Klimchik, A.; Magid, E. Humanoid robot kinematic calibration using industrial manipulator. In Proceedings of the 2017 International Conference on Mechanical, System and Control Engineering (ICMSC), St. Petersburg, Russia, 19–21 May 2017; pp. 184–189.
4. Sarajchi, M.; Al-Hares, M.K.; Sirlantzis, K. Wearable Lower-Limb Exoskeleton for Children With Cerebral Palsy: A Systematic Review of Mechanical Design, Actuation Type, Control Strategy, and Clinical Evaluation. *IEEE Trans. Neural Syst. Rehabil. Eng.* **2021**, *29*, 2695–2720. [CrossRef] [PubMed]
5. Rehman, I.U.; Javed, S.B.; Chaudhry, A.M.; Azam, M.R.; Uppal, A.A. Model-based dynamic sliding mode control and adaptive Kalman filter design for boiler-turbine energy conversion system. *J. Process Control* **2022**, *116*, 221–233. [CrossRef]
6. Mechali, O.; Xu, L.M.; Xie, X.M.; Iqbal, J. Theory and practice for autonomous formation flight of quadrotors via distributed robust sliding mode control protocol with fixed-time stability guarantee. *Control Eng. Pract.* **2022**, *123*, 105150. [CrossRef]

7.	Bano, S.; Azam, M.R.; Uppal, A.A.; Javed, S.B.; Bhatti, A.I. Robust p53 recovery using chattering free sliding mode control and a gain-scheduled modified Utkin observer. *J. Theor. Biol.* **2021**, *532*, 110914. [CrossRef]
8.	Qin, L.; Liang, L.; Liu, F.; Jin, Z. The application of adaptive backstepping sliding mode for hybrid humanoid robot arm trajectory tracking control. In Proceedings of the 2013 International Conference on Advanced Mechatronic Systems, Luoyang, China, 25–27 September 2013; pp. 730–735.
9.	Kuan, J.Y.; Huang, H.P. Independent joint dynamic sliding mode control of a humanoid robot arm. In Proceedings of the 2008 IEEE International Conference on Robotics and Biomimetics, Dali, China, 6–8 December 2019; pp. 602–607.
10.	Sanchez-Magos, M.; Ballesteros-Escamilla, M.; Cruz-Ortiz, D.; Salgado, I.; Chairez, I. Terminal sliding mode control of a virtual humanoid robot. In Proceedings of the 2019 6th International Conference on Control, Decision and Information Technologies (CoDIT), Paris, France, 23–26 April 2019; pp. 726–731.
11.	Ganjefar, S.; Sarajchi, M.H.; Hoseini, S.M.; Shao, Z. Lambert W Function Controller Design for Teleoperation Systems. *Int. J. Precis. Eng. Manuf.* **2019**, *20*, 101–110. [CrossRef]
12.	Gao, W.; Hung, J.C. Variable structure control of nonlinear systems: A new approach. *IEEE Trans. Ind. Electron.* **1993**, *40*, 45–55.
13.	Ullah, M.I.; Ajwad, S.A.; Irfan, M.; Iqbal, J. Non-linear Control Law for Articulated Serial Manipulators: Simulation Augmented with Hardware Implementation. *Elektron. Elektrotechnika* **2016**, *22*, 3–7. [CrossRef]
14.	Utkin, V. Variable structure systems with sliding modes. *IEEE Trans. Autom. Control* **1977**, *22*, 212–220. [CrossRef]
15.	Fuh, C.C. Variable-thickness boundary layers for sliding mode control. *J. Mar. Sci. Technol.* **2008**, *16*, 7. [CrossRef]
16.	Young, K.D.; Utkin, V.; Ozguner, U.A. control engineer's guide to sliding mode control. *IEEE Trans. Control Syst. Technol.* **1999**, *7*, 328–342. [CrossRef]
17.	Soheil, G.; Mohammad, H.S.; Mahmoud Hoseini, S. Teleoperation Systems Design Using Singular Perturbation Method and Sliding Mode Controllers. *J. Dyn. Syst. Meas. Control* **2014**, *136*, 051005.
18.	Chen, Y.-Q.; Wei, Y.-H.; Wang, Y. On 2 types of robust reaching laws. *Int. J. Robust Nonlinear Control* **2018**, *28*, 2651–2667. [CrossRef]
19.	Giap, V.-N.; Huang, S.-C. Effectiveness of fuzzy sliding mode control boundary layer based on uncertainty and disturbance compensator on suspension active magnetic bearing system. *Meas. Control* **2020**, *53*, 934–942. [CrossRef]
20.	Tayebi-Haghighi, S.; Piltan, F.; Kim, J.M. Robust Composite High-Order Super-Twisting Sliding Mode Control of Robot Manipulators. *Robotics* **2018**, *6*, 13. [CrossRef]
21.	Floquet, T.; Barbot, J.P.; Perruquetti, W. Higher order sliding mode stabilization for a class of non holomic perturbed system. *Automatica* **2003**, *39*, 1077–1083. [CrossRef]
22.	Davila, J.; Fridman, L.; Levant, A. Second-order sliding-mode observer for mechanical systems. *IEEE Trans. Autom. Control* **2005**, *50*, 1785–1789. [CrossRef]
23.	Davila, J.; Fridman, L.; Poznyak, A. Observation and identification of mechanical systems via second order sliding modes. In *International Workshop on Variable Structure Systems*; IEEE: New York, NY, USA, 2006; pp. 232–237.
24.	Guzmán, E.; Moreno, J.A. Super-twisting observer for second-order systems with time-varying coefficient. *IET Control Theory Appl.* **2015**, *9*, 553–562. [CrossRef]
25.	Mu, C.; Sun, C. A new finite time convergence condition for super-twisting observer based on Lyapunov analysis. *Asian J. Control* **2015**, *17*, 1050–1060. [CrossRef]
26.	Tenoch, G.; Jaime, A.M.; Fridman, L. Variable Gain Super-Twisting Sliding Mode Control. *IEEE Trans. Autom. Control* **2012**, *57*, 2100–2105.
27.	Utkin, V. On convergence time and disturbance rejection of super-twisting control. *IEEE Trans. Autom. Control* **2013**, *58*, 2013–2017. [CrossRef]
28.	Lochan, K.; Singh, J.P.; Roy, B.K. Adaptive time-varying super-twisting global SMC for projective synchronisation of flexible manipulator. *Nonlinear Dyn.* **2018**, *93*, 2071–2088. [CrossRef]
29.	Nasim, U.; Yasir, M.; Jawad, A.; Amjad, A.; Jamshed, I. UAVS-UGV Leader Follower Formation Using Adaptive Non-Singular Terminal Super Twisting Sliding Mode Control. *IEEE Access* **2021**, *5*, 3081483.
30.	Moreno, J.A.; Osorio, M. A Lyapunov approach to second-order sliding mode controllers and observers. In Proceedings of the 47th IEEE Conference on Decision and Control, Cancún, Mexico, 9–11 December 2008; pp. 2856–2861.
31.	Kali, Y.; Saad, M.; Benjelloun, K.; Khairallah, C. Super-twisting algorithm with time delay estimation for uncertain robot manipulators. *Nonlinear Dyn.* **2018**, *93*, 557–569. [CrossRef]
32.	Yin, C.; Chen, Y.; Zhong, S.-M. Fractional-order sliding mode based extremum seeking control of a class of nonlinear systems. *Automatica* **2014**, *50*, 3173–3181. [CrossRef]
33.	Chen, D.; Zhang, J.D.; Li, Z.K. A Novel Fixed-Time Trajectory Tracking Strategy of Unmanned Surface Vessel Based on the Fractional Sliding Mode Control Method. *Electronics* **2022**, *11*, 726. [CrossRef]
34.	Maaruf, M.; Khalid, M. Global sliding-mode control with fractional-order terms for the robust optimal operation of a hybrid renewable microgrid with battery energy storage. *Electronics* **2021**, *11*, 88. [CrossRef]
35.	Zhu, L.K.; Chen, X.R.; Qi, X.; Zhang, J. Research on Fractional-Order Global Fast Terminal Sliding Mode Control of MDF Continuous Hot-Pressing Position Servo System Based on Adaptive RBF Neural Network. *Electronics* **2022**, *11*, 1117. [CrossRef]
36.	Aghababa, M.P. A fractional-order controller for vibration suppression of uncertain structures. *ISA Trans.* **2013**, *52*, 881–887. [CrossRef]

37. Cuong, H.M.; Dong, H.Q.; Trieu, P.V.; Tuan, L.A. Adaptive fractional-order terminal sliding mode control of rubber-tired gantry cranes with uncertainties and unknown disturbances. *Mech. Syst. Signal Proc.* **2021**, *154*, 107601. [CrossRef]
38. Wang, Y.; Yan, F.; Zhu, K.; Chen, B.; Wu, H. A new practical robust control of cable-driven manipulators using time-delay estimation. *Int. J. Robust Nonlinear Control.* **2019**, *29*, 3405–3425. [CrossRef]
39. Duc, T.M.; Hoa, N.V.; Dao, T.P. Adaptive Fuzzy Fractional-Order Nonsingular Terminal Sliding Mode Control for a Class of Second-Order Nonlinear Systems. *J. Comput. Nonlinear Dyn.* **2018**, *13*, 031004. [CrossRef]
40. Tuan, L.A. Neural observer and adaptive fractional-order backstepping fast-terminal sliding-mode control of RTG cranes. *IEEE Trans. Ind. Electron.* **2021**, *68*, 434–442. [CrossRef]
41. Nojavanzadeh, D.; Badamchizadeh, M. Adaptive Fractional-order Non-singular Fast Terminal Sliding Mode Control for Robot Manipulators. *IET Control Theory Appl.* **2016**, *10*, 1565–1572. [CrossRef]
42. Su, L.; Guo, X.; Ji, Y.; Tian, Y. Tracking control of cable-driven manipulator with adaptive fractional-order nonsingular fast terminal sliding mode control. *J. Vib. Control* **2021**, *27*, 2482–2493. [CrossRef]
43. Wang, Y.Y.; Peng, J.W.; Zhu, K.W.; Chen, B.; Wu, H.T. Adaptive PID-fractional-order nonsingular terminal sliding mode control for cable-driven manipulators using time-delay estimation. *Int. J. Syst. Sci.* **2020**, *51*, 3118–3133. [CrossRef]
44. Wang, Y.Y.; Gu, L.Y.; Xu, Y.H.; Cao, X.X. Practical tracking control of robot manipulators with continuous fractional-order nonsingular terminal sliding mode. *IEEE Trans. Ind. Electron.* **2016**, *63*, 6194–6204. [CrossRef]
45. Wang, Y.Y.; Chen, J.W.; Yan, F.; Zhu, K.W.; Chen, B.; Wang, Y. Adaptive super-twisting fractional-order nonsingular terminal sliding mode control of cable-driven manipulators. *ISA Trans.* **2018**, *75*, 163–180. [CrossRef]
46. Zhang, Y.Y.; Yang, X.H.; Wei, P.; Liu, P.X. Fractional-order adaptive non-singular fast terminal sliding mode control with time delay estimation for robotic manipulators. *IET Control Theory Appl.* **2020**, *14*, 2556–2565. [CrossRef]
47. Hu, R.; Deng, H.; Zhang, Y. Novel dynamic-sliding-mode-manifold-based continuous fractional-order nonsingular terminal sliding mode control for a class of second-order nonlinear systems. *IEEE Access* **2020**, *8*, 19820–19829. [CrossRef]
48. Kilbas, A.; Srivastava, H.M.; Trujillo, J.J. Fractional integrals and fractional derivatives. In *Theory Applications of Fractional Differential Equations*; Elsevier: Amsterdam, The Netherlands, 2006; pp. 69–133.

*Article*

# IM Fed by Three-Level Inverter under DTC Strategy Combined with Sliding Mode Theory

**Salma Jnayah *, Intissar Moussa and Adel Khedher**

Université de Sousse, Ecole Nationale d'Ingénieurs de Sousse, LATIS-Laboratory of Advanced Technology and Intelligent Systems, 4023 Sousse, Tunisie
* Correspondence: salmajnayah@gmail.com; Tel.: +216-20-671-166

**Abstract:** The classical direct torque control (CDTC) of the induction motor (IM) drive is characterized by high ripples in the stator flux and the electromagnetic torque waveforms due to the use of hysteresis comparators. Furthermore, the motor speed in this control strategy is ensured through a proportional integral (PI) regulator, due to its simple structure. Nonetheless, this controller is sensitive to load disturbances. Hence, it is not robust against parameter variance, which can degrade the motor performance. To overcome this deficiency, many endeavors have been conducted in the literature to ensure a high dynamic response of the motor in all speed ranges, with minimum flux and torque undulations. Thus, the DTC of an IM associated with a three-level inverter based on sliding mode (SM) flux, torque and speed controllers was adopted to substitute the hysteresis comparators and the traditional PI regulator, since the SM speed controller is able to prevail against external disturbances. The second contribution of this manuscript is to develop the proposed DTC_SM approach using the Xilinx System Generator (XSG) in order to implement it on a field programmable gate array (FPGA) Virtex 5 on account of its ability to adopt parallel processing. The hardware co-simulation results verify clearly the merits of the suggested modified DTC strategy.

**Keywords:** DTC of induction motor; three-level inverter; hysteresis comparators; ripple minimization; sliding mode; robust control; Xilinx System Generator (XSG); FPGA

**Citation:** Jnayah, S.; Moussa, I.; Khedher, A. IM Fed by Three-Level Inverter under DTC Strategy Combined with Sliding Mode Theory. *Electronics* **2022**, *11*, 3656. https://doi.org/10.3390/electronics11223656

Academic Editors: Jamshed Iqbal, Ali Arshad Uppal and Muhammad Rizwan Azam

Received: 17 October 2022
Accepted: 4 November 2022
Published: 9 November 2022

**Publisher's Note:** MDPI stays neutral with regard to jurisdictional claims in published maps and institutional affiliations.



## 1. Introduction

Induction motors (IM) are nowadays the most used machines in the industrial area due to their reasonable cost, good performance, simple structure and plain control [1]. Therefore, numerous control strategies have been proposed to satisfy the industry's growing demand. Within this trend, the direct torque control (DTC) approach, initially introduced by Takahashi and Depenbrock in the middle of the 1980s [2], has gained extensive interest. It has been touted as the first field-oriented control competitor since it is less sensitive to machine parameter variations and ensures the precise and fast dynamics of the torque. The basic concept of this strategy is to control, independently and directly, the electromagnetic torque and the stator flux through two hysteresis comparators. However, the latter systematize the motor into a variable switching frequency range, which definitely engenders fluctuations in flux and torque waveforms [3]. Among the several prevalent methods developed to enhance the CDTC strategy is the use of the space vector modulation (SVM) technique [4–6]. Nevertheless, the DTC-SVM algorithm is somewhat complex when compared to CDTC, and it is sensitive to the machine parameter variations and external disturbances [3]. A further improvement in the motor torque response is attained by using intelligent techniques such as fuzzy logic and neural networks. However, the switching frequency in this approach remains variable, which does not satisfy the requirement for torque and flux ripple reduction. To enhance the CDTC performance, another method is suggested, built on multilevel inverters instead of the standard two-level one. This scheme offers a greater number of active voltage vectors able to minimize the torque ripples [7]. Other attempts to improve the DTC strategy have been elaborated, notably combining the DTC with predictive control [8], where the optimized switching state is selected in a manner in which torque ripples are significantly minimized. However, the structure of the algorithm is mathematically complex because it needs the discretization of the power converter.

In most aforementioned cases, flux and torque ripples have been noticeably decreased but the robustness of the motor control is neglected. Thence, solutions adopting the sliding mode (SM) approach become a promising alternative. In fact, it ensures not only a fast dynamic response, stability and robustness against parameter variation, but also reduced flux and torque undulation, as well as a low current total harmonic distortion (THD) rate.

For this reason, unlike the previously published work, the main contribution and novelty of this paper is to combine several approaches that ensure the improvement of the classical DTC control to enable the IM to operate with minimum flux and torque ripples in all speed ranges. The proposed DTC-SM method in this paper is established by substituting the conventional two-level inverter feeding the motor with a three-level one. Indeed, the semiconductor switches in the three-level neutral point clamped (NPC) inverter are subjected to a lower voltage change rate (dv/dt), which causes the harmonic losses on the motor side to be significantly lower. Furthermore, the flux and torque hysteresis comparators in the suggested DTC strategy are replaced with sliding mode controllers. Consequently, and through this work, torque and flux oscillations are notably reduced. Likewise, the tuning of the PI speed controller is delicate because its parameters are strongly related to the IM parameters. It is also sensitive to any load torque disturbance, which can arguably degrade the machine performance. As an auspicious solution, this research work seeks to replace the classical PI speed controller with a sliding mode one in view of its robustness against external disturbances [9]. The second part of this manuscript is devoted to implementing the developed control strategy in a digital platform.

Ordinarily, two hardware design solutions are exploited when controlling the IM drive. The first one involves digital signal processors (DSPs) and microcontrollers (STM32, etc.) [10,11]. The second category is the field programmable gate array (FPGA). Nowadays, the first software family is almost excluded because of the sequential treatment of its processors, and then they became able to cope with the market needs. In fact, the processing speed of DSP and μcontrollers relies on the complexity of the implemented control system, which can heavily increase the computational development. Moreover, the DSP computing speed imposes a limit for the inverter switching frequency that must not be exceeded. Accordingly, the control algorithm's performance is affected [12]. To remedy the above-mentioned limitations, the FPGA board is proposed to guarantee faster executions [13]. Indeed, this software target is chosen as a promising alternative to implement the DTC strategy with a shorter processing time due to its parallel execution, flexibility and large computing capability. To configure an FPGA, various digital design languages are deployed, namely Verilog and the Very-High-speed integrated circuits Hardware Description Language (VHDL). However, programming Verilog and VHDL requires extensive skills and a large amount of development time. To address this dilemma, the Xilinx System Generator (XSG) tool is used in this paper to generate automatically the bitstream file required for FPGA configuration, and so this toolbox enables designers to test and validate the effectiveness of the control algorithm without deteriorating the real system [14]. This method is called "hardware co-simulation" or "hardware in the loop" (HIL).

In this work, the DTC-SM of the IM drive fed by a three-level NPC inverter is developed. This control strategy is based on three sliding mode controllers dedicated to torque, flux and motor speed, instead of hysteresis and PI regulators. To validate the performance of the proposed method, hardware-in-the-loop simulations are carried out between MATLAB/Simulink and FPGA Virtex 5.

The previously published research works either deal with modeling the standard DTC-SM strategy (based on a two-level inverter) with XSG or with implementing the DTC-

SM strategy on an FPGA using VHDL code, which require advanced programming skills. The literature has not yet taken into account the HIL simulation of DTC-SM based on a three-level inverter using the XSG toolbox.

The rest of this paper is organized as follows. Section 2 is devoted to developing the DTC strategy of the IM drive controlled with a three-level inverter. The sliding mode flux, torque and speed controllers are detailed in Section 3. A hardware co-simulation of the modified DTC approach implemented in the FPGA Virtex 5 board is presented in Section 4. Finally, the main conclusions are drawn in the Section 5. Table 1 provides a summary of the main contributions of this study in comparison to other publications. The proposed improved DTC method, the type of motor, the type of inverter driving the motor, the current THD rate, the digital platform for hardware implementation and the programming language were the six criteria considered to conduct this brief review.

**Table 1.** Review of some proposed methods to improve DTC strategy.

| Ref. | Proposed Method | Motor Type | Converter Type | Current THD | Support | Programming Language |
|---|---|---|---|---|---|---|
| [10] | DTC based on twelve sectors and modified hysteresis controllers | Three-phase IM | Two-level inverter | 19.92% | DSP TMS320F28069M | C |
| [15] | DTC based on a specific selection of the inverter switching states | Five-phase IM | Three-level five-phase inverter | 11.59% | DSP TMS320F28377S | C |
| [16] | DTC based on intelligent technique | Three-phase IM | Two-level inverter | 5.882% | FPGA Virtex 5 | XSG |
| [17] | DTC with sliding mode controller and observer | Four-phase switched reluctance motor | Asymmetrical half-bridge | - | dsPACE 1401 | C |
| [18] | Supertwisting sliding mode DTC | Three-phase IM | Two-level inverter | - | TMS320F28335 eZdsp platform | C++ |
| This work | DTC based on sliding mode theory | Three-phase IM | Three-level NPC inverter | 4.19% | FPGA Virtex 5 | XSG |

## 2. DTC of IM Associated with Three-Level NPC Inverter

### 2.1. IM Modeling

The first step in designing the DTC strategy is to build the mathematical model of the IM in the stationary reference frame ($\alpha$, $\beta$), as introduced in Equation (1) [19]:

$$\begin{cases} \frac{d}{dt}X = [A]X + [B]U \\ Y = CX \end{cases} \tag{1}$$

where X is the state vector defined as $X = \begin{bmatrix} i_{s\alpha} \\ i_{s\beta} \\ \varphi_{s\alpha} \\ \varphi_{s\beta} \end{bmatrix}$.

U and Y are the state control and output vectors determined as $U = \begin{bmatrix} V_{s\alpha} \\ V_{s\beta} \end{bmatrix}$, $Y = \begin{bmatrix} i_{s\alpha} \\ i_{s\beta} \end{bmatrix}$.

[A], [B] and [C] are three matrices given in the following form:

$$[A] = \begin{pmatrix} -(\frac{R_s}{\sigma R_r} + \frac{R_r}{\sigma L_r}) & -\omega & \frac{R_r}{\sigma L_s L_r} & \frac{\omega}{\sigma L_s} \\ \omega & -(\frac{R_s}{\sigma R_r} + \frac{R_r}{\sigma L_r}) & \frac{\omega}{\sigma L_s} & \frac{R_r}{\sigma L_s L_r} \\ -R_s & 0 & 0 & 0 \\ 0 & -R_s & 0 & 0 \end{pmatrix} ; [B] = \begin{bmatrix} \frac{1}{\sigma L_s} & 0 \\ 0 & \frac{1}{\sigma L_s} \\ 1 & 0 \\ 0 & 1 \end{bmatrix} ; [C] = \begin{bmatrix} 1 & 0 \\ 0 & 1 \\ 0 & 0 \\ 0 & 0 \end{bmatrix}$$

$R_s$, $R_r$ and $\omega$ are the stator resistance, the rotor resistance and the motor speed. Likewise, $L_s$ and $L_r$ are the stator and rotor inductances. $\sigma = 1 - \frac{M_{sr}^2}{L_s L_r}$ is the Blondel coefficient, where $M_{sr}$ represents the mutual inductance.

The equation describing the mechanical behavior of the IM is established hereafter [20]

$$J \frac{d}{dt}\Omega = T_{em} - f\Omega - T_l \tag{2}$$

where $T_{em}$, $T_l$, $J$ and $f$ are, respectively, the electromagnetic torque, the load torque, the inertia moment and the viscous friction coefficient.

The motor's parameters are presented in Appendix A.

### 2.2. DTC Principle

The fundamental concept of the DTC approach, as originally introduced in the 1980s, is based on the direct and independent control of both stator flux and electromagnetic torque through selecting a suitable voltage vector that should be applied to the inverter switches [21]. The basic scheme of this control strategy is illustrated in Figure 1.

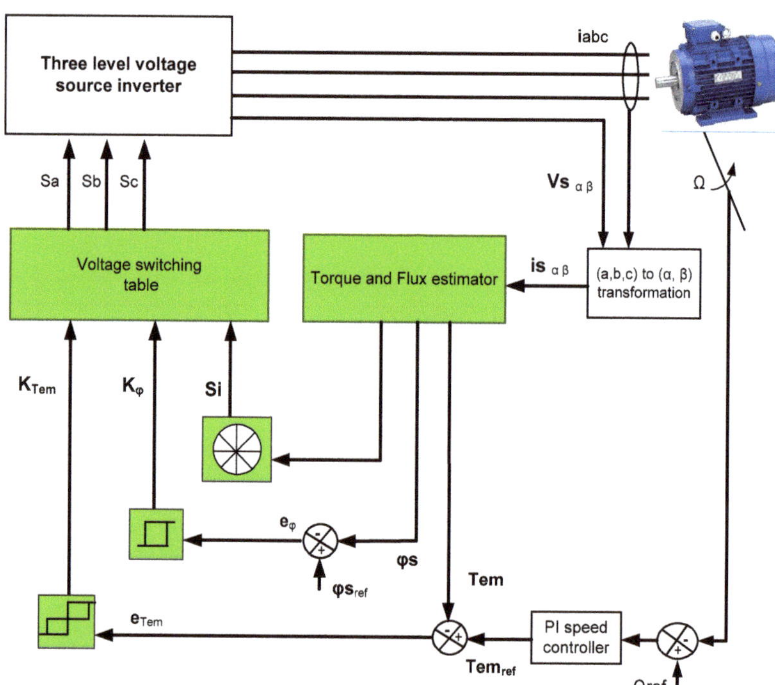


**Figure 1.** DTC structure of IM fed with three-level inverter.

Referring to Figure 1, the DTC strategy is mainly composed of three blocks.

Torque and flux estimator: The torque can be estimated based on Equation (3):

$$T_{em} = \frac{3}{2} n_p \left( \varphi_{s\alpha} \, i_{s\beta} - \varphi_{s\beta} \, i_{s\alpha} \right) \tag{3}$$

where $n_p$ is the pair pole number.

Likewise, the stator flux components are estimated as given in Equation (4):

$$\begin{cases} \varphi_{s\alpha} = \int \left( V_{s\alpha} - R_s \, i_{s\alpha} \right) dt \\ \varphi_{s\beta} = \int \left( V_{s\beta} - R_s \, i_{s\beta} \right) dt \end{cases} \tag{4}$$

Therefore, the module of the stator flux as well as its corresponding angle are evaluated as

$$\begin{cases} |\varphi_s| = \sqrt{\varphi_{s\alpha}^2 + \varphi_{s\beta}^2} \\ \theta_s = \arg(\varphi_s) \end{cases} \tag{5}$$

Torque and flux hysteresis comparators: Once the torque and stator flux are estimated, they are compared to their reference values. Indeed, the obtained errors present the inputs of two hysteresis controllers. These latter produce, in their outputs, two logical rules $K_\varphi$ and $K_{Tem}$.

Selecting table: The output voltage vector of the three-level inverter, given in Equation (6), is chosen based on a switching table. This table needs as inputs $K_\varphi$, $K_{Tem}$ and the sector where the stator flux is located [22].

$$V_s = \sqrt{\frac{2}{3}} V_{dc}(S_a + S_b.e^{j\frac{2\pi}{3}} + S_c.e^{j-\frac{2\pi}{3}}) \tag{6}$$

$S_a$, $S_b$ and $S_c$ are the inverter switching states.

The paper [23] provides a thorough explanation of the DTC control approach for additional details.

### 2.3. Three-Level NPC Inverter

Continuous developments in the field of power electronics have prompted researchers to use three-level inverter topologies in controlling IM drives. The increase in level number can resolve efficiently torque and flux ripples, as well as current distortion, in the DTC strategy when compared to conventional ones [24,25].

Figure 2 illustrates the general scheme of the three-level NPC inverter. The number of switching states in this case is $3^3 = 27$. Among them are three zero vectors and 24 active vectors.

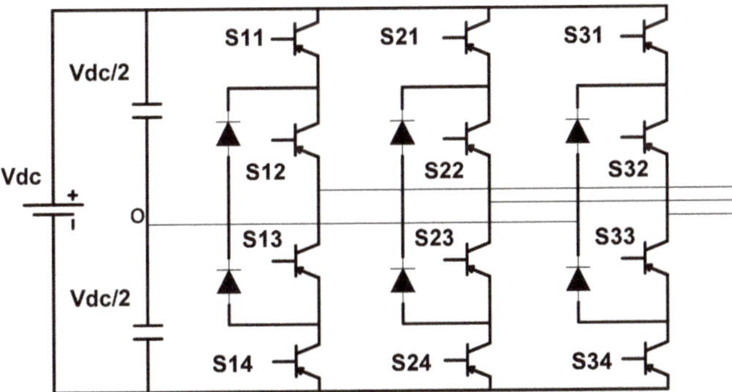


**Figure 2.** Three-level NPC inverter structure.

## 3. Proposed DTC Based on Sliding Mode Approach

In the DTC of the three-level inverter-fed IM drive, ripples are notably reduced when compared to the classical control approach. However, the voltage vector selection in this strategy is mainly based on the hysteresis controllers, which are responsible for torque and stator flux undulations. Furthermore, the speed control loop is settled by a PI regulator, known by its sensitivity to parameter variations. To surmount the above-mentioned deficiencies, the stator flux, torque and motor speed are controlled by sliding mode functions [26]. Indeed, the strength characteristics of the SM regulators are a fast response, robustness against parameter uncertainties and simple hardware implementation [27–30]. The design of SMC hinges on two steps. The first one is the determination of the switching surface and the second one relies on designing the control law in such a manner as to maintain the system trajectory toward the sliding surface [31].

The SM structure, shown in Figure 3, is the most commonly used one for electrical machine drives owing to the relay function, which is considered as a suitable control for power electronic converters.

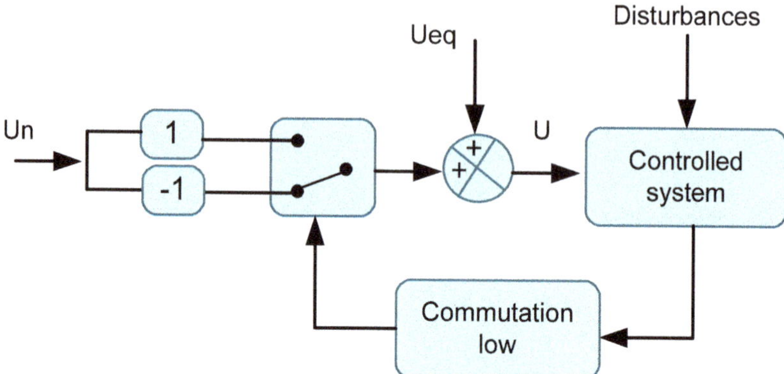


**Figure 3.** SM control structure.

The control law expression is provided by

$$U = U_{eq} + U_n \tag{7}$$

where $U_{eq}$ is the equivalent command, which is a continuous function used to maintain the controlled variable on the sliding surface. It is obtained when the invariance conditions of the sliding surface are satisfied.

$$\begin{cases} S = 0 \\ \frac{d}{dt}S = 0 \end{cases} \tag{8}$$

$U_n$ is the switching term based on the sign function of the sliding surface.

### 3.1. Speed Integral Sliding Mode Controller

The speed regulation loop is designed to generate the electromagnetic toque reference and to provide a fast response and good dynamics. The speed tracking error considered in some research works is defined as

$$e_\Omega = \Omega_{ref} - \Omega \tag{9}$$

where $\Omega_{ref}$ denotes the motor speed reference.

To provide a faster dynamic response with good accuracy, an integral sliding surface is suggested to establish the control law, as formulated in Equation (10):

$$S_\Omega = e_\Omega + \lambda_\Omega \int e_\Omega \, dt \tag{10}$$

$\lambda_\Omega$ is a positive constant.

Its derivative is therefore

$$\dot{S}_\Omega = \dot{\Omega}_{ref} - \dot{\Omega} + \lambda(\Omega_{ref} - \Omega) \tag{11}$$

When substituting Equation (2) into (10), and assuming that there is convergence to the sliding surface ($\dot{S}_\Omega = 0$), we obtain the equation below:

$$\frac{1}{J}Tem_{eq} = \dot{\Omega}_{ref} + \frac{1}{J}T_l + \frac{f}{J}\Omega + \lambda_\Omega(\Omega_{ref} - \Omega) \tag{12}$$

Based on Equation (7), the expression of the reference torque is written as

$$\text{Tem}_{\text{ref}} = \underbrace{J\left(\dot{\Omega}_{\text{ref}} + \frac{1}{J}T_l + \frac{f}{J}\Omega + \lambda_\Omega(\Omega_{\text{ref}} - \Omega)\right)}_{\text{Tem}_{\text{eq}}} + \underbrace{K_\Omega|S_\Omega|\text{sign}(S_\Omega)}_{\text{Tem}_n} \qquad (13)$$

$K_\Omega$ is a positive gain, $0 < \alpha < 1$, and sign(.) denotes the sign function.

The diagram corresponding to the implementation of this speed regulation loop based on the variable structure control is given in Figure 4.

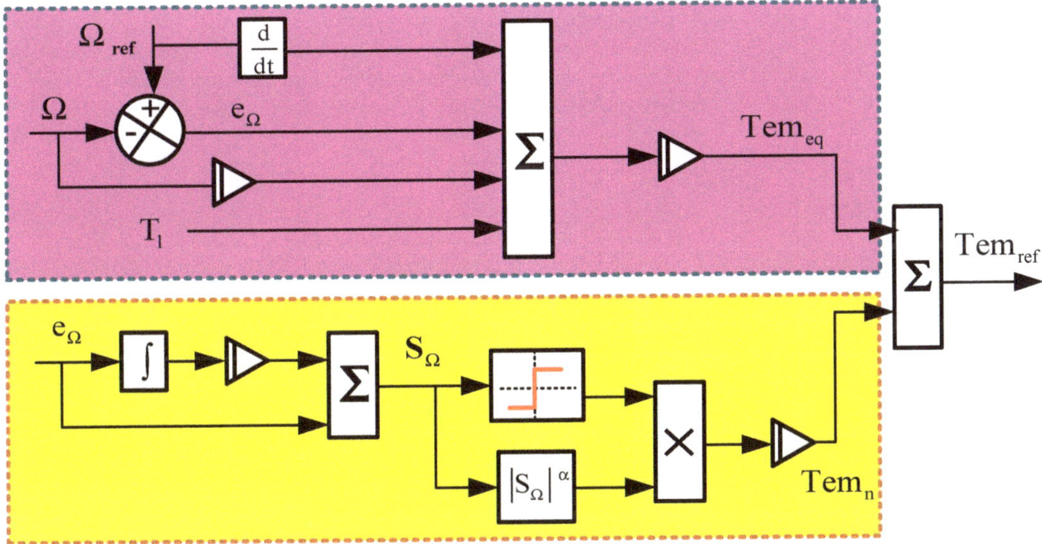



**Figure 4.** Reference electromagnetic torque generated by the SM speed control loop.

*Stability proof:* Once the design of the sliding control laws is settled, it is necessary to verify the system's stability, which can be assessed in terms of the Lyapunov function as follows:

$$V = \frac{1}{2}S^2 \qquad (14)$$

If the derivative of this function is defined as negative, then the system trajectory is driven to the sliding mode surface and remains on it,

$$\dot{V} = S\dot{S} < 0 \qquad (15)$$

Then, replacing expression (12) in (13), we obtain the following equation:

$$\dot{S} = -\frac{k_\Omega}{J}|S_\Omega|\text{sign}\,(S_\Omega) < 0 \qquad (16)$$

Subsequently, the stability condition is approved.

### 3.2. Sliding Mode PI Controller of Flux and Torque

Using Equations (3) and (4), and assuming that the direction of the stator flux vector is aligned with the d component, which means that $\varphi_{sq} = 0$, we can write the following equation:

$$\begin{cases} V_s = V_{sd} + jV_{sq} = R_s i_s + \frac{d}{dt}\varphi_s + j\omega_{\varphi s}\varphi_s = \underbrace{R_s i_{sd} + \frac{d}{dt}\varphi_s}_{V_{sd}} + \underbrace{R_s i_{sq} + \omega_{\varphi s}\varphi_s}_{V_{sq=0}} \\ T_{em} = \frac{3}{2}p\,\varphi_s i_{sq} = \frac{3}{2}p\,\varphi_s \frac{(V_{sq} - \omega_{\varphi s}\varphi_s)}{R_s} \end{cases} \tag{17}$$

where $\omega_{\varphi s}$ is the angular speed of the stator flux.

According to the aforementioned equations, flux and torque control can be accomplished using the direct and quadrature stator voltage components, respectively ($V_{sd}$ and $V_{sq}$). The two selected sliding surfaces are $S_\varphi$ and $S_{Tem}$. While $S_{Tem}$ represents the sliding surface of the electromagnetic torque, $S_\varphi$ is defined for the stator flux error to control the direct component of voltage $V_s$.

Equation (18) provides the corresponding sliding surfaces, given below:

$$\begin{cases} S_{Tem} = e_{Tem} + k_{Tem}\int e_{Tem}\,dt \\ S_\varphi = e_\varphi + k_\varphi \int e_\varphi\,dt \end{cases} \tag{18}$$

$k_{Tem}$ and $k_\varphi$ are two positive constants. $e_{Tem}$ and $e_\varphi$ are, respectively, the torque and flux errors, defined as

$$\begin{cases} e_{Tem} = Tem_{ref} - Tem \\ e_\varphi = \varphi_{ref} - \varphi \end{cases} \tag{19}$$

Hence, we proceed to write the following equation:

$$\begin{cases} V_{sd} = \left(K_{p\varphi s} + \frac{K_{I\varphi s}}{S}\right)sat(S_1) \\ V_{sq} = \left(K_{pTem} + \frac{K_{ITem}}{S}\right)sat(S_2) + \omega_{\varphi s} \end{cases} \tag{20}$$

where $K_{p\varphi s}$, $K_{pTem}$, $K_{I\,\varphi s}$ and $K_{I\,Tem}$ are the PI regulator gains that must be properly chosen to satisfy the condition of stability using the Lyapunov criterion.

The proposed flux and torque variable structure controller (VSC) is depicted in Figure 5.

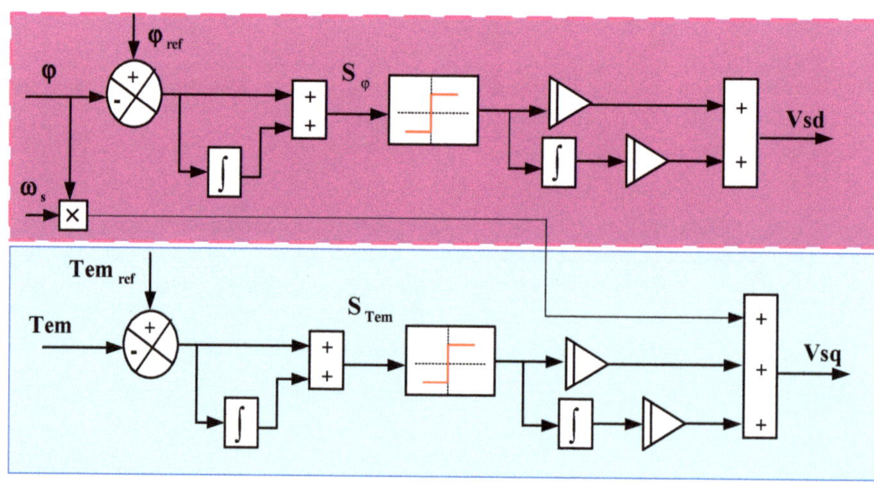


**Figure 5.** SM-PI flux and torque regulation.

The major problem of the SM controllers is the chattering phenomenon, due to the discontinuous nature of the control law caused by an infinite commutation around the sliding surface [26]. To solve this problem, the sign function is replaced by a smoother one (Sat function), as illustrated in Figure 6.

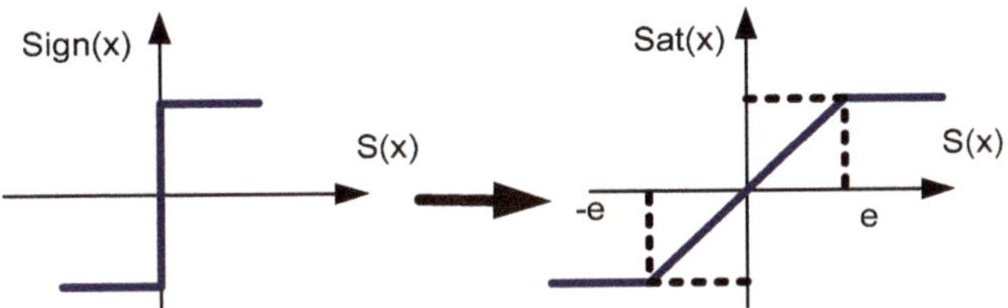


**Figure 6.** Commutation functions.

*3.3. Simulation Results*

To substantiate the effectiveness of the proposed DTC-SM control strategy and evaluate the designed control system's dynamic response, we have modeled it in Matlab/Simulink software. The applied reference speed profile of the given system is depicted in Figure 7.

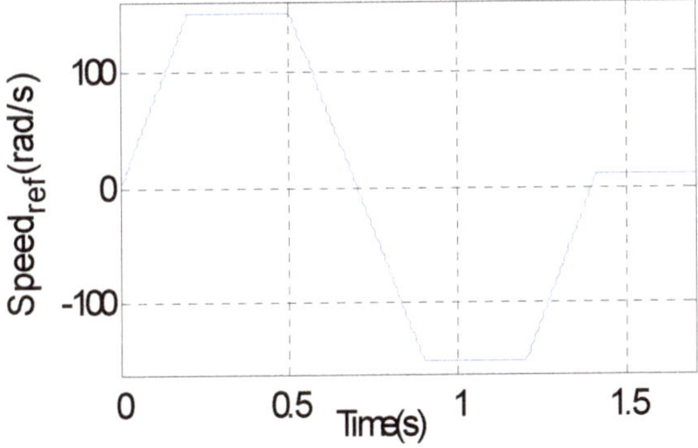


**Figure 7.** Reference speed.

To test the robustness of the suggested control approach, sudden electromagnetic load torques are applied in each speed range, as given in Figure 8.

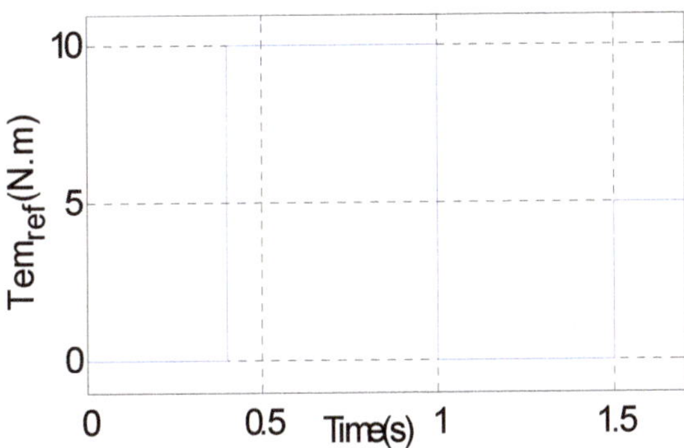


**Figure 8.** Load torque application.

In the startup, the machine is not loaded until t = 0.4 s. Then, from t = 0.4 s to t = 1 s, we apply a load torque of 10 N·m, which represents the motor's nominal electromagnetic torque. From t = 1 s to t = 1.5 s, the IM operates again without any load. Finally, a load torque of 5 N·m is suddenly applied at a low speed range.

The following figures illustrate the comparative analysis of the DTC of the three-level inverter-fed IM (based on hysteresis flux and torque controllers and PI speed regulator) and DTC_SM (based on SM torque, flux and speed controllers).

Figure 9a,b illustrate, respectively, the stator flux trajectory in the ($\alpha$, $\beta$) frame. It can be noted that when using both strategies, a fully circular trajectory was guaranteed. This demonstrates that the flux vector amplitude is kept constant throughout all operation stages. However, compared to the DTC technique, the suggested DTC-SM delivers a finer and smoother trajectory, which confirms the reliability of this control approach. Figure 9c,d represent magnified views of the stator current waveforms of the IM controlled with the DTC and DTC-SM strategies, respectively. It is clear that in the case of the DTC-SM, the harmonics content was less important, with THDi = 4.19%, in comparison to the case of DTC, which was characterized by THDi = 26.68%. The torque responses and the stator current modules of DTC (red line) and DTC-SM (blue line) with the same load scenario are illustrated in Figure 9e,f. It can be clearly perceived that the proposed strategy has reduced torque and current ripple levels when compared to the DTC control method. In fact, the torque ripple range of the DTC technique is approximately $\Delta_{Tem}$ = 0.4 N·m, while that of the proposed DTC-SM is neglected, which confirms the effectiveness of the suggested control technique in terms of torque and current ripple reduction.

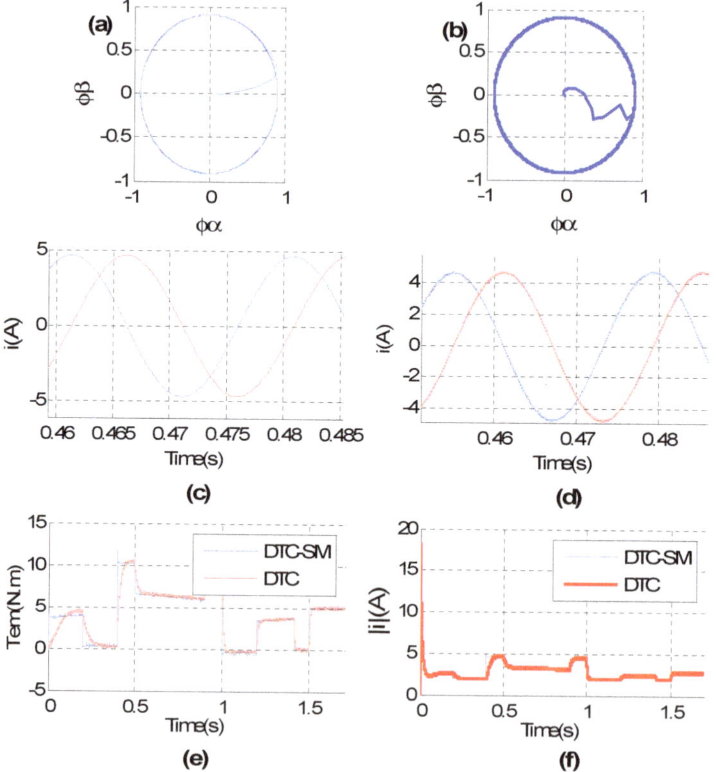

**Figure 9.** Simulation results of IM performance when controlled with DTC and DTC-SM techniques: (**a**) flux trajectory under DTC strategy; (**b**) flux trajectory under DTC-SM strategy; (**c**) magnified view of the stator current under DTC strategy; (**d**) magnified view of the stator with DTC-SM technique; (**e**) torque responses under DTC and DTC-SM strategies; (**f**) stator current magnitude under both strategies.

The figure exhibiting the motor speed with the DTC of the three-level inverter-fed IM drive and DTC based on sliding mode theory, with a change in load level during operation, is illustrated in Figure 10.

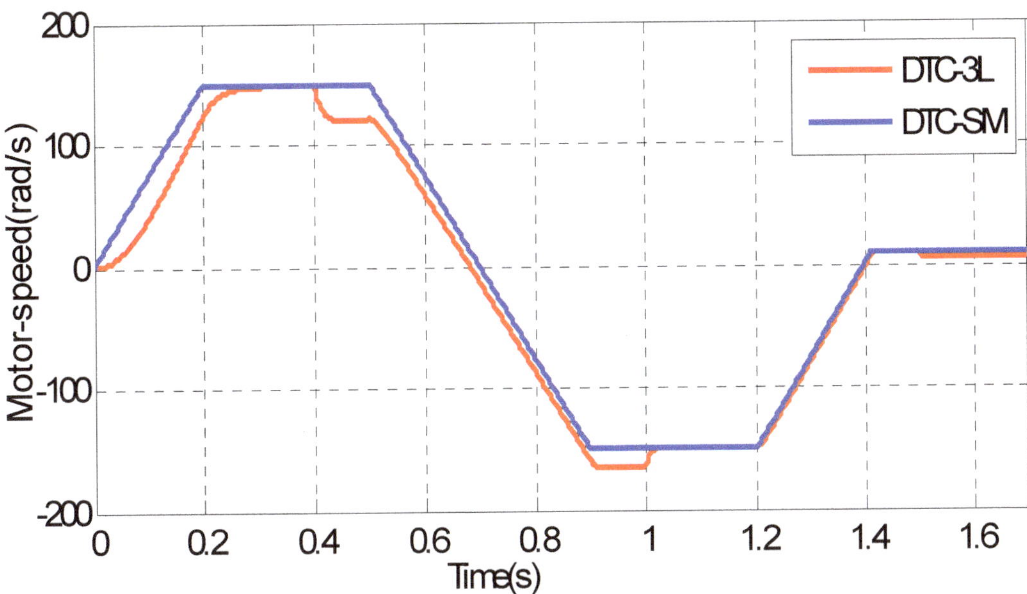


**Figure 10.** Motor speed with DTC and DTC-SM techniques under load torque variations.

The speed responses of the machine with both developed control techniques converge towards their reference values. However, during the load torque's sudden application (Figure 8), we recorded speed drops, which are given in Table 2.

**Table 2.** Speed drop of the IM under both control strategies in different operating ranges.

|  | At t = 0.4 s | At t = 1 s | At t = 1.5 s |
| --- | --- | --- | --- |
| Speed drop with DTC strategy under load variation | 29.7 rad/s | 13.8 rad/s | 4.8 rad/s |
| Speed drop with DTC-SM strategy under load variation | 0.4 rad/s | 0.4 rad/s | 0.07 rad/s |

In the case of the DTC strategy, the motor speed under external load introduction of 10 N·m at t = 0.4 s displays a drop of approximately 29.7 rad/s, while the speed curve in DTC-SM is not significantly affected by the load torque disturbance (very slight drop of 0.4 rad/s). Likewise, when a load torque of 5 N·m is applied at a low speed range at t = 1.5 s, the speed drop with DTC is around 4.8 rad/s, while in DTC-SM, the speed curve closely follows the reference value.

It is obvious from Figure 10 and Table 2 that, in various operating ranges, when the load torque scenario is applied, as shown in Figure 8, the speed curve generated by the IM under the DTC-SM approach thoroughly tracks the reference curve with a disregarded disturbance, before regaining immediately its reference value. Furthermore, it can also be observed that there exists faster dynamics and better speed reference tracking with the suggested control technique. Therefore, the proposed DTC-SM method has better speed robustness against load perturbation and a better dynamic response when compared to the DTC method.

## 4. HIL Simulation of DTC-SM Strategy

The first step to implement the developed control strategy on the FPGA board is to design it under the Matlab/Simulink environment using the XSG toolbox. Thus, this section is devoted to describing the software configuration of the suggested approach based on XSG blocks. Nonetheless, many operations are not directly available, mainly the signal generators and trigonometric functions. To address this constraint, we can mathematically approximate the desired expression by exploiting the toolbox's basic elements [14].

The co-simulation method is used to validate the control strategy performance and optimize the time required for experimental test execution. The fundamental step of the HIL consists of generating a JTAG block that substitutes the previously designed control architecture. This block is automatically created between the "gateway in" and "gateway out" resources. Concerning the data type, we used the fixed point format instead of the floating point, because it requires a longer execution time, especially for complex mathematical functions. The diagram in Figure 11 explains the approach to hardware co-simulation.

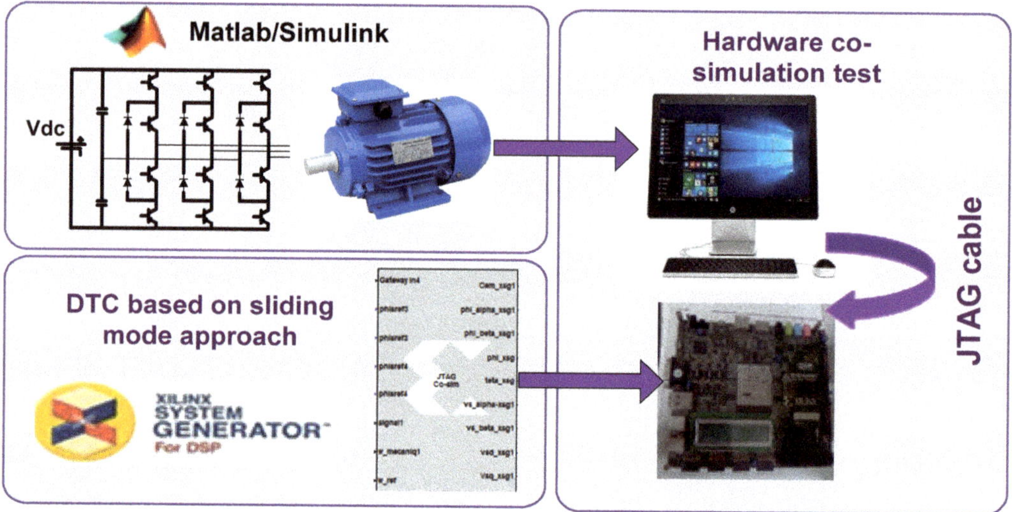


**Figure 11.** Synoptic diagram of co-simulation test.

### 4.1. XSG Design of DTC Based on Sliding Mode Approach of IM Supplied by Three-Level Inverter

Exploiting the basic elements of the XSG toolbox, the DTC of the IM built on the sliding mode approach is as depicted in Figure 12. By comparing the design functionality to the reference model behavior displayed in the Simulink interface, this library enables us to validate the design functionality via digital simulation. The communication between the proposed architecture under XSG and the association of the IM/inverter is conferred through the "gateway in" and "gateway out" blocks.

The architecture of the coordinate transformation $(d, q)$ to $(\alpha, \beta)$ frame is illustrated in Figure 13. This block is based on trigonometric functions that are not directly available in the XSG toolbox library. Therefore, we have resorted to an ROM unit. The accuracy of the generated value depends on the internal parameters, such as the size of the ROM, the step size and the gain coefficient in the input module.

Likewise, to settle the angle of the reference stator vector as well as its module, the arc-tangent function existing in the CORDIC block of the XSG library produces a computation error, which leads to inaccurate results. To address this deficiency, we chose to use ROM modules. The flux angle and module designed using the XSG toolbox are outlined in Figure 14.

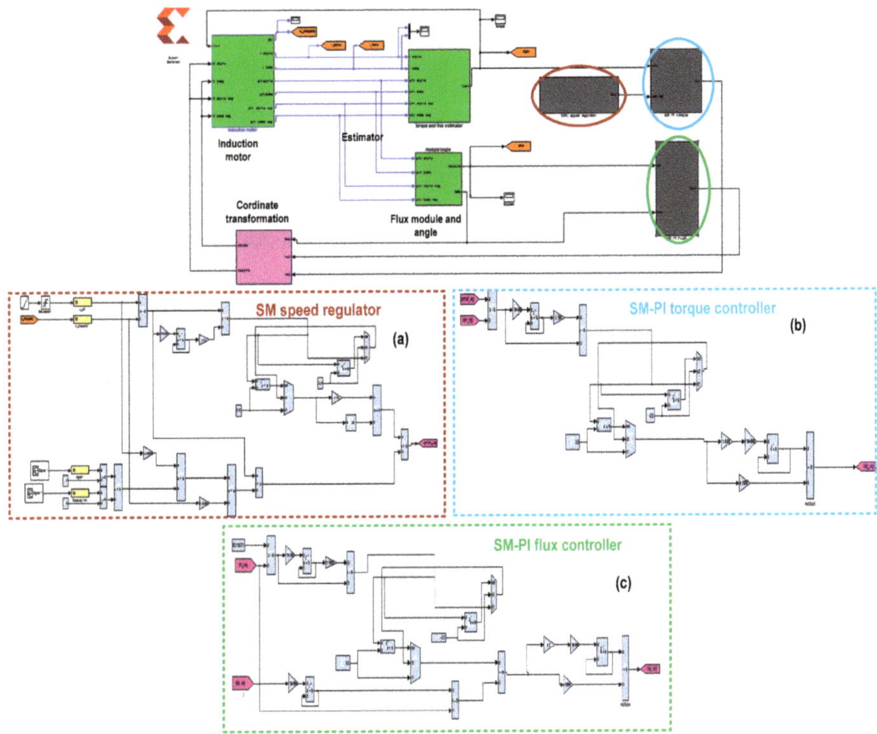

**Figure 12.** DTC-SM modeling using XSG toolbox: (**a**) speed regulation loop using SM controller; (**b**) SM-PI torque regulator; (**c**) SM-PI controller dedicated to stator flux.

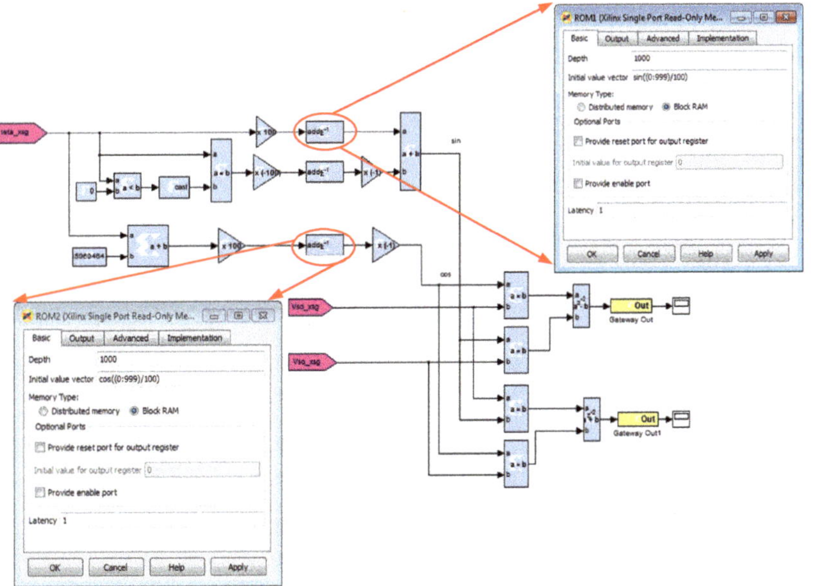

**Figure 13.** XSG modeling of the coordinate transformation $(d, q)$ to $(\alpha, \beta)$ using ROM blocks.

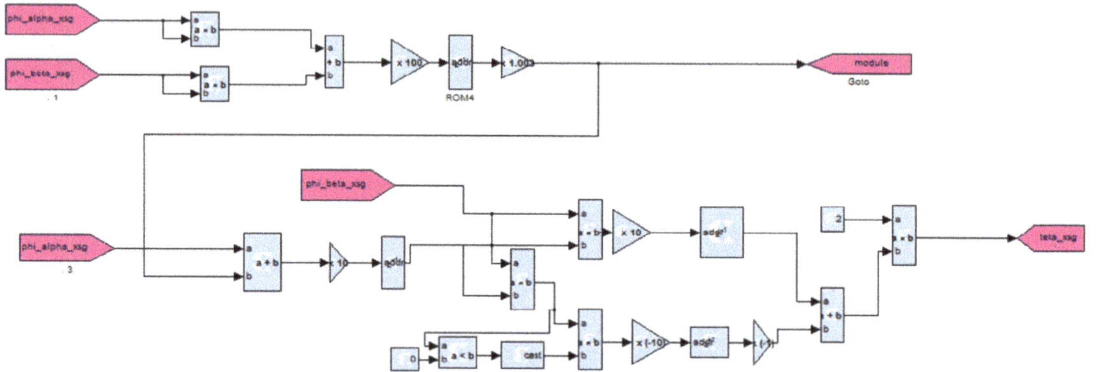

**Figure 14.** Designed flux angle and magnitude using elementary XSG blocks based on mathematical estimations of the unavailable functions in the Xilinx library.

*4.2. Software in-the-Loop Simulation Results*

To show the performance of the DTC-SM, a simulation study is carried out using the XSG toolbox, giving the following results depicted in Figure 15, which presents the motor speed, the electromagnetic torque and the stator flux magnitude.

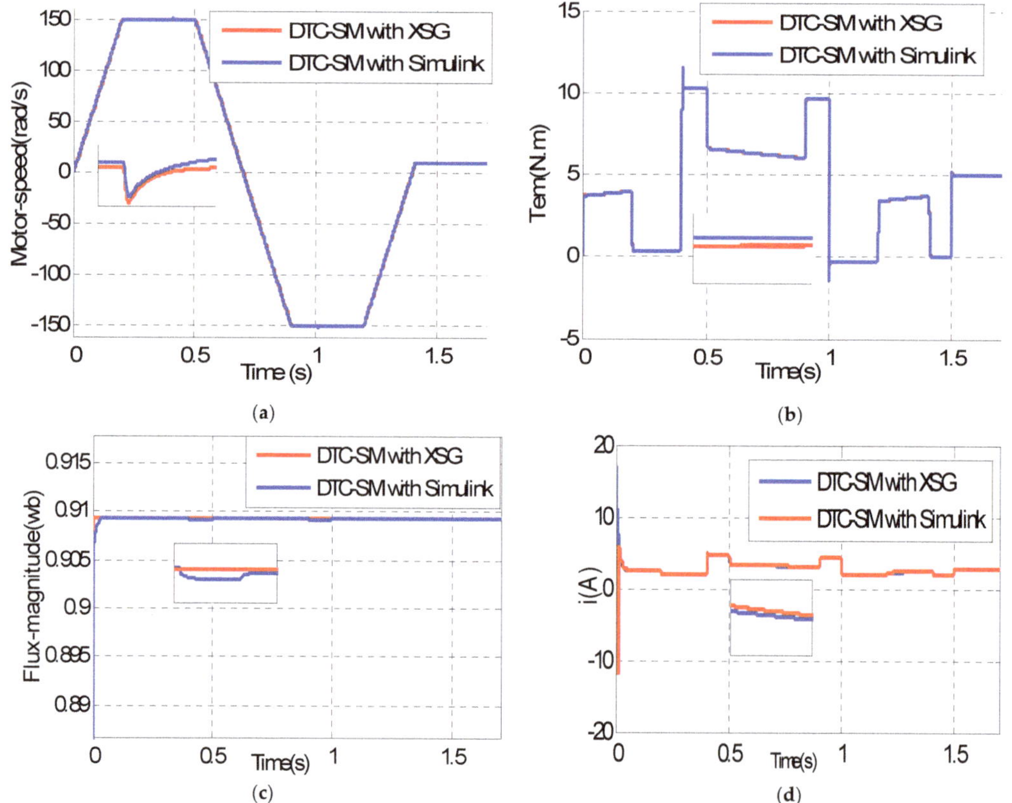

**Figure 15.** Evolution of the (**a**) motor speed; (**b**) electromagnetic torque; (**c**) stator flux module with both Simulink and XSG models; and (**d**) current module with XSG and Simulink.

The same scenario of load torque variation and reference speed evolution, described in Figures 8 and 9, is applied. As can be observed, the speed profiles in Figure 15a follow the reference value quite closely, with only a few minor deviations at the moment of load torque application. Figure 15b–d show the evolution of the electromagnetic torque, the stator flux module and the current magnitude under the DTC-SM strategy for both the Simulink and XSG models. It can be claimed that the obtained results using XSG tools are very similar to those obtained by numerical simulation in terms of speed tracking, electromagnetic torque response and stator flux and current aspects, which confirms the reliability and effectiveness of the XSG modeling in designing architectures on FPGA.

The error rates between quantities modeled in Simulink and those modeled with XSG tools are presented in Table 3. These promising results allow us to proceed further with the hardware implementation of the suggested control technique.

**Table 3.** Error rate collection between quantities modeled in Simulink and those modeled with XSG.

|  | Speed | Torque | Flux | Current |
|---|---|---|---|---|
| Errors between Simulink and XSG | 0.066% | 1% | 0.022% | 0.69% |

The next step is to proceed to the experimental validation of the proposed control strategy. Hence, the test bench conceived in our research laboratory, based on a three-phase IM fed with a three-level NPC inverter and controlled through an FPGA board in closed loop mode, is presented in Figure 16.

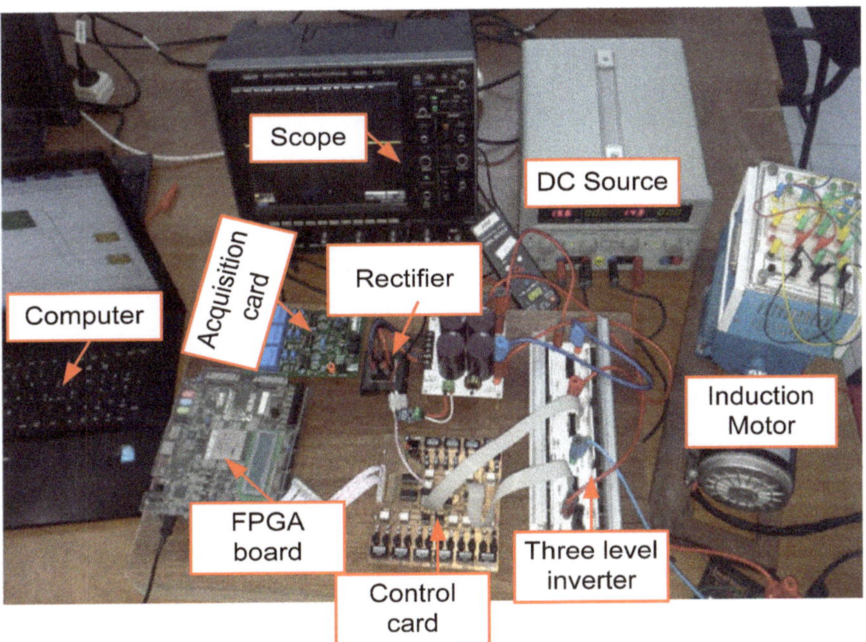


**Figure 16.** Experimental test bench.

## 5. Conclusions

This study proposed a DTC strategy based on sliding mode theory for a high-performance IM controlled with a three-level NPC inverter. The suggested control paradigm is designed by combining the SM algorithm dedicated to flux, torque and speed loops with the merits of the multi-level inverter. Hence, the control structure solves not only the problem of high flux and torque ripples caused by the discrete nature of hysteresis regulators, but also improves the robustness and stability of the system. Additionally, in various speed operating ranges, the suggested DTC-SM strategy's robustness against load torque perturbation has been extensively demonstrated.

Once the conceived control system is verified through simulations, the XSG modeling enables the implementation of the whole design by loading the generated bitstream file on the FPGA board and, thus, a hardware co-simulation is launched.

Overall, the simulation results of this study offer a very appealing and prospective control technique for high-performance AC drives. For future studies, the proposed methodology will be tested on an experimental setup. Moreover, the robust control strategy can be further improved by other variants, such as using a sliding mode observer to ensure sensorless control, or exploiting the DTC-SM in electric vehicle traction applications, as well as in solar- and PV-powered water pumping systems.

**Author Contributions:** Conceptualization, S.J. and A.K.; Methodology, S.J.; Software, S.J.; Validation, S.J. and A.K.; Investigation, S.J. and I.M.; Writing—original draft preparation, S.J.; Writing—review and editing, S.J. and A.K.; Visualization, S.J.; Supervision, A.K. All authors have read and agreed to the published version of the manuscript.

**Funding:** This research received no external funding.

**Conflicts of Interest:** The authors declare no conflict of interest.

## Appendix A

**Table A1.** The three-phase induction motor's parameters in SI units, used for simulations, are listed below.

| | |
|---|---|
| Rated power | 1.5 kW |
| Rated speed | 1435 tr/min |
| Rated frequency | 50 Hz |
| Rated current | 5.5/3.2 A |
| Number of pole pairs | 2 |
| Stator resistance | 5.72 $\Omega$ |
| Rotor resistance | 4.28 $\Omega$ |
| Stator inductance | 0.464 H |
| Rotor inductance | 0.464 H |
| Mutual inductance | 0.44 H |
| Moment of inertia | 0.0049 kg·m² |
| Viscous friction coefficient | 0.002 |

## References

1. Bocker, J.; Mathapati, S. State of the art of induction motor control. In Proceedings of the International Electric Machines & Drives Conference, Antalya, Turkey, 3–5 May 2007.
2. Takahashi, I.; Noguchi, T. A new quick-response and high-efficiency control strategy of an induction motor. *IEEE Trans. Ind. Appl.* **1986**, *5*, 820–827. [CrossRef]
3. Krim, S.; Gdaim, S.; Mtibaa, A.; Mimouni, M.F. Contribution of the FPGAs for complex control algorithms: Sensorless DTFC with an EKF of an induction motor. *Int. J. Autom. Comput.* **2019**, *16*, 226–237. [CrossRef]
4. Costa, B.L.G.; Graciola, C.L.; Angélico, B.A.; Goedtel, A.; Castoldi, M.F.; de Andrade Pereira, W.C. A practical framework for tuning DTC-SVM drive of three-phase induction motors. *Control Eng. Pract.* **2019**, *88*, 119–127. [CrossRef]
5. El Ouanjli, N.; Derouich, A.; El Ghzizal, A.; Motahhir, S.; Chebabhi, A.; El Mourabit, Y.; Taoussi, M. Modern improvement techniques of direct torque control for induction motor drives-a review. *Prot. Control Mod. Power Syst.* **2019**, *4*, 1–12. [CrossRef]
6. Vaezi, S.A.; Iman-Eini, H.; Razi, R. A new space vector modulation technique for reducing switching losses in induction motor DTC-SVM scheme. In Proceedings of the 10th International Power Electronics, Drive Systems and Technologies Conference (PEDSTC), Shiraz, Iran, 12–14 February 2019.

7. Sedaghati, F.; Latifi, S.H. Application of a three-phase multilevel inverter for DTC based induction motor drive. In Proceedings of the 9th Annual Power Electronics, Drives Systems and Technologies Conference (PEDSTC), Tehran, Iran, 13–15 February 2018.

8. Li, Y.H.; Wu, T.X.; Zhai, D.W.; Zhao, C.H.; Zhou, Y.F.; Qin, Y.G.; Su, J.S.; Qin, H. Hybrid Decision Based on DNN and DTC for Model Predictive Torque Control of PMSM. *Symmetry* **2022**, *14*, 693. [CrossRef]

9. Dendouga, A. A comparative study between the PI and SM controllers used by nonlinear control of induction motor fed by SVM matrix converter. *IETE J. Res.* **2022**, *68*, 3019–3029. [CrossRef]

10. El Haissouf, M.; El Haroussi, M.; Ba-Razzouk, A. DSP In the Loop implementation of an enhanced Direct Torque Control for Induction Motor drive. In Proceedings of the 2nd International Conference on Innovative Research in Applied Science, Engineering and Technology (IRASET), Meknes, Morocco, 3–4 March 2022; IEEE: New York, NY, USA, 2022.

11. Soliman, A.I.; Vedadi, M.; Kahia, B.; Abdelrahem, M.; Kennel, R. Flexible Test Bench Arrangement and Particular Implementation of Three Level IGBT Based VSI for Self-Sensing Model Predictive Control of Induction Motor. In Proceedings of the 21st Workshop on Control and Modeling for Power Electronics (COMPEL), Aalborg, Denmark, 09–12 November 2020; IEEE: New York, NY, USA, 2020.

12. Lakka, M.; Koutroulis, E.; Dollas, A. Development of an FPGA-based SPWM generator for high switching frequency DC/AC inverters. *IEEE Trans. Power Electron.* **2013**, *29*, 356–365. [CrossRef]

13. Alouane, A.; Rhouma, A.B.; Khedher, A. FPGA implementation of a new DTC strategy dedicated to delta inverter-fed BLDC motor drives. *Electr. Power Compon. Syst.* **2018**, *46*, 688–700. [CrossRef]

14. Moussa, I.; Khedher, A. Software in-the-Loop Simulation of an Advanced SVM Technique for 2$\phi$-Inverter Control Fed a TPIM as Wind Turbine Emulator. *Electronics* **2022**, *11*, 187. [CrossRef]

15. Tatte, Y.; Aware, M.; Khadse, C. Torque ripple reduction in three-level five-phase inverter-fed five-phase induction motor. *Electr. Eng.* **2022**, 1–13. [CrossRef]

16. Krim, S.; Gdaim, S.; Mtibaa, A.; Mimouni, M.F. Control with high performances based DTC strategy: FPGA implementation and experimental validation. *EPE J.* **2019**, *29*, 82–98. [CrossRef]

17. Sun, X.; Wu, J.; Lei, G.; Guo, Y.; Zhu, J. Torque ripple reduction of SRM drive using improved direct torque control with sliding mode controller and observer. *IEEE Trans. Ind. Electron.* **2020**, *68*, 9334–9345. [CrossRef]

18. Lascu, C.; Argeseanu, A.; Blaabjerg, F. Supertwisting sliding-mode direct torque and flux control of induction machine drives. *IEEE Trans. Power Electron.* **2019**, *35*, 5057–5065. [CrossRef]

19. Maes, J.; Melkebeek, J.A. Speed-sensorless direct torque control of induction motors using an adaptive flux observer. *IEEE Trans. Ind. Appl.* **2000**, *36*, 778–785. [CrossRef]

20. Martins, C.A.; Roboam, X.; Meynard, T.A.; Carvalho, A.S. Switching frequency imposition and ripple reduction in DTC drives by using a multilevel converter. *IEEE Trans. Power Electron.* **2002**, *17*, 286–297. [CrossRef]

21. Jnayah, S.; Khedher, A. DTC of Induction Motor Drives Fed by Two and Three-Level Inverter: Modeling and Simulation. In Proceedings of the 19th International Conference on Sciences and Techniques of Automatic Control and Computer Engineering (STA), Sousse, Tunisia, 24–26 March 2019; IEEE: New York, NY, USA, 2019.

22. Jnayah, S.; Khedher, A. Fuzzy-Self-Tuning PI Speed Regulator for DTC of Three-Level Inverter fed IM. In Proceedings of the 17th International Multi-Conference on Systems, Signals & Devices (SSD), Monastir, Tunisia, 20–23 July 2020; IEEE: New York, NY, USA, 2021.

23. Jnayah, S.; Khedher, A. Improvement of DTC Performance of Three Level Inverter Fed IM Drive with High Gain Flux Observer. In Proceedings of the International Conference on Digital Technologies and Applications (ICDTA), Fez, Morocco, 28–30 January; Springer: Cham, Switzerland, 2022.

24. Tatte, Y.N.; Aware, M.V. Torque ripple and harmonic current reduction in a three-level inverter-fed direct-torque-controlled five-phase induction motor. *IEEE Trans. Ind. Electron.* **2017**, *64*, 5265–5275. [CrossRef]

25. Tatte, Y. Torque ripple minimization with modified comparator in DTC based three-level five-phase inverter fed five-phase induction motor. *IET Power Electron.* **2021**, *14*, 1713–1723. [CrossRef]

26. Ammar, A.; Bourek, A.; Benakcha, A. Robust SVM-direct torque control of induction motor based on sliding mode controller and sliding mode observer. *Front. Energy* **2020**, *14*, 836–849. [CrossRef]

27. Zerzeri, M.; Jallali, F.; Khedher, A. A Robust Nonlinear Control Based on SMC Approach for Doubly-Fed Induction Motor Drives Used in Electric Vehicles. In Proceedings of the of the International Conference on Signal, Control and Communication (SCC), Hammamet, Tunisia, 16–18 December 2019; IEEE: New York, NY, USA, 2020.

28. Ferhat, N.; Aounallah, T.; Essounbouli, N.; Hamzaoui, A.; Bouchafaa, F. Fractional Adaptive Fuzzy Sliding Mode Control Algorithm for Permanent Magnet Synchronous Generators. In Proceedings of the International Conference on Control, Automation and Diagnosis (ICCAD), Grenoble, France, 3–5 November 2021; IEEE: New York, NY, USA, 2021.

29. Ullah, M.I.; Ajwad, S.A.; Irfan, M.; Iqbal, J. Non-linear control law for articulated serial manipulators: Simulation augmented with hardware implementation. *Elektron. Ir. Elektrotechnika* **2016**, *22*, 3–7. [CrossRef]

30. Iqbal, J.; Ullah, M.I.; Khan, A.A.; Irfan, M. Towards sophisticated control of robotic manipulators: An experimental study on a pseudo-industrial arm. *Strojniški vestnik. J. Mech. Eng.* **2015**, *61*, 465–470. [CrossRef]

31. Benkahla, M.; Taleb, R.; Boudjema, Z. A new robust control using adaptive fuzzy sliding mode control for a DFIG supplied by a 19-level inverter with less number of switches. *Электротехника И Электромеханика* **2018**, *4*, 11–19. [CrossRef]

*electronics*

*Article*

# Adaptive Chattering-Free PID Sliding Mode Control for Tracking Problem of Uncertain Dynamical Systems

**Yufei Liang, Dong Zhang, Guodong Li and Tao Wu ***

Key Laboratory of Intelligent Equipment Technology for Harsh Environments in Shanxi Province,
North University of China, Taiyuan 030051, China
* Correspondence: b1902009@st.nuc.edu.cn

**Abstract:** Aiming at the trajectory tracking problem with unknown uncertainties, a novel controller composed of proportional-integral-differential sliding mode surface (PIDSM) and variable gain hyperbolic reaching law is proposed. A PID-type sliding mode surface with an inverse hyperbolic integral terminal sliding mode term is proposed, which has the advantages of global convergence of integral sliding mode (ISM) and finite time convergence of terminal sliding mode (TSM), and the control effect is significantly improved. Then, a variable gain hyperbolic approach law is proposed to solve the sliding mode approaching velocity problem. The variable gain term can guarantee different approaching velocities at different distances from the sliding mode surface, and the chattering problem is eliminated by using a hyperbolic function instead of the switching function. The redesign of the sliding mode surface and the reaching law ensures the robustness and tracking accuracy of the uncertain system. Adaptive estimation can compensate for uncertain disturbance terms in nonlinear systems, and the combination with sliding mode control further improves the tracking accuracy and robustness of the system. Finally, the Lyapunov stability principle is used for stability analysis, and the simulation study verifies that the proposed control scheme has the advantages of fast response, strong robustness, and high tracking accuracy.

**Keywords:** sliding mode control; hyperbolic reach law; variable gain; adaptive; trajectory tracking

**Citation:** Liang, Y.; Zhang, D.; Li, G.; Wu, T. Adaptive Chattering-Free PID Sliding Mode Control for Tracking Problem of Uncertain Dynamical Systems. *Electronics* **2022**, *11*, 3499. https://doi.org/10.3390/electronics11213499

Academic Editor: Erdal Kayacan

Received: 3 October 2022
Accepted: 27 October 2022
Published: 28 October 2022

**Publisher's Note:** MDPI stays neutral with regard to jurisdictional claims in published maps and institutional affiliations.

## 1. Introduction

Most of the controlled systems at work are second-order nonlinear systems, which will be affected by internal uncertainties and external disturbances, resulting in a decline in output performance [1]. In order to improve the control performance, many effective control schemes have been proposed, including adaptive control [2,3], optimal control [4–6], backstepping control [7,8], intelligent control [9,10], model predictive control (MPC) [11,12], and sliding mode control [13–15], etc. Because of the advantages of good transient performance, strong robustness to parameter changes, and insensitivity to disturbances, sliding mode control has been widely used in practical work. The basic idea of sliding mode control is to establish a sliding mode manifold, which abstracts the controlled system into a form that keeps moving on the sliding mode surface. However, traditional linear sliding mode surfaces rely on high gain switching terms to ensure asymptotic convergence, resulting in chattering. In order to reduce the output chattering and steady-state error, the sliding mode surface and the reaching law are redesigned respectively. In order to solve the lumped disturbance term in the nonlinear system, the disturbance term must be compensated with the disturbance estimation method. The disturbance estimation methods mainly include adaptive estimation [16,17], observer estimation [18], and neural network approximation [19]. Compared with the other two estimation methods, adaptive estimation achieves a better estimation effect with fewer computing resources.

Many scholars have their ideas on the design of sliding surfaces. TSM and ISM are the most widely used. TSM can guarantee finite-time convergence of the control system, but the fractional power term will have singularities. Therefore, a non-singular terminal sliding mode control (NTSM) scheme was proposed [20]. NTSM control has the advantages of limited time convergence, avoidance of singular points, and has a faster convergence rate when far away from the origin, while ISM has the advantage of global convergence. In [21], a robust adaptive integral terminal sliding mode (AITSM) control scheme with an uncertainty observer for electronic throttle systems is proposed, which uses a novel integral terminal sliding surface to ensure the closed-loop system is forced to start from the sliding surface, thereby ensuring good error convergence and tracking accuracy. To ensure that the ISM surface can quickly converge to zero in a finite time without singularity problems and control input is continuous in [22], a second-order integral terminal sliding mode (ITSM) surface combining an ISM surface and a fast non-singular integral terminal sliding mode surface is proposed. A recursive structure ITSM consisting of a non-singular terminal sliding mode and a fractional power integral terminal sliding mode is proposed in [23], where the approach control input is obtained in the form of an integral rather than a signal function, and the system state is obtained from the sliding surface starts and then converges to the origin in a finite time. The combination of terminal sliding mode and integral sliding mode can obtain a better control effect. NTSM with the inverse hyperbolic tangent function was proposed in [24], which guarantees the system state to converge to the origin in a finite time and guarantees a faster convergence rate. A practical TSM constructed based on a hyperbolic tangent function nested with terminal attractors is proposed in [25], which guarantees a Lipschitz continuous but steep slope generalized velocity at the origin. Therefore, terminal sliding mode with hyperbolic function can effectively improve the control effect. This research innovatively proposes a PID-type integral terminal sliding mode manifold with an inverse hyperbolic tangent function.

The important role of the sliding surface in the controller has been outlined above, and the reaching law plays a crucial role in reaching the sliding surface velocity and reducing the chattering effect. In [26], an exponential reaching law for the nonlinear multivariable coupled system of a permanent magnet synchronous motor is proposed, which improves the dynamic response performance and robustness of the permanent magnet synchronous motor control system and reflects the effect of the reaching law on the important role of the control system. In [27], a reaching law including the system state variables and the sliding mode surface $s$ is proposed. There are two different representations in the reaching process, which improves the reaching speed of the system, which improves the reaching speed of the system and show that the method of introducing system state variables into the reaching law is effective. In [28], the hyperbolic reaching law proposed for the sluggish transient behavior of quadrotors eliminates buffeting while guaranteeing a dynamic response. In [29], a continuous reaching law using two hyperbolic functions with similar rates of change and opposite amplitudes was proposed, namely, the inverse hyperbolic sine function and the hyperbolic tangent function, which ensured the fast convergence of the algorithm and no chattering characteristics. In [30], a reaching law based on an inverse hyperbolic function and a variable gain term is proposed, and a sliding mode manifold is introduced in the variable gain term; when the system state converges to the surface, better control can be achieved due to the reduction of the gain term. This paper proposes a hyperbolic approach law with variable gain similar to [30], which has different approach speeds according to the distance from the sliding mode surface and eliminates chattering.

Aiming at the tracking problem of uncertain control systems, we innovatively propose a control scheme based on a PID-type sliding surface with an inverse hyperbolic integral terminal term and a variable gain hyperbolic reaching law that can obtain a good tracking effect under uncertain conditions. The main contributions of this paper are as follows: (1) Combining the advantages of ISM and TSM, this study innovatively proposes a PID-type sliding mode surface with an inverse hyperbolic integral terminal term, which can achieve a better tracking effect while ensuring finite time convergence. (2) The variable gain hyperbolic reaching law is proposed, in which the variable gain term can achieve the purpose of the variable speed approach according to the distance from the sliding mode surface and $\mathrm{asinh}(\cdot)$ is a continuous function, the continuity of the sliding mode control can be guaranteed. The two work together to eliminate the high-frequency unmodeled dynamics of the system caused by the traditional reaching law relying on high-gain switching. (3) The adaptive estimation method can effectively estimate and compensate for the lumped disturbance when the upper bound of uncertainty and external disturbance is unknown and consumes less computational resources.

The rest of the paper is organized as follows: Section 2 states the problem. In Section 3, the proposed PID-type sliding mode surface and variable-gain hyperbolic reaching law are presented, and their stability is proved according to the Lyapunov theorem. In Section 4, we generalize the PIDSM method to N-DOF robots. Numerical simulation results are presented in Section 5. Finally, the work in this paper is summarized in Section 6.

## 2. Problem Statement

The following second-order nonlinear dynamic model shown below is considered in this paper:

$$\begin{aligned} y &= x_1 \\ \dot{x}_1 &= x_2 \\ \dot{x}_2 &= f(x) + \Delta f(x) + b(x)u(t) + \Delta b(x)u(t) + d(t) \end{aligned} \tag{1}$$

where $x_1$ and $x_2$ are the state variables of the system, $y$ is the output signal of the system, $f(x)$ and $b(x)$ are nonlinear functions $x$ that are closely related to the control system itself, $\Delta f(x)$ is the uncertainty of the system, $d(t)$ is the external disturbances, and $u(t)$ is the control input of the system.

Define the tracking error $e_1 = y - y_d$ and where $y_d$ is the reference signal. Then, substitute into the nonlinear system (1) and take the derivative to obtain the following:

$$\begin{aligned} \dot{e}_1 &= \dot{y} - \dot{y}_d = e_2 \\ \dot{e}_2 &= F(x) + b(x)u(t) + D(x,t) \end{aligned} \tag{2}$$

where $F(x) = f(x) - \ddot{y}_d$, $D(x,t) = \Delta f(x) + \Delta b(x)u(t) + d(t)$ is the lumped disturbance of the system. In order to enable the controller to achieve accurate tracking in the presence of uncertainties and external disturbances in the controlled system, the following assumptions are made.

**Assumption 1.** *The reference signal $y_d$ must be a twice continuously differentiable function with respect to $t$.*

**Assumption 2.** *The lumped disturbance $D(x,t)$ has an upper bounded by a positive function at all times and one has:*

$$|D(x,t)| = |\Delta f(x) + d(t)| \leq \delta \tag{3}$$

*where we assume that the lumped disturbance upper bound $\delta$ is a function containing only position and velocity measurements, the following conclusions can be drawn [31]:*

$$\delta = a_0 + a_1|x_1| + a_2|x_2|^2 \tag{4}$$

*where $a_0$, $a_1$ and $a_2$ are unknown positive constants.*

**Remark 1.** *According to the actual working conditions of the controlled system, the above assumptions are very reasonable because the maximum charge rate of external disturbance and system disturbance limits the value of the parameters.*

**Notation**: $\mathrm{sign}(\cdot)$ is the symbolic function. For a variable vector $x = [x_1, \cdots, x_n]^T$ is a real vector and define the fractional power of vectors as follows:

$$x^\alpha = [x_1^\alpha, \cdots, x_n^\alpha]^T \tag{5}$$

## 3. Adaptive Chattering-Free PIDSM Controller Design

The design of a sliding mode controller usually includes the following three main contents: (1) the design of the sliding surface; (2) the design of equivalent control law; (3) the design of switching control law.

### 3.1. PIDSM Controller Design

Firstly, to track a given desired signal of the controlled system (1), a novel PID-type sliding surface containing a hyperbolic function is designed as follows:

$$s = e_2 + k_1 e_1 + k_2 \int e_1^\alpha \arctan(e_1) d\sigma \tag{6}$$

where $k_1$ and $k_2$ are positive constants, $\alpha \geq 1$ and $\arctan(\cdot)$ is the arctangent function.

By observing (6), we find that if the initial condition of the integral term is chosen as $-1/k_2(e_2(0) + k_1 e_1(0))$, the sliding variable $s$ will start from the sliding surface $s = 0$. This characteristic of ISM can effectively eliminate the arrival phase of the controlled system and improve the response speed of the controller.

The time derivative of the sliding surface (6) has the following:

$$\dot{s} = \dot{e}_2 + k_1 e_2 + k_2 e_1^\alpha \arctan(e_1) \tag{7}$$

Substituting (2) into (7), one has the following:

$$\dot{s} = F(x) + b(x)u(t) + k_1 e_2 + k_2 e_1^\alpha \arctan(e_1) \tag{8}$$

Then design the equivalent control law $u_{eq}(t)$. $\dot{s} = 0$ is a necessary condition for the state trajectory to remain on the sliding surface. So, we can obtain the equivalent control law by (8) as follows:

$$u_{eq}(t) = -b(x)^{-1}(F(x) + k_1 e_2 + k_2 e_1^\alpha \arctan(e_1)) \tag{9}$$

The equivalent control law can guarantee that the system maintains the motion state on the sliding surface without uncertainty and external disturbance. However, there are bound to be factors such as uncertainty and external disturbances in practical work. To meet the actual work situation, the following approach law is designed:

$$\dot{s} = \frac{k_3 |s| \operatorname{asinh}(bs)}{\left(\varepsilon + \left(1 + \frac{1}{|y|} - \varepsilon\right)e^{-\delta|s|}\right)} \tag{10}$$

where $k_3 > 0, \delta > 0, 0 < \varepsilon < 1, b = k_2/k_1$ and $y$ is the system state.

We can obtain the switching control law by (10), one has the following:

$$u_{sw}(t) = -b(x)^{-1}\left(\frac{k_3 |s| \operatorname{asinh}(bs)}{\left(\varepsilon + \left(1 + \frac{1}{|y|} - \varepsilon\right)e^{-\delta|s|}\right)} + \delta \operatorname{asinh}(s)\right) \tag{11}$$

Finally, the control law is designed as follows:

$$u(t) = u_{eq}(t) + u_{sw}(t) \tag{12}$$

For the brevity of the paper, the notion of time for all given variables is omitted in the following. The main results of the designed adaptive chatter-free PID sliding mode control are summarized as Theorem 1.

**Theorem 1.** *Consider a nonlinear uncertain system (1) where the lumped disturbance satisfies constraint (4). If the APIDSM surface is set to (6) and the control law is designed to be (12), the system state converges to the sliding surface in a finite time.*

**Proof.** Choose the Lyapunov function as follows:

$$V = \frac{1}{2}s^2 \tag{13}$$

Differentiating $V$ with respect to time and substituting in (7), one has the following:

$$\dot{V} = s\left(\dot{e}_2 + k_1 e_2 + k_2 e_1^\alpha \arctan(e_1)\right) \tag{14}$$

By substituting the error dynamics (2) and control law (12) into (14), one has the following:

$$\dot{V} = s\left(D - \frac{k_3|s|\mathrm{asinh}(bs)}{\left(\varepsilon + \left(1 + \frac{1}{|y|} - \varepsilon\right)e^{-\delta|s|}\right)} - \delta\mathrm{sign}(s)\right) \tag{15}$$

It is easy to verify that

$$\dot{V} \le -\frac{k_3\mathrm{asinh}(bs)\mathrm{sign}(s)}{\left(\varepsilon + \left(1 + \frac{1}{|y|} - \varepsilon\right)e^{-\delta|s|}\right)}s^2 + (|D| - \delta)|s| \tag{16}$$

According to (13), we can obtain the following:

$$\dot{V} \le -\rho_1 V - \rho_2 V^{\frac{1}{2}} \le 0 \tag{17}$$

where $\rho_1 = \dfrac{2k_3\mathrm{asinh}(bs)\mathrm{sign}(s)}{\left(\varepsilon + \left(1 + \frac{1}{|y|} - \varepsilon\right)e^{-\delta|s|}\right)} > 0$ and $\rho_2 = \delta - |D| > 0$.

According to the Lyapunov stability theorem, the state of the system converges asymptotically to the sliding surface $s = 0$.

To prove that this convergence occurs in finite time, it can be obtained by a simple operation from (13) as follows:

$$\frac{dV}{dt} \le -\rho_1 V - \rho_2 V^{\frac{1}{2}} \tag{18}$$

One has the following:

$$dt \le \frac{-dV}{\rho_1 V + \rho_2 V^{\frac{1}{2}}} = \frac{-2dV^{\frac{1}{2}}}{\rho_1 V^{\frac{1}{2}} + \rho_2} \tag{19}$$

Assuming that the arrival time of the initial error $e \ne 0$ to $e = 0$ is $t$, integrating both sides of (19) can be obtained as follows:

$$\int_0^t dt \le \int_{V(0)}^{V(t)} \frac{-2dV^{\frac{1}{2}}}{\rho_1 V^{\frac{1}{2}} + \rho_2} \tag{20}$$

After simple calculation, we can obtain the following:

$$t \le \frac{2}{\rho_1} \ln\left(\frac{\rho_1 V(0)^{\frac{1}{2}} + \rho_2}{\rho_2}\right) \tag{21}$$

Therefore, according to the Lyapunov stability criterion, the proposed PIDSM surface can converge to zero in finite time $t$. At the same time, the output tracking error of the system can also converge to zero in finite time. Completes the proof. $\square$

**Remark 2.** *The sliding-mode reaching law (10) consists of a variable gain term and an inverse hyperbolic sine function, which can ensure the variable-speed reaching process and is chattering-free. The system state $x_1$ is introduced into the variable gain term, which can achieve the purpose of the variable speed approaching the sliding mode surface. The inverse hyperbolic sine function replaces the traditional sign function* $\mathrm{sign}(\cdot)$*, which ensures the continuity of the control law. The two work together to achieve a chatter-free effect. Figure 1 shows a simplified reaching law* $\dot{s} = -k|s|\mathrm{asinh}(bs)$ *to express the effect of variable gain.*

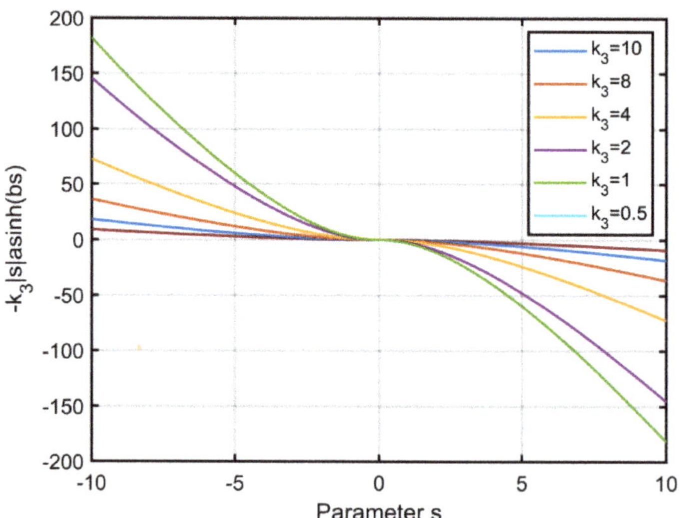



**Figure 1.** Plot of $-k_3|s|\mathrm{asinh}(bs)$ with $b = 0.3$.

### 3.2. Adaptive PIDSM (APIDSM) Controller Design

In this section, an adaptive estimation method is used to estimate the upper bound of the lumped disturbance. The designed APIDSM control block diagram is shown in Figure 2.

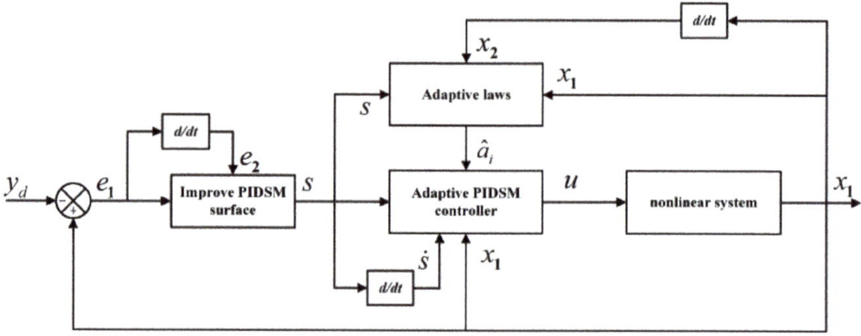



**Figure 2.** Control diagram of the designed controller.

The switching control law (11) can be redesigned as follows:

$$u_{asw} = -b^{-1}\left( \frac{k_3|s|\mathrm{asinh}(bs)}{\left(\varepsilon + \left(1 + \frac{1}{|y|} - \varepsilon\right)e^{-\delta|s|}\right)} + \hat{\delta}\mathrm{sign}(s) \right) \tag{22}$$

where $\hat{\delta}$ can update the estimate of $D(x,t)$ by the following adaptive law:

$$\hat{\delta} = \hat{a}_0 + \hat{a}_1 |x_1| + \hat{a}_2 |x_2|^2 \tag{23}$$

and where $\hat{a}_0$, $\hat{a}_1$ and $\hat{a}_2$ are respectively as follows [32]:

$$\begin{aligned}
\dot{\hat{a}}_0 &= |s| \\
\dot{\hat{a}}_1 &= |s||x_1| \\
\dot{\hat{a}}_2 &= |s||x_2|^2
\end{aligned} \tag{24}$$

Hence, the control law (12) is rewritten as follows:

$$u = u_{eq} + u_{asw} \tag{25}$$

**Theorem 2.** *Consider system (1), where the upper bound of lumped disturbance (4) is assumed to be completely unknown. Using the control law of (25) can ensure that the system state converges to the sliding surface.*

**Proof.** Choose the Lyapunov function as follows:

$$V = \frac{1}{2}s^2 + \tilde{a}_0^2 + \tilde{a}_1^2 + \tilde{a}_2^2 \tag{26}$$

where $\tilde{a}_i = \hat{a}_i - a_i$ is the adaption error, $i = 0, 1, 2$. The time derivative of $V$ and substituting in (2) and (7) to obtain the following:

$$\begin{aligned}
\dot{V} &= s\dot{s} + \tilde{a}_0\dot{\hat{a}}_0 + \tilde{a}_1\dot{\hat{a}}_1 + \tilde{a}_2\dot{\hat{a}}_2 \\
&= s\left(\dot{e}_2 + k_1 e_2 + k_2 e_1^{\alpha}\arctan(e)\right) + \tilde{a}_0\dot{\hat{a}}_0 + \tilde{a}_1\dot{\hat{a}}_1 + \tilde{a}_2\dot{\hat{a}}_2 \\
&= s\left(F + bu + D + k_1 e_2 + k_2 e_1^{\alpha}\arctan(e)\right) + \tilde{a}_0\dot{\hat{a}}_0 + \tilde{a}_1\dot{\hat{a}}_1 + \tilde{a}_2\dot{\hat{a}}_2
\end{aligned} \tag{27}$$

Substitute in (25) and it is easy to verify the following:

$$\begin{aligned}
\dot{V} &= s\left(D - \frac{k_3|s|\,\mathrm{asinh}(bs)}{\left(\varepsilon + \left(1 + \frac{1}{|y|} - \varepsilon\right)e^{-\delta|s|}\right)} - \hat{\delta}\mathrm{sign}(s)\right) + \tilde{a}_0\dot{\hat{a}}_0 + \tilde{a}_1\dot{\hat{a}}_1 + \tilde{a}_2\dot{\hat{a}}_2 \\
&\leq -\frac{k_3\,\mathrm{asinh}(bs)\mathrm{sign}(s)}{\left(\varepsilon + \left(1 + \frac{1}{|y|} - \varepsilon\right)e^{-\delta|s|}\right)}s^2 + |D||s| - \left(\hat{a}_0 + \hat{a}_1|x_1| + \hat{a}_2|x_2|^2\right)|s| + (\hat{a}_0 - a_0)\dot{\hat{a}}_0 \\
&\quad + (\hat{a}_1 - a_1)\dot{\hat{a}}_1 + (\hat{a}_2 - a_2)\dot{\hat{a}}_2
\end{aligned}$$

We can obtain the following:

$$\dot{V} \leq -\frac{k_3\,\mathrm{asinh}(bs)\mathrm{sign}(s)}{\left(\varepsilon + \left(1 + \frac{1}{|y|} - \varepsilon\right)e^{-\delta|s|}\right)}s^2 - \left(\left(a_0 + a_1|x_1| + a_2|x_2|^2\right) - |D|\right)|s| \leq 0 \tag{28}$$

By the Lyapunov stability theorem, the state of the system converges asymptotically to the sliding surface $s = 0$. Completes the proof. $\square$

**Remark 3.** *The upper bound value of the lumped disturbance is estimated according to the Lyapunov stability theorem, so no prior knowledge is required. Although we can determine the upper bound value according to work experience, due to the variability of the actual situation and different operating conditions, the upper bound value is time-varying, which reduces the robustness of the controlled system. The adaptive estimation method successfully eliminates the selection process of the upper bound of uncertainty, which ensures the control output's high precision and the controller's robustness.*

**Remark 4.** *Since the parameter estimates (23) are adaptively tuned in the Lyapunov sense, they do not necessarily converge to the true value. However, according to the proof of (28), when the sliding mode variable $s = 0$, the adaptive value will remain unchanged to ensure the finite-time convergence of the tracking error and achieve the robustness of the controller.*

**Remark 5.** *During the working process, due to the uncertainty and measurement noise in the system, the sliding variable $s$ appears to chatter near the equilibrium point, which will be $a_i$ overestimated. To mitigate this problem, the dead zone technique is often employed, as shown below:*

$$
\begin{aligned}
\dot{\hat{a}}_0 &= \begin{cases} |s| & \text{if } |s| \geq \varepsilon \\ 0 & \text{if } |s| < \varepsilon \end{cases} \\
\dot{\hat{a}}_1 &= \begin{cases} |s||x_1| & \text{if } |s| \geq \varepsilon \\ 0 & \text{if } |s| < \varepsilon \end{cases} \\
\dot{\hat{a}}_2 &= \begin{cases} |s||x_2|^2 & \text{if } |s| \geq \varepsilon \\ 0 & \text{if } |s| < \varepsilon \end{cases}
\end{aligned}
\tag{29}
$$

*where $\varepsilon$ is a small positive constant. Obviously, when $s$ is in range, $\hat{a}_0$, $\hat{a}_1$ and $\hat{a}_2$ will not continue to increase but keep their current values. The above method can ensure that as long as the control system reaches the true sliding mode, it can converge to the neighborhood of $s$ in a finite time, and the steady-state accuracy of $s$ is given. Additionally, we can easily show that finite-time convergence is guaranteed since the threshold (28) above is still satisfied.*

**Remark 6.** *To better suppress chatter, we use saturation function to replace $\mathrm{sign}(\cdot)$ in (22). The saturation function is the following:*

$$
\mathrm{sat}(x) = \begin{cases} \frac{x}{\zeta}, & |x| < \zeta \\ \mathrm{sign}(x), & |x| \geq \zeta \end{cases}
\tag{30}
$$

*where $\zeta > 0$ is a constant and select $\zeta = 0.2$. $\zeta$ value is determined by considering the tracking accuracy and smoothness of the control output.*

### 3.3. Control Parameters Selection

In the actual work process, it is impossible to keep the performance of the controlled system in the best state at the same time. It is necessary to consider various influencing factors such as control input, command smoothness, and noise at the same time to obtain a compromise value. We give the following parameter selection guidelines for the proposed controller by trial and error:

1. Selections of $\alpha$, $k_1$ and $k_2$: The parameters $\alpha$, $k_1$, $k_2$ directly affect the dynamics of the sliding surface $s$ in (6). An appropriate increase in $\alpha$, $k_1$ and $k_2$ can effectively improve the tracking accuracy and response speed, but an excessively large value will cause it to develop in the opposite direction;
2. Selections of $k_3$ and $\delta$: The parameters $k_3$ directly influence the approach speed of sliding mode control, which has a positive correlation. The larger parameter $\delta$ can improve the smoothness of the control output and the approach speed of sliding mode control;
3. Selections of $\varepsilon$: First, the parameter $\varepsilon$ is used to speed up the approach speed when it is closer to the sliding surface, and a smaller value can speed up the approach speed. Second, the parameter $\varepsilon$ is quite important for the adaptation control law (29) and the value of $\varepsilon$ directly affects the tracking accuracy of the controller.

## 4. Robotic Manipulators

It is worth noting that the robot system can be represented in the form of (2), so the following generalization is made to the robot system as the research example object. To this end, the robot model is transformed in the form of (2) through appropriate coordinate transformations.

The $n$-joint rigid robot manipulators model is as follows:

$$M(q)\ddot{q} + C(q,\dot{q})\dot{q} + G(q) = \tau + \tau_{d1} \tag{31}$$

where $q, \dot{q}, \ddot{q} \in R^n$ are the joint's position, velocity, and acceleration, respectively. $M(q) \in R^{n \times n}$ is the positive inertia matrix, $C(q,\dot{q}) \in R^{n \times n}$ denotes the centrifugal-Coriolis matrix, $G(q) \in R^{n \times n}$ is the gravity term, $\tau_{d1} \in R^n$ is the external disturbances and $\|\tau_{d1}\| \leq d_m$ with $d_m > 0$, $\tau \in R^n$ is the input torque.

In practical engineering applications, it is impossible to obtain a completely accurate robot dynamics model. Therefore, the above dynamic quantities can be expressed as follows [20]:

$$\begin{aligned} M(q) &= M_0(q) + \Delta M(q) \\ C(q,\dot{q}) &= C_0(q,\dot{q}) + \Delta C(q,\dot{q}) \\ G(q) &= G_0(q) + \Delta G(q) \end{aligned} \tag{32}$$

where $M_0(q), C_0(q,\dot{q})$ and $G_0(q)$ denote the nominal parts, $\Delta M(q), \Delta C(q,\dot{q})$ and $\Delta G(q)$ denote the model uncertainties.

According to (32), the dynamic equation of the robotic manipulator is replaced as follows:

$$M_0(q)\ddot{q} + C_0(q,\dot{q})\dot{q} + G_0(q) = \tau + \tau_d \tag{33}$$

where $\tau_d = \tau_{d1} - \Delta M(q) - \Delta C(q,\dot{q}) - \Delta G(q)$ is the lumped disturbance, which includes the uncertainty in (32), external disturbances and measurement noise.

Let the tracking error denoted by $e, \dot{e} \in R^n$ are defined as follows:

$$\begin{cases} e = q - q_d \\ \dot{e} = \dot{q} - \dot{q}_d \end{cases} \tag{34}$$

where $q_d$ is the reference trajectory.

Equation (34) is used for system (33), one has the following:

$$\begin{cases} \dot{e}_1 = e_2 = \dot{q} - \dot{q}_d \\ \dot{e}_2 = F(q,\dot{q}) + M_0^{-1}(q)\tau + \tau_d \end{cases} \tag{35}$$

where $F(q,\dot{q}) = -M_0^{-1}(q)\left(C_0(q,\dot{q})\dot{q} + G_0(q)\right) - \ddot{q}_d$. By Assumption 2, the lumped disturbance $\tau_d$ has an upper bounded, and one has the following:

$$\|\tau_d\| \leq a_0 + a_1\|q\| + a_2\|\dot{q}\|^2 \tag{36}$$

where $a_0$, $a_1$ and $a_2 > 0$ are unknown positive constants. $q$ and $\dot{q}$ are bounded.

It is easy to verify that the error dynamic of (35) is the form of (2). Hence, the APIDSM controller proposed in Section 3 can be applied to n-DOF robotic manipulators. The proof is similar to Theorem 2, so it is omitted here.

## 5. Simulation Results

To verify the effectiveness of the proposed control algorithm, we use a two-link manipulator, as shown in Figure 3.

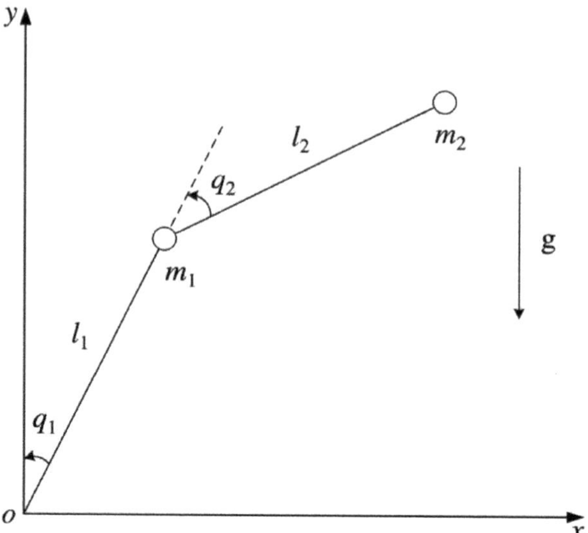



**Figure 3.** The robotic manipulator system.

The dynamics of the two-link robotic manipulator in (33) is given by the following:

$$M(q) = \begin{bmatrix} f_1 + 2f_2\cos(q_2) & f_3 + f_2\cos(q_2) \\ f_3 + f_2\cos(q_2) & f_4 \end{bmatrix} \tag{37}$$

$$C(q,\dot{q}) = \begin{bmatrix} -f_2\sin(q_2)\dot{q}_1 & -2f_2\sin(q_2)\dot{q}_1 \\ 0 & f_2\sin(q_2)\dot{q}_2 \end{bmatrix} \tag{38}$$

$$G(q) = \begin{bmatrix} f_5\cos(q_1) + f_6\cos(q_1+q_2) \\ f_6\cos(q_1+q_2) \end{bmatrix} \tag{39}$$

where

$$\begin{aligned} f_1 &= (m_1+m_2)l_1^2 + m_2 l_2^2 + J_1 \\ f_2 &= m_2 l_1 l_2 \\ f_3 &= m_2 l_2^2 \\ f_4 &= f_3 + J_2 \\ f_5 &= (m_1+m_2)l_1 g \\ f_6 &= m_2 l_2 g \end{aligned} \tag{40}$$

where $q = [q_1, q_2]^T$ is the position of the robot manipulator joints 1 and 2, respectively. $\tau = [\tau_1, \tau_2]^T$ is the input torque, and we ignore friction between joints. Due to the limited output torque in the actual situation, so set the upper limit of actuator output to $\tau_{max} = [150, 150]^T$. Table 1 shows the values of the physical parameters used in the simulation analysis of the robotic manipulator.

The terms $M_0(q)$, $C_0(q,\dot{q})$ and $G_0(q)$ used in the proposed controller are given by the known $\hat{m}_1$ and $\hat{m}_2$. The initial position and velocity are set as $q(0) = [0.5, 1.5]^T$ rad and $\dot{q}(0) = [0,0]^T$ rad/s. The expected signal is as follows:

$$q_d = \begin{bmatrix} 1.25 - \frac{7}{5}e^{-t} + \frac{7}{20}e^{-4t} \\ 1.25 + e^{-t} - \frac{1}{4}e^{-4t} \end{bmatrix} \tag{41}$$

**Table 1.** Physical parameters of the robotic manipulator used in the simulation.

| Symbols | Definition | Values |
|---|---|---|
| $\hat{m}_1$ | The nominal mass of link 1 | 0.4 kg |
| $\hat{m}_2$ | The nominal mass of link 2 | 1.2 kg |
| $m_1$ | Mass of link 1 | 0.5 kg |
| $m_2$ | Mass of link 2 | 1.5 kg |
| $l_1$ | Length of link 1 | 1 m |
| $l_2$ | Length of link 2 | 0.8 m |
| $J_1$ | Moment of inertia of link 1 | 5 kgm$^2$ |
| $J_2$ | Moment of inertia of link 2 | 5 kgm$^2$ |
| g | Acceleration due to gravity | 9.81 m/s$^2$ |

To show the robustness of the proposed control scheme, a time-varying external disturbance is introduced into the system as follows:

$$\tau_{d1} = \begin{bmatrix} 2\sin(t) + 0.5\sin(200\pi t) \\ \cos(2t) + 0.5\sin(200\pi t) \end{bmatrix} \tag{42}$$

and the effect of sudden load variation is introduced, assuming that the robot picks up a 2 kg weight at 2 s, namely, the mass of link 2 is the increasing from 1.2 kg to 3.2 kg after $t \geq 2$ s.

### 5.1. Performance Evaluation of the APIDSMC

Considering the APIDSM controller described in Section 3, the control law of the two-link manipulator is as follows:

$$\tau = \tau_{eq} + \tau_{asw} \tag{43}$$

where $\tau_{eq}$ is the equivalent control. $\tau_{sw}$ is the switching control. $\tau_{eq}$ and $\tau_{asw}$ are, respectively, as follows:

$$\tau_{eq} = -M_0(q)\left(k_1 e_2 + k_2 e_1^\alpha \arctan(e_1) + F(q,\dot{q})\right) \tag{44}$$

$$\tau_{asw} = -M_0(q)\left( \frac{k_3|s|\,\mathrm{asinh}(bs)}{\left(\varepsilon + \left(1 + \frac{1}{|y|} - \varepsilon\right)e^{-\delta|s|}\right)} + \left(\hat{a}_0 + \hat{a}_1\|q\| + \hat{a}_2\|\dot{q}\|^2\right)\mathrm{sat}(s) \right) \tag{45}$$

where $k_i = \mathrm{diag}(k_{i1}, \cdots, k_{in}), i = 1, 2$ is a positive definite matrix, $\hat{a}_0$, $\hat{a}_1$ and $\hat{a}_2$ are the real-time estimated value of $a_0$, $a_1$ and $a_2$.

The proposed controller parameters are chosen as $k_1 = \mathrm{diag}(12, 12)$, $k_2 = \mathrm{diag}(25, 15)$, $\alpha = 4$, $k_3 = (15, 10)$, $\varepsilon = 0.1$ and $\delta = 12$. The initial values of the adaptive update law are $\hat{a}_0(0) = 10$, $\hat{a}_1(0) = 5$ and $\hat{a}_2(0) = 0$. The effectiveness of the proposed controller is verified by simulation under the conditions of introducing external disturbances and payload changes.

Figures 4–9 show the effect of the controlled system on tracking performance under time-varying external disturbances (42) and sudden load changes. It is clear from Figures 4–6 that the control performance does not deteriorate when the upper bound of the external disturbance of the system is unknown and the load changes. The control torque shown in Figure 7 is smooth and chattering-free, indicating that the system can still guarantee excellent tracking performance under the influence of lumped disturbances, proving the effectiveness of the proposed controller. The adaptive estimates shown in Figure 8 all converge to a constant value, indicating that the estimates are adaptively adjusted until the sliding variable converges to zero (Figure 9).

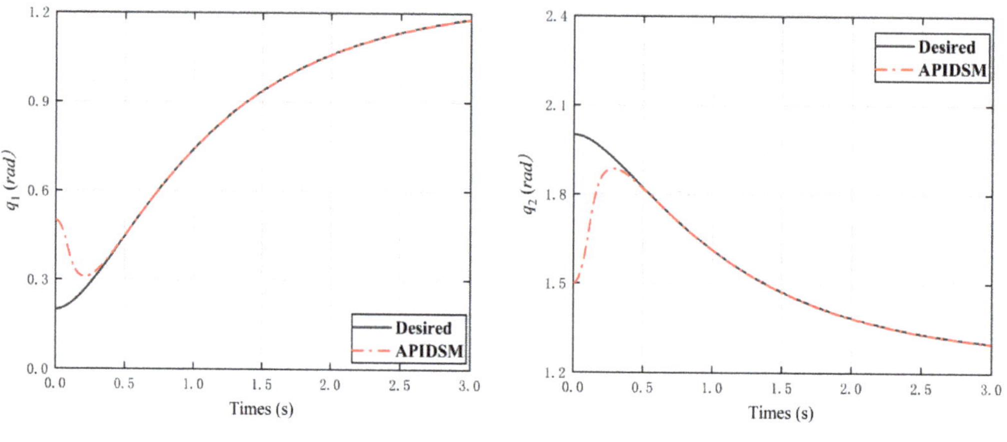

**Figure 4.** Position tracking results of Joint 1 and Joint 2.

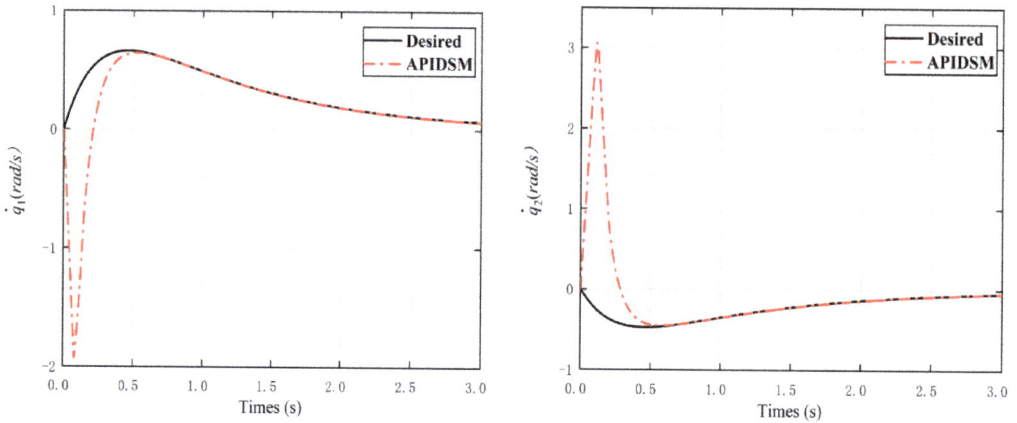

**Figure 5.** Velocity tracking results of Joint 1 and Joint 2.

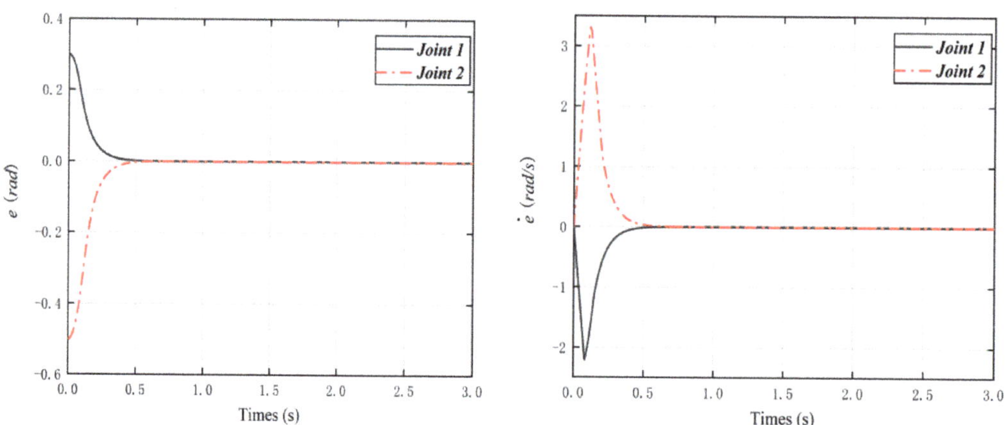

**Figure 6.** Position tracking error and velocity tracking error of Joint 1 and Joint 2.

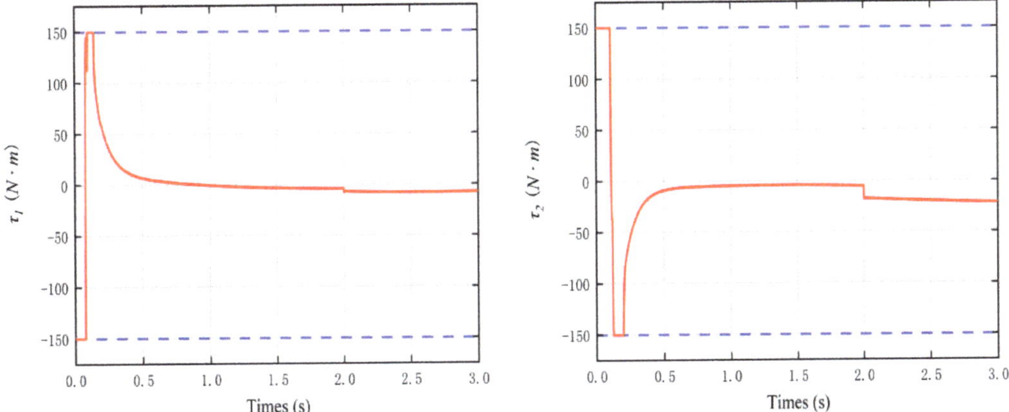

**Figure 7.** Position tracking error and velocity tracking error of Joint 1 and Joint 2.

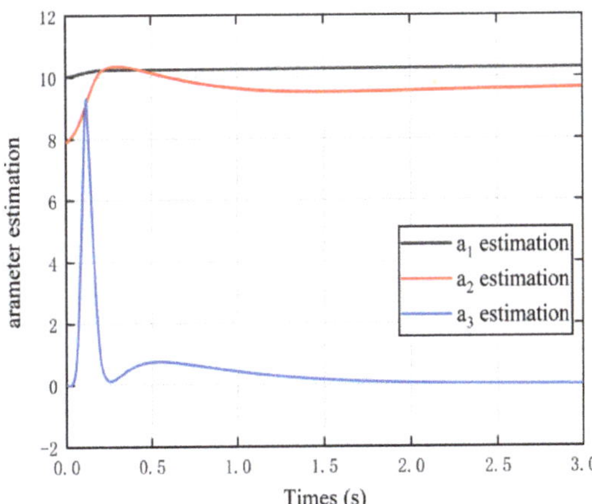

**Figure 8.** Parameter estimation.

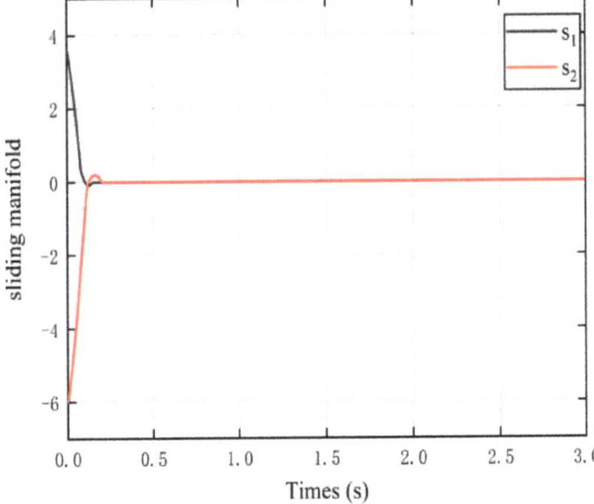

**Figure 9.** Sliding manifold.

*5.2. Comparative Study*

To better verify the effectiveness of APIDSMC, we compare the two methods of ITSM [13] and adaptive nonsingular fast terminal sliding mode control (ANFTSM) [33]. To obtain convincing results, all simulations are performed under the same simulation conditions as the proposed control algorithm.

The sliding surface $s$ and the control law $\tau$ of ITSM are the following:

$$s = \dot{e} + \int_0^t \left( K_1 \text{sign}^\alpha(e_1) + K_2 \text{sign}^\beta(e_2) \right) d\sigma \tag{46}$$

$$\tau = \tau_{eq} + \tau_{asw} \tag{47}$$

where

$$\tau_{eq} = M_0(q)\ddot{q}_d + C_0(q,\dot{q})\dot{q} + G_0(q) - M_0(q)\left( k_1 \text{sign}^\alpha(e_1) + k_2 \text{sign}^\beta(e_2) \right)$$

$$\tau_{sw} = -\frac{s}{\|s\|(1-\delta)}\left( k_0 + a_0 + a_2\|\dot{q}\|^2 + \delta\|\tau\| \right)$$

and where $0 < \alpha < 1$, $\beta = 2\alpha/(\alpha+1)$, $K_1, K_2 = R^{n \times n}$ are constant positive-definite diagonal matrices and $k_i$ are positive constants.

The sliding surface $s$, the control law $\tau$ and the adaptive law of ANFTSM are the following:

$$s = e + k_1|e_1|^\alpha \text{sgn}(e_1) + k_2|e_2|^\beta \text{sgn}(e_2) \tag{48}$$

$$\tau = \tau_{eq} + \tau_{asw} \tag{49}$$

where

$$\tau_{eq} = M_0(q)\ddot{q}_d + C_0(q,\dot{q})\dot{q} + G_0(q) - \frac{M_0(q)}{\beta k_2}|e_2|^{2-\beta}\left( 1 + \alpha k_1|e_1|^{\alpha-1}\text{sgn}(e_2) \right)$$

$$\tau_{sw} = -M_0(q)\left( ks + \left( \hat{a}_0 + \hat{a}_1\|q\| + \hat{a}_2\|\dot{q}\|^2 + \xi \right)\text{sgn}(s) \right)$$

$$\dot{\hat{a}}_0 = \lambda_0\|s\|\|e_2\|^{\beta-1}$$

$$\dot{\hat{a}}_1 = \lambda_1\|s\|\|e_2\|^{\beta-1}\|q\|$$

$$\dot{\hat{a}}_2 = \lambda_2\|s\|\|e_2\|^{\beta-1}\|\dot{q}\|^2$$

where $1 < \beta < 2$, $\alpha > \beta$ and $k_1, k_2, \xi, \lambda_i$ are positive constants.

Table 2 shows the controller parameters selected for ITSM and ANFTSM. The tracking performance comparison effect is shown in Figures 10–13. It is observed from Figures 10–12 that the proposed controller has a faster convergence speed in position and velocity tracking. In addition, it can be seen from Figure 13 that the proposed controller has no chatter with the ITSM controller, but the ANFTSM has larger chatter.

**Table 2.** Controller parameters of ITSM and ANFTSM.

| Controller | Parameters |
|---|---|
| ITSM | $\alpha = 0.5$, $K_1 = \text{diag}(9,10)$, $K_2 = \text{diag}(6,7)$, $k_0 = 1$, $a_0 = 12$, $a_2 = 2.8$, $\delta = 0.3793$ |
| ANFTSM | $\alpha = 2$, $\beta = 5/3$, $k_1 = 1$, $k_2 = 1$, $k = 50$, $\xi = 0.5$, $\lambda_0 = \lambda_1 = \lambda_2 = 1$ |

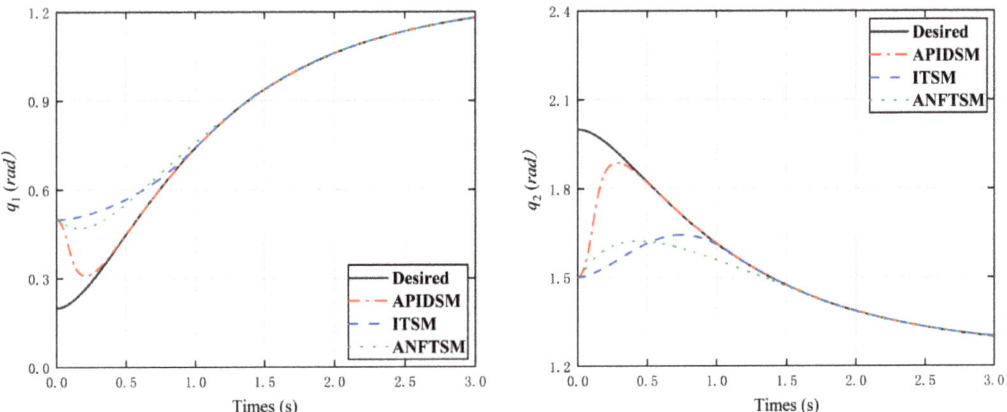

**Figure 10.** Comparison of position tracking results between controllers of Joint 1 and Joint 2.

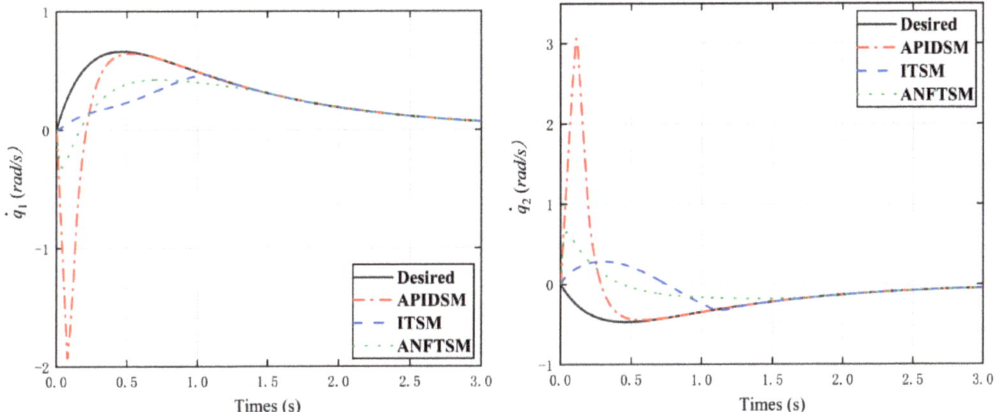

**Figure 11.** Comparison of velocity tracking results between controllers of Joint 1 and Joint 2.

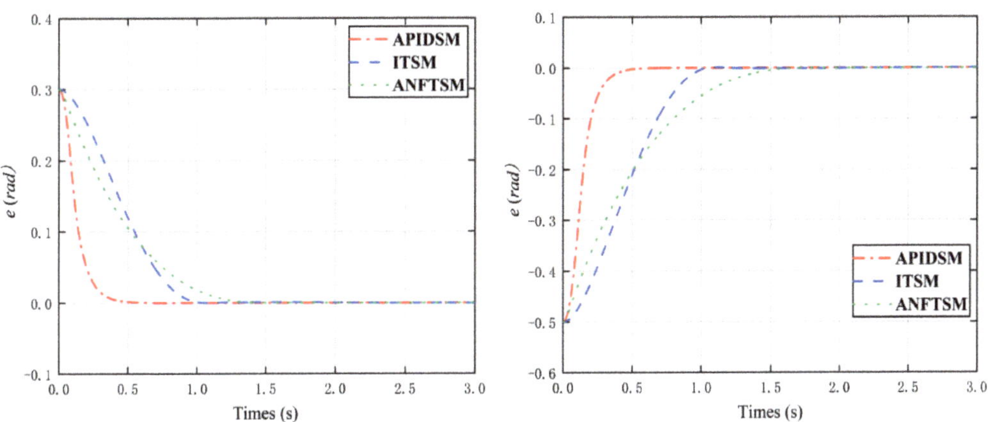

**Figure 12.** Comparison of position tracking error results between controllers of Joint 1 and Joint 2.

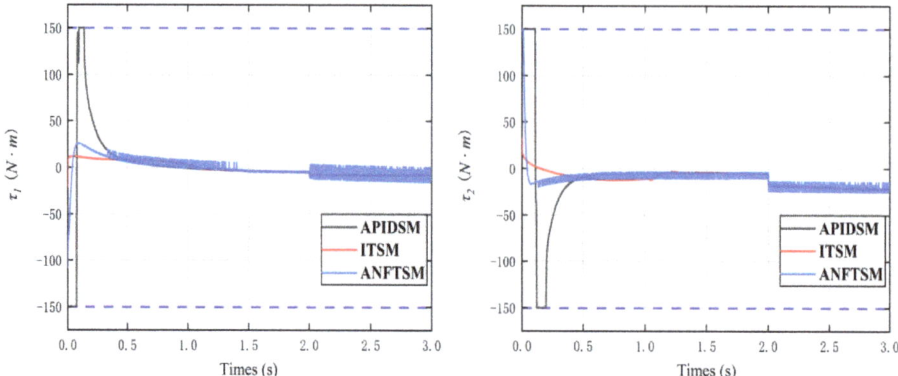

**Figure 13.** Comparison of input torque results between controllers of Joint 1 and Joint 2.

To obtain the compared results more clearly, the integration of the absolute value of the error (IAE) and the integral of the time multiplied by the absolute value of the error (ITAE) were introduced for quantitative analysis [34]. Table 3 shows the comparison results between the performance indicators IAE and ITAE. These are defined as follows:

$$IAE = \int_0^{t_f} |e_1| dt \tag{50}$$

$$ITAE = \int_0^{t_f} t \cdot |e_1| dt \tag{51}$$

where $t_f$ is the total running time.

**Table 3.** Comparison of performance indicators.

| Controllers | Joints | IAE | ITAE |
|---|---|---|---|
| APIDSM | Joint 1 | 0.03921 | 0.00393 |
|  | Joint 2 | 0.07269 | 0.007806 |
| ITSM | Joint 1 | 0.135 | 0.03919 |
|  | Joint 2 | 0.2306 | 0.06861 |
| ANFTSM | Joint 1 | 0.1272 | 0.04142 |
|  | Joint 2 | 0.2415 | 0.09188 |

It can be seen from Table 3 that the proposed control method has lower IAE and ITAE values compared with the comparison method. Therefore, the simulation results can prove that the proposed control method can ensure faster-tracking speed, faster convergence speed, strong robustness, lower chattering, and adaptive ability. The effectiveness of the proposed APIDSM controller is verified.

## 6. Conclusions

In this paper, a sliding mode controller composed of a PID-type sliding mode surface with an arctangent hyperbolic function and a variable gain hyperbolic reaching law is proposed for trajectory tracking under uncertainty. Especially in the case of external time-varying interference and variable load, it can still maintain a good tracking effect. Therefore, the proposed controller still maintains continuity, smoothness, and chatter-free output without knowing the upper bound of the lumped disturbance. To verify the effectiveness of the proposed controller, a two-link robot is simulated under the condition of external time-varying disturbance and load variation, and the effectiveness of the proposed control method is verified by comparing it with other control methods.

**Author Contributions:** Conceptualization, Y.L. and T.W.; methodology, G.L.; software, D.Z.; validation, Y.L., G.L. and T.W; formal analysis, D.Z.; investigation, D.Z.; resources, G.L.; data curation, G.L.; writing—original draft preparation, Y.L.; writing—review and editing, Y.L.; visualization, D.Z.; supervision, T.W.; project administration, T.W. All authors have read and agreed to the published version of the manuscript.

**Funding:** This work was supported by The Central Government Guided Local Special Foundation, grant number YDZX20191400002765; The National Natural Science Foundation of China, grant number 52075503 and 51875532.

**Data Availability Statement:** The data that support the findings of this study are available from the corresponding author upon reasonable request.

**Acknowledgments:** We acknowledge academic editor for his careful revision of the language and grammatical structures in this article.

**Conflicts of Interest:** The authors declare no conflict of interest.

## References

1. Wang, Y.; Xie, L.; de Souza, C.E. Robust control of a class of uncertain nonlinear systems. *Syst. Control Lett.* **1992**, *19*, 139–149. [CrossRef]
2. Rincon, L.; Coronado, E.; Hendra, H.; Phan, J.; Zainalkefli, Z.; Venture, G. Expressive states with a robot arm using adaptive fuzzy and robust predictive controllers. In Proceedings of the 2018 3rd International Conference on Control and Robotics Engineering (ICCRE), Nagoya, Japan, 20 April 2018; pp. 11–15.
3. Fukao, T.; Nakagawa, H.; Adachi, N. Adaptive tracking control of a nonholonomic mobile robot. *IEEE Trans. Robot. Autom.* **2000**, *16*, 609–615. [CrossRef]
4. Nguyen, N.D.K.; Chu, B.L.; Nguyen, T.T. Optimal control for the stable walking gait of a biped robot. In Proceedings of the 2017 14th International Conference on Ubiquitous Robots and Ambient Intelligence (URAI), Jeju, South Korea, 28 June–1 July 2017; pp. 309–313.
5. Wang, J.; Xin, M. Distributed optimal cooperative tracking control of multiple autonomous robots. *Robot. Auton. Syst.* **2012**, *60*, 572–583. [CrossRef]
6. Verscheure, D.; Demeulenaere, B.; Swevers, J.; De Schutter, J.; Diehl, M. Time-energy optimal path tracking for robots: A numerically efficient op-timization approach. In Proceedings of the 2008 10th IEEE International Workshop on Advanced Motion Control, Trento, Italy, 26–28 March 2008; pp. 727–732.
7. Lu, E.; Ma, Z.; Li, Y.; Xu, L.; Tang, Z. Adaptive backstepping control of tracked robot running trajectory based on realtime slip parameter estimation. *Int. J. Agric. Biol. Eng.* **2020**, *13*, 178–187. [CrossRef]
8. Jiang, Z.H.; Shinohara, K. Workspace trajectory tracking control of flexible joint robots based on backstepping method. In Proceedings of the 2016 IEEE Region 10 Conference (TENCON), Singapore, 22–25 November 2016; pp. 3473–3476.
9. Liu, Y.; Jing, H.; Liu, X.; Lv, Y. An improved hybrid error control path tracking intelligent algorithm for omnidirectional AGV on ROS. *Int. J. Comput. Appl. Technol.* **2020**, *64*, 115–125. [CrossRef]
10. Li, Z.F.; Li, J.T.; Li, X.F.; Yang, Y.J.; Xiao, J.; Xu, B.W. Intelligent tracking obstacle avoidance wheel robot based on arduino. *Procedia Comput. Sci.* **2020**, *166*, 274–278. [CrossRef]
11. Bai, G.; Liu, L.; Meng, Y.; Luo, W.; Gu, Q.; Wang, J. Path Tracking of Wheeled Mobile Robots Based on Dynamic Prediction Model. *IEEE Access* **2019**, *7*, 39690–39701. [CrossRef]
12. Dai, L.; Yu, Y.; Zhai, D.-H.; Huang, T.; Xia, Y. Robust Model Predictive Tracking Control for Robot Manipulators with Disturbances. *IEEE Trans. Ind. Electron.* **2020**, *68*, 4288–4297. [CrossRef]
13. Su, Y.; Zheng, C. A new nonsingular integral terminal sliding mode control for robot manipulators. *Int. J. Syst. Sci.* **2020**, *51*, 1418–1428. [CrossRef]
14. Tayebi-Haghighi, S.; Piltan, F.; Kim, J.M. Robust composite high-order super-twisting sliding mode control of robot manipula-tors. *Robotics* **2018**, *7*, 13. [CrossRef]
15. Su, C.-Y.; Stepanenko, Y. Adaptive sliding mode control of robot manipulators: General sliding manifold case. *Automatica* **1994**, *30*, 1497–1500. [CrossRef]
16. Cheng, C.C.; Hsu, S.C. Design of adaptive sliding mode controllers for a class of perturbed fractional-order nonlinear systems. *Nonlinear Dyn.* **2019**, *98*, 1355–1363. [CrossRef]
17. Shi, K.; Yin, X.; Jiang, L.; Liu, Y.; Hu, Y.; Wen, H. Perturbation Estimation Based Nonlinear Adaptive Power Decoupling Control for DFIG Wind Turbine. *IEEE Trans. Power Electron.* **2019**, *35*, 319–333. [CrossRef]
18. Kwon, S.J.; Chung, W.K. Combined synthesis of state estimator and perturbation observer. *J. Dyn. Syst. Meas. Control* **2003**, *125*, 19–26. [CrossRef]
19. Song, Q.; Spall, J.C.; Soh, Y.C.; Ni, J. Robust neural network tracking controller using simultaneous perturbation stochastic ap-proximation. *IEEE Trans. Neural Netw.* **2008**, *19*, 817–835. [CrossRef]

20. Feng, Y.; Yu, X.; Man, Z. Non-singular terminal sliding mode control of rigid manipulators. *Automatica* **2002**, *38*, 2159–2167. [CrossRef]
21. Wang, H.; Li, Z.; Jin, X.; Huang, Y.; Kong, H.; Yu, M.; Ping, Z.; Sun, Z. Adaptive Integral Terminal Sliding Mode Control for Automobile Electronic Throttle via an Uncertainty Observer and Experimental Validation. *IEEE Trans. Veh. Technol.* **2018**, *67*, 8129–8143. [CrossRef]
22. Hao, S.; Hu, L.; Liu, P.X. Second-order adaptive integral terminal sliding mode approach to tracking control of robotic manipulators. *IET Control Theory Appl.* **2021**, *15*, 2145–2157. [CrossRef]
23. Shao, K. Nested adaptive integral terminal sliding mode control for high-order uncertain nonlinear systems. *Int. J. Robust Nonlinear Control* **2021**, *31*, 6668–6680. [CrossRef]
24. Zhai, J.; Xu, G. A Novel Non-Singular Terminal Sliding Mode Trajectory Tracking Control for Robotic Manipulators. *IEEE Trans. Circuits Syst. II Express Briefs* **2020**, *68*, 391–395. [CrossRef]
25. Dong, H.; Yang, X.; Gao, H.; Yu, X. Practical Terminal Sliding Mode Control and its Applications in Servo Systems. *IEEE Trans. Ind. Electron.* **2022**, *70*, 752–761. [CrossRef]
26. Wang, A.; Jia, X.; Dong, S. A New Exponential Reaching Law of Sliding Mode Control to Improve Performance of Permanent Magnet Synchronous Motor. *IEEE Trans. Magn.* **2013**, *49*, 2409–2412. [CrossRef]
27. Wang, Y.; Feng, Y.; Zhang, X.; Liang, J. A new reaching law for anti-disturbance sliding-mode control of PMSM speed regulation system. *IEEE Trans. Power Electron.* **2019**, *35*, 4117–4126. [CrossRef]
28. Xu, L.; Shao, X.; Zhang, W. USDE-based continuous sliding mode control for quadrotor attitude regulation: Method and ap-plication. *IEEE Access* **2021**, *9*, 64153–64164. [CrossRef]
29. Tao, L.; Chen, Q.; Nan, Y.; Wu, C. Double Hyperbolic Reaching Law with Chattering-Free and Fast Convergence. *IEEE Access* **2018**, *6*, 27717–27725. [CrossRef]
30. Asad, M.; Bhatti, A.; Iqbal, S.; Asfia, Y. A smooth integral sliding mode controller and disturbance estimator design. *Int. J. Control Autom. Syst.* **2015**, *13*, 1326–1336. [CrossRef]
31. Zhihong, M.; Yu, X. Adaptive terminal sliding mode tracking control for rigid robotic manipulators with uncertain dynamics. *JSME Int. J. Ser. C* **1997**, *40*, 493–502. [CrossRef]
32. Yi, S.; Zhai, J. Adaptive second-order fast nonsingular terminal sliding mode control for robotic manipulators. *ISA Trans.* **2019**, *90*, 41–51. [CrossRef]
33. Boukattaya, M.; Mezghani, N.; Damak, T. Adaptive nonsingular fast terminal sliding-mode control for the tracking problem of uncertain dynamical systems. *ISA Trans.* **2018**, *77*, 1–19. [CrossRef]
34. Li, T.H.S.; Huang, Y.C. MIMO adaptive fuzzy terminal sliding-mode controller for robotic manipulators. *Inf. Sci.* **2010**, *180*, 4641–4660. [CrossRef]

*Article*

# Fuel Cell Voltage Regulation Using Dynamic Integral Sliding Mode Control

Amina Yasin [1,*], Abdul Rehman Yasin [2], Muhammad Bilal Saqib [2], Saba Zia [2], Robina Nazir [1], Ridab Adlan Elamin Abdalla [1] and Shaherbano Bajwa [1]

[1] Department of Basic Sciences, Preparatory Year Deanship, King Faisal University, Al Hofuf 31982, Saudi Arabia
[2] Department of Electrical Engineering, The University of Lahore, Lahore 54000, Pakistan
* Correspondence: ayasin@kfu.edu.sa; Tel.: +966-5-6986-5050

**Abstract:** Fuel cells guarantee ecological ways of electricity production by promising zero emissions. Proton exchange membrane fuel cells (PEMFCs) are considered one of the safest methods, with a low operating temperature and maximum conversion efficiency. In order to harness the full potential of PEMFC, it is imperative to ensure the membrane's safety through appropriate control strategies. However, most of the strategies focus on fuel economy along with viable fuel cell life, but they do not assure constant output voltage characteristics. A comprehensive design to regulate and boost the output voltages of PEMFC under varying load conditions is addressed with dynamic integral sliding mode control (DISMC) by combining the properties of both the dynamic and integral SMC. The proposed system outperforms in robustness against parametric uncertainties and eliminates the reaching phase along with assured stability. A hardware test rig consisting of a portable PEMFC is connected to the power converter using the proposed technique that regulates voltage for varying loads and power conditions. The results are compared with a proportional integral (PI) based system. Both simulation and hardware results are provided to validate the proposed technique. The experimental results show improvements of 35.4%, 34% and 50% in the rise time, settling time and robustness, respectively.

**Keywords:** fuel cell; clean energy; power electronics; voltage regulation; sliding mode control

**Citation:** Yasin, A.; Yasin, A.R.; Saqib, M.B.; Zia, S.; Riaz, M.; Nazir, R.; Abdalla, R.A.E.; Bajwa, S. Fuel Cell Voltage Regulation Using Dynamic Integral Sliding Mode Control. *Electronics* **2022**, *11*, 2922. https://doi.org/10.3390/electronics11182922

Academic Editors: Jamshed Iqbal, Ali Arshad Uppal and Muhammad Rizwan Azam

Received: 3 August 2022
Accepted: 11 September 2022
Published: 15 September 2022

**Publisher's Note:** MDPI stays neutral with regard to jurisdictional claims in published maps and institutional affiliations.



## 1. Introduction

Fuel cells (FC) are regarded as ecologically clean electrochemical devices which perform highly efficient energy conversion with almost zero emissions [1–3]. They are also merited as renewable energy sources, owing to their use of hydrogen and air as fuel. A FC is constructed with varied configurations among which proton exchange membrane fuel cells (PEMFC) are mostly found in commercial vehicles on account of their vibration resilience, comparatively low operating temperature and noise-suppressing attributes [4,5].

PEMFC operation is inherently nonlinear and varying load conditions affect its maximum efficiency; therefore, various control strategies have been put forward to ensure favorable operating conditions for fuel cells. The appropriate control strategy also warrants an enhanced lifetime of the fuel cell by avoiding fuel starvation within the cell. These approaches have mostly been evaluated with respect to fuel utilization (hydrogen and oxygen), humidity control within the cell, optimal cell operating temperature, fuel pressure, and maximum power point parameters [6–10].

Fuel cell control becomes a challenging problem, especially when dealing with time-delayed systems. An adaptive control technique, extremum seeking (ES), based on the real-time tuning of parameters of nonlinear systems was investigated in [11]. To optimize fuel usage in terms of flow rate, the research community used model predictive control (MPC) due to its robustness against disturbance within defined limits [12–14]. The management of temperature stress was also ascribed to feedforward MPC in PEMFC [15]. However, MPC bounds are narrow, and its complex implementation requires more execution time [16]. The oxygen flow rate from air with appropriate pressure in PEMFC was also implemented using fuzzy, proportional integral derivative (PID) controllers or with an amalgam of both fuzzy and PID controllers; possessing the capability of reducing steady state error and oscillations, along with eliminating the need for a precise mathematical model [17–20]. Although the combination of the two strategies presents a promising outcome, the limitations of fuzzy controller in terms of initial assumptions and need for constant system upgrading do not qualify it as a reliable solution [21]. Variations of the sliding mode control (SMC) approach, such as SMC based on the super twisting algorithm (STA) and second-order SMC implementations, have also been utilized for improving performance in handling fuel starvation, as a consequence of load variations and parametric uncertainties, as well as maximum power point tracking (MPPT) algorithms [22,23]. The authors of [24,25] used a fast track integral SMC for regulating the fuel-excess ratio, and first- and-second order SMC variations of the super twisting algorithm to analyze performance of various PEMFC parameters, respectively.

However, the control strategies mentioned above only promise fuel economy and viable fuel cell life and do not guarantee constant output voltage or current characteristics, thus requiring the integration of the optimal power converter with PEMFC [26]. PEMFC is considered an ideal candidate for electric vehicles (EV); however, its output voltage needs to be boosted to meet the electrical requirements of EVs. In this regard, ref. [27] worked on the cell voltage stability of a fuel cell stack during the aging process under dynamic load conditions. Similarly, ref. [28] provided a comparative analysis of PID and MPC techniques for controlling the dynamics of voltage and temperature of PEMFC, and [29] worked on the modified whale optimization algorithm to achieve a stable voltage curve. An ES technique was utilized by [30] for PEMFC control with respect to constant current and hydrogen influx. State-of-the-art research in the control domain classifies SMC as the best candidate for nonlinear systems with unknown disturbances and load uncertainties, along with the improved transient response [31–33]. However, it is only true if implemented through the discontinuous control law, where the switching frequency approaches infinity.

A more accurate approach was discussed in [34,35], where a hysteresis comparator replaced the discontinuous sign function, resulting in a finite switching frequency. The associated drawback is a variable switching frequency that depends on the states of the system [36]. Moreover, the electromagnetic compatibility (EMC) issues resulting through the variable frequency pose additional design challenges [37–39]. Therefore, it is necessary to achieve a fixed switching frequency in SMC because power converters are composed of reactive components, and their accurate design is frequency dependent, thereby making it an active area for research [40–42]. In this domain, some stimulating work was done by means of equivalent control via filter extraction [42], full bridge converters with phase shifts [43], buck converters with interleaving [39], controlling the current of DC/AC machines [44], cascaded converters [45], and DC power converters [46]. Furthermore, Repecho et al. in [47] provided a technique to fix the frequency in the hysteresis band of the SMC.

One way of fixing the frequency utilizes an adaptive controller, where the width of the hysteresis band is adjusted as proposed in [38,39,42] and [48]. However, information about the plant parameters is needed for a precise design. Using additional sensors for parametric detection adds to the system cost, thereby limiting its utilization. Repecho et al. exploited an external signal for a fixed-rate SMC; however, the addition of new hardware and the underlying condition of requiring the system time constant to be large enough as compared to the switching frequency are its weaknesses.

Another technique, called zero average dynamics (ZAD), was utilized in [49,50], where the goal was to find a duty cycle that corresponds to the zero average value of the dynamics of the selected sliding manifold. This allows the switching rate of the steady state to be configured as desired, resulting in a performance average which is near the ideal behavior. However, the use of the algorithm is limited by the need to perform complex calculations, thereby requiring a powerful real-time processor.

Yet another method which uses the first derivative of the required value corresponding to the switching manifold provides a direct equivalent control implementation, and is mentioned by the researchers in [51–54]. However, reduced robustness and increased system order are its weaknesses, as the discontinuous function is not implemented directly. Ref. [55] used a smooth second-order SMC for robustness, but it suffers from slow convergence. Moreover, another design to achieve a fixed switching rate is by using pulse-width modulation (PWM), as stated by Abrishamifar et al. [56] and Ye et al. [57]. It provides a trade-off between robustness and chattering as the control law morphs into a limiting layer.

Researchers in [58–60] achieved a fixed switching rate in SMC by means of an extra control loop, based on proportional-integral (PI) control, to administer the hysteresis width. It requires slower dynamics of the PI loop to achieve interference-free regulation of the voltage and current, and is hindered by the need for additional calculations for this extra layer of control.

Scientists in [61] utilized the integral sliding-mode controller (ISMC) for converter voltage stability. This technique enforces the sliding mode through the entire response since it eliminates the reaching phase, thereby ensuring robustness from the beginning. In traditional first-order SMC, the robustness of the system is compromised during the reaching phase. The system becomes parametric invariant only when the sliding phase is established. This issue is resolved using integral SMC, where the reaching phase is eliminated, and the system enters the sliding mode from the very initial phase, thus ensuring robustness throughout the control. A double integral SMC was proposed by [62] which provides a constant operational switching frequency with improved robustness and true parametric independence, but it suffers from the higher complexity of implementation.

A dynamic integral SMC for power converters was proposed by [63] which is robust against parametric uncertainties and other perturbations, and the technique itself performs better than ISMC [64]. It eliminates the reaching phase, thereby canceling out the matched disturbances. These control techniques of power converters are summarized in Table 1.

**Table 1.** Converter control techniques summary.

| Ref # | Control Technique | Relative Cost | Convergence Speed | Parametric Independence | Complexity |
|---|---|---|---|---|---|
| [63] | Dynamic Integral SMC[1] | Moderate | Fast | Independent | Low |
| [38,39] | Parabolic Modulated SMC | Low | Fast | Dependent | Moderate |
| [42] | Equivalent Control via Filter Extraction | Low | Real-time | Independent | Low |
| [62] | Double Integral SMC | Moderate | Fast | Independent | High |
| [45,51–54,56,57] | Fixed-Frequency via PWM[2] | Low | Real-time | Dependent | Moderate |
| [43,61] | Fixed-Frequency Integral SMC | Low | Real-time | Dependent | Moderate |
| [60] | Fixed-Frequency Boundary Control | Low | Real-time | Independent | High |
| [58] | Fixed-Frequency Mixed-signal Hysteresis Control | High | Moderate | Independent | High |
| [44,59] | Discrete-Time SMC | High | Moderate | Independent | High |
| [55] | 2nd Order SMC | High | Moderate | Independent | Moderate |
| [49,50] | Zero Average Dynamics | High | Fast | Independent | High |
| [46] | Adaptive Backstepping SMC | High | Moderate | Independent | High |
| [34,35,47,48] | Hysteresis Band SMC * | Moderate | Real-time | Independent | Low |

[1] Sliding mode control (SMC). [2] Pulse-width modulation (PWM). * Provides variable switching frequency.

In this work, a portable PEMFC is connected to a DC-DC power converter circuit so that the dynamics of both systems can be observed and controlled in a series cascade operation. In this regard, a control loop is developed to administer the partial pressure of fuel gases within the PEMFC, while the main controller regulates the boosted output of the fuel cell using the dynamic integral SMC (DISMC). To the best of our knowledge, this is the first time a PEMFC is integrated with a DISMC controlled power converter. The proposed technique provides the constant voltage at the output of the converter despite the uncertain load conditions and varying input voltage of PEMFC.

The rest of the work is organized such that Sections 2 and 3 explain the aggregated system and its mathematical modeling, respectively. Section 4 presents the hardware setup that is used for the experimentation. Results and relevant discussion are provided in Section 5, and finally, the research is concluded in Section 6.

## 2. System Description

A PEMFC-based power system consists of a fuel cell and a DC-DC boost converter with a high output capacitance. A fuel cell is composed of bipolar plates with a gas diffusion mechanism and chemical exchange membranes. The hydrogen reacts with the oxygen in the presence of a catalyst, resulting in DC power with water and heat as byproducts of the chemical reaction.

The open circuit voltage of a single bipolar plate of a fuel cell stack is around 1 V, which drops significantly when a load is connected. This drop is due to the internal resistance of the cell, resulting from the construction of the electrode and the membrane. Hence, a fuel cell represents a typical power source, having low voltage and high current characteristics. In order to utilize the fuel cell in power systems, it is necessary to boost the voltages before applying to the load, which is achieved using a PWM-type DC-DC boost converter. It operates on the principle of storing electrical energy in a switching inductor that transfers it to a capacitor, resulting in boosted output voltage. Therefore, a DC-DC power converter is an integral part of this power system.

Figure 1 shows the key components of the system consisting of a controller for fuel regulation and DC-DC boost converter. The associated control strategies are developed, and finally a new technique based on DISMC is proposed to provide a constant output voltage. It is ensured that the chemical reactants are fully humidified before entering the fuel cell, and the flow rate of hydrogen is directly proportional to the flow rate of air; therefore, by controlling the blower voltage, the flow rates of both hydrogen and air are controlled. The heat produced during the exothermic reaction is removed by a fan-based air-cooled mechanism. Thus, the model for a fuel cell can be reduced to a voltage source, having unregulated voltage, which is regulated by the control circuitry illustrated in Figure 2.

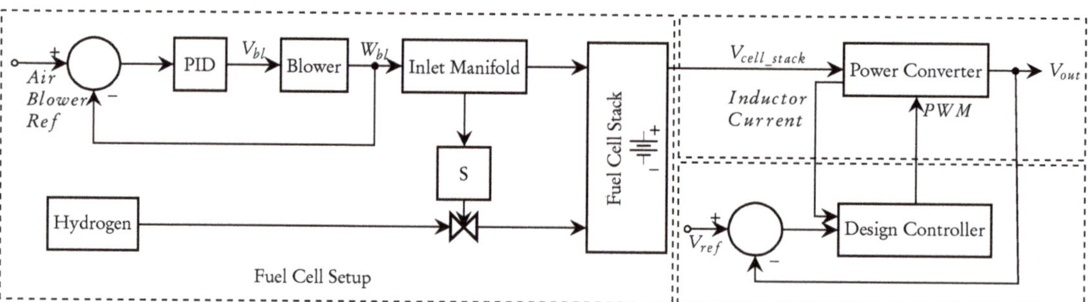


**Figure 1.** Fuel cell setup with power control circuitry and its control system.

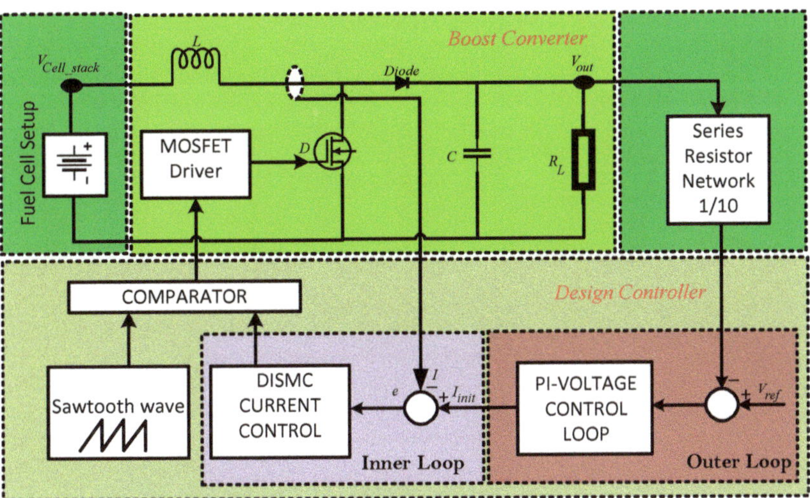



**Figure 2.** Block diagram of power control circuitry.

## 3. Mathematical Model

The mathematical modeling of the system is described in this section.

### 3.1. PEMFC Stack Voltage

PEMFC model is built upon founding electrochemical equations, which are detailed in [4,65–67]. Figure 1 represents the system level block diagram that shows a fuel cell setup, where a PID controller ensures an optimum blow rate, $W_{bl}$, by controlling the blower voltage, $V_{bl}$, for maximum efficiency. The unregulated voltage of the fuel cell stack, $V_{cell\_stack}$, is fed to the power converter system, which is responsible for its regulation. $V_{cell\_stack}$ is defined as the difference between the FC thermodynamic potential, $V_{thermo}$, and voltage losses outlined by its polarization curve, $V_{polarization}$ for the total number of cells in a stack, $\eta_{total}$. The polarization curve elucidates the variations in voltage behavior of the stack against the applied current density.

$$V_{cell\_stack} = \left( V_{thermo} - V_{polarization} \right) \times \eta_{total} \tag{1}$$

Thermodynamic potential is the theoretical open circuit voltage drop which is principally based on Nernst expression with $T_{cell\_stack}$ representing the stack operating temperature. The Nernst potential specified at standard temperature readings is defined by the equation below:

$$V_{thermo} = 1.229 - \left[ 8.5 \times 10^{-4} \cdot (T_{cell\_stack} - 298.15) + \frac{RT_{cell\_stack}}{2F} \cdot \ln\left( H_{2_{pressure}} \cdot O_{2_{pressure}}^{1/2} \right) \right] \tag{2}$$

where $H_{2_{pressure}}$ and $O_{2_{pressure}}^{1/2}$ represent the partial pressures for hydrogen and oxygen, respectively, $R$ stands for the universal gas constant, while $F$ denotes Faraday's constant.

The potential drop by $V_{polarization}$ is defined by

$$V_{polarization} = V_{activation} + V_{resistive} + V_{mass\_concerntration} \tag{3}$$

Among the three potential drops, $V_{activation}$ accounts for the major voltage loss and is given by the equation below:

$$V_{activation} = \kappa_1 - \kappa_2 T_{cell\_stack} + \kappa_3 T_{cell\_stack}\left(\ln\left(O_{2_{concentration}}\right)\right) + \kappa_4 T_{cell\_stack}\ln(\rho_i) \quad (4)$$

where $\rho_i$ is the current density, and $\kappa_j$ is the constants with $j$ ranging from 1 to 4. The values of $\kappa_1$, $\kappa_2$, $\kappa_3$ and $\kappa_4$ are $-0.9514$, $0.00312$, $-0.000187$ and $0.000074$, respectively. The $O_{2_{concentration}}$ represents the oxygen gas concentration and is defined by the equation below:

$$O_{2_{concentration}} = \frac{O_{2_{pressure}}^{1/2}}{5.08 \times 10^6 \cdot e^{\frac{498}{T_{cell\_stack}}}} \quad (5)$$

Impedance in the ion and electron flow accounts for resistive or ohmic polarization and is stated by the equations below:

$$V_{resistive} = \rho_i \times \Omega_{resistive} \quad (6)$$

$$\Omega_{resistive} = \Omega_{ionic} + \Omega_{elect} + \Omega_{contact} \quad (7)$$

where $\Omega_{ionic}$ is the ionic resistance of the electrolyte, $\Omega_{elect}$ defines the electrical resistance of cell plates, and $\Omega_{contact}$ is the contact resistance of the electrodes. For PEMFC, $\Omega_{ionic}$ accounts for almost all of the resistive polarization due to specific membrane characteristics.

Polarization exposition for reactant consumption by electrodes generating concentration gradient is given by the $V_{mass\_concentration}$ equation below:

$$V_{mass\_concentration} = \frac{RT_{cell\_stack}}{nF} \ln\left(\frac{1}{1 - \frac{\rho_i}{\rho_{i_l}}}\right) \quad (8)$$

where $n = 4$ represents two electrons for each hydrogen molecule and $\rho_{i_l}$ is the limiting current density.

It is important to mention that including the dynamics of the fuel cell in Equation (1) can have an impact on the efficiency of the fuel cell. In case of the proposed technique where a PID controller is used in the first stage, there will be no significant improvement, as the controllers are experimentally tuned for the best performance. However, there will be a significant impact on the efficiency in cases where advanced nonlinear controllers are used at the fuel cell efficiency stage.

### 3.2. Power Converter

Dynamics of a boost converter designed on first principles of knowledge are defined by Equations (9) and (10), where $V_{cell\_stack}$ is the output voltage of PEMFC and $V_{out}$ is the stable and uniform voltage of the boost converter.

$$\dot{I} = -(1 - u)\frac{1}{L} \cdot V_{out} + \frac{1}{L} \cdot V_{cell\_stack} \quad (9)$$

$$\dot{V}_{out} = (1 - u)\frac{1}{C} \cdot I - \frac{1}{RC} \cdot V_{out}. \quad (10)$$

where $u = \{0, 1\}$ represents the switching signal, also termed as control input, $L$ is the coil inductance, $C$ is the storage capacitance, and $R_L$ is the load resistance defined by Equation (11).

$$R_L = R_{in} + \Delta R \quad (11)$$

$R_{in}$ represents the initial resistance value, whereas $\Delta R$ is the uncertain change in the resistance.

Owing to Equations (9) and (10), the system exhibits non-minimum phase behavior, forcing it to be implemented in a two-loop fashion, wherein the inner loop controls the current based on DISMC, while the outer loop controls the voltage with the PI-based approach.

### 3.3. Hybrid Model

The composite and dynamic approach for the two-loop implementation of a hybrid boost converter is approached. The control signal $\hat{u}$ consists of two parts: the first part is an analog signal, $\hat{u}_0$, designed using the pole-placement technique, while $\hat{u}_1$ is the discontinuous part used to cancel out unknown disturbances, and is presented below.

$$\hat{u} = \hat{u}_0 + \hat{u}_1 \tag{12}$$

The current error, $e$, is defined as $e = I_{init} - I$, where $I_{init}$ is the reference current. The continuous part of control Equation (12) is described as

$$\hat{u}_0 = -\left(\eta e + k \int e \, dt\right) \tag{13}$$

where $\eta$ and $k$ are positive constants representing the gain of the controller. By taking the time derivative of error, $e$, and by substituting the value of $\dot{I}$ from Equation (9), $\dot{e}$ becomes Equation (14):

$$\dot{e} = \dot{I}_{init} + \hat{u}\frac{1}{L}V_{out} - \frac{1}{L}V_{cell\_stack} \tag{14}$$

and $\hat{u} = 1 - u$. With this elaboration, the robust integral sliding surface is generated by the equation framework given in Equations (15) to (20):

$$s = x(e) + z \tag{15}$$

where $x(e)$ represents the systems equations and is defined by Equation (16):

$$x(e) = \dot{e} + \lambda e \tag{16}$$

Taking the derivative of Equation (16),

$$\dot{x}(e) = \ddot{e} + \lambda \dot{e} \tag{17}$$

$z$ is the integral component in Equation (15) and its derivative is defined by Equation (18).

$$\dot{z} = -\lambda \dot{e} - \hat{u}_0 \tag{18}$$

with the initial condition specified by $z(0) = -x(e)$. Taking the derivative of Equation (15) with respect to time gives Equation (19).

$$\dot{s} = \dot{x}(e) + \dot{z}(t) \tag{19}$$

and substituting Equations (17) and (18) into Equation (19) produces Equation (20).

$$\dot{s} = \ddot{e} - \hat{u}_0 \tag{20}$$

To design the discontinuous part, we take the derivative of Equation (14), and incorporate the system dynamics:

$$\ddot{e} = \ddot{I}_{init} + \hat{u} \cdot \frac{1}{L}V_{out} - \frac{1}{L}V_{cell\_stack} + \frac{\hat{u}^2}{LC}I + \frac{\hat{u}}{R_{in}LC}V_{out} \tag{21}$$

and by substituting Equation (21) into (20), we obtain Equation (22)

$$\dot{s} = \ddot{I}_{init} + \hat{u} \cdot \frac{1}{L}V_{out} - \frac{1}{L}V_{cell\_stack} + \frac{\hat{u}^2}{LC}I + \frac{\hat{u}}{R_{in}LC}V_{out} - \hat{u}_0 \tag{22}$$

By the existence of the sliding mode condition $-D \times sign(s) = \dot{s}$ and using the equation of $\dot{u}$, Equation (22) can be rewritten as

$$-D \times sign(s) = \ddot{I}_{init} + (\dot{\hat{u}_0} + \dot{\hat{u}_1}) \cdot \frac{1}{L} V_{out} - \frac{1}{L} V_{cell\_stack} + \frac{\hat{u}^2}{LC} I + \frac{\hat{u}}{R_{in}LC} V_{out} - \dot{\hat{u}_0} \quad (23)$$

where duty cycle, $D \in \Re^+$. The discontinuous control $\dot{\hat{u}_1}$ can now be extracted by Equation (23) as

$$\dot{\hat{u}_1} = -\frac{L}{V_{out}} \left[ D \times sign(s) + \ddot{I}_{init} - \dot{\hat{u}_0}\left(1 + \frac{V_{out}}{L}\right) - \frac{1}{L} V_{cell\_stack} + \frac{\hat{u}^2}{LC} I + \frac{\hat{u}}{R_{in}LC} V_{out} \right] \quad (24)$$

The entire control law specifying the complete PWM modulated signal $\dot{u}$ can now be rewritten using Equations (13) and (24) to give Equation (25)

$$\dot{u} = -\left( \eta e + k \int e\, dt \right) - \frac{L}{V_{out}} \left[ D \times sign(s) + \ddot{I}_{init} - \dot{\hat{u}_0}\left(1 + \frac{V_{out}}{L}\right) - \frac{1}{L} V_{cell\_stack} + \frac{\hat{u}^2}{LC} I + \frac{\hat{u}}{R_{in}LC} V_{out} \right] \quad (25)$$

*3.4. Lyapunov based Existence Condition*

For the existing model, the Lyapunov stability can be defined as

$$V(x,t) = \frac{1}{2} s^2 \quad (26)$$

By taking the derivative of Equation (26), we obtain Equation (27).

$$\dot{V}(x,t) = s\dot{s} \quad (27)$$

Substituting Equation (22) and existence condition

$$\dot{V}(x,t) = s\left( \ddot{I}_{init} + \dot{\hat{u}} \cdot \frac{1}{L} V_{out} - \frac{1}{L} V_{cell\_stack} + \frac{\hat{u}^2}{LC} I + \frac{\hat{u}}{(R_{in} - \Delta R)LC} V_{out} - \dot{\hat{u}_0} \right) \quad (28)$$

Equation (28) can be further elaborated as

$$\dot{V}(x,t) = -s\left[ D \times sign(s) - \frac{\Delta R V_{out} \hat{u}}{R_{in}(R_{in} - \Delta R)LC} \right] \quad (29)$$

where $\Delta R \leq \mu$, the ultimate value of the load resistance uncertainty. Hence, Equation (29) can be redefined as

$$\dot{V}(x,t) \leq -|s|\left[ D - \frac{\mu V_{out} \hat{u}}{R_{in}(R_{in} - \mu)LC} \right] \quad (30)$$

Thus, when $D > \dfrac{\mu V_{out} \hat{u}}{R_{in}(R_{in} - \mu)LC} + \theta$, the time-domain derivative of the Lyapunov function becomes negative definite, where $\theta \in \Re^+$ and $\mu = 72$, proving $s = 0$ in a finite time.

## 4. Hardware Setup

A 100 W, 24 cell portable fuel cell device from Horizon Fuel Cell Technologies bearing the model number MT08M965-100 was employed. Its detailed specifications are provided in Table 2.

Table 3 shows the boost converter's parameters, and Figure 3 shows the experimental setup. The electronic switch comprises IRF540 power MOSFET, having a channel resistance of 0.06 Ω and can sustain up to 20 A of continuous drain current. The converter operates at a switching frequency of 32 kHz, which is selected based on a trade-off between the switching losses and chattering amplitude. The output of the converter is fed to the control circuitry, through a series resistor network having an attenuation of 10, as shown in Figure 2.

**Table 2.** Specifications of the Horizon Fuel Cell Technologies' produced fuel cell used in this research.

| Parameter | Description |
| --- | --- |
| Type | Portable |
| Operating Temperature | 65 °C |
| Volume | $15 \times 10.9 \times 9.4\,\text{cm}^3$ |
| Weight | 90 g |
| Oxident Supply | Open Cathode |

**Table 3.** Boost Converter Specification.

| Specification | Representation | Value |
| --- | --- | --- |
| Input Voltage | $V_{cell\_stack}$ | 11−19 V |
| Output voltage | $V_{out}$ | 32 V |
| Capacitance | $C$ | 2200 µF |
| Coil inductance | $L$ | 110 µH |
| Switching Frequency | $F$ | 32 kHz |
| Load Resistor | $R_{init}$ | 82 Ω |
| Variation in Load | $\Delta R$ | 0−47 Ω |

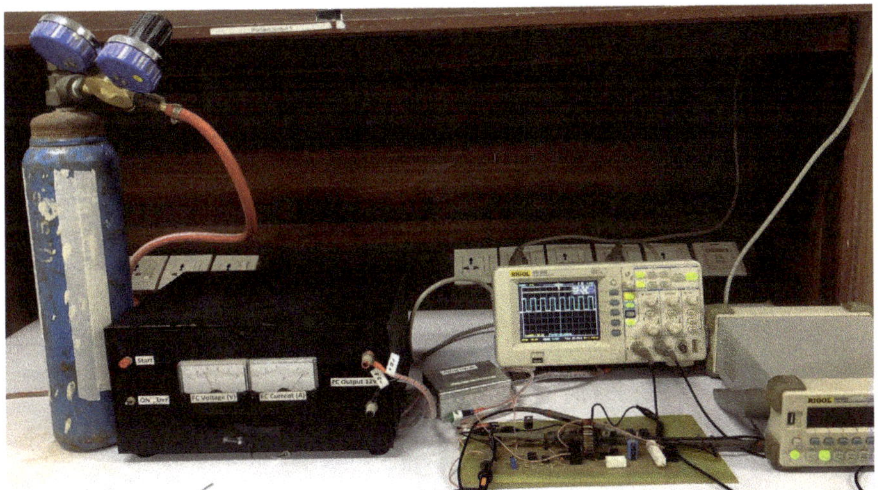

**Figure 3.** Hardware Setup.

Experiments show that the equivalent resistance of the MOSFET drain should be low enough to quickly discharge the capacitance of the body diode; a higher value adversely affects the device's turn-off process. A 47 kΩ resistor is placed between the MOSFET gate and source to ensure the discharge of the gate capacitance. Another resistor with a value of 0.47 Ω is positioned in series with the inductor current to measure its instantaneous value. According to Ohm's law, the current flowing through an inductor is $I_L = V_{me}/R_{me} = 2.1\,V_{me}$. Hence, the voltage measured at $R_{me}$ is multiplied by a factor of 2.1 to obtain the exact value of the inductor current. All hardware-related wave forms in this work were captured with a 70 MHz Rigol oscilloscope at the rate of 1 G $^{sample}/_s$. Furthermore, the power supply that runs the control circuitry is accurate to 0.1 V.

## 5. Results and Discussion

A complete schematic of the system is shown in Figure 4. The controller is implemented using Artix-7 family FPGA with model XC7A35T-ICPG236C. The voltages from the PEMFC are fed to the FPGA. The 12-bit, 1 MSPS, on-board Xilinx analog-to-digital converter (XADC) interface operates at input values between 0 and 1 V, thus requiring a series network of resistive elements to convert the input levels to the desired values. The inductor current is also provided to this XADC. A moving average filter is applied to the samples after every 10 μs. The extracted PWM signal with the required duty cycle is fed to the DC-DC boost converter implemented with the help of discrete components. Since FPGA implements the controller in the digital domain, all the fractional values are handled using signed fixed-point number representation.

The system is also simulated to validate the mathematical model and results. The simulations were performed using Runge–Kutta (ODE4) solver, having a fixed step size of $10^{-8}$ in the Simulink environment. The results suggest that the simulation outputs resemble quite accurately the experimental results. It is observed that the unmodeled dynamics of the system as well as the uncertainties of the experimental setup do have an impact, causing a slight variation between the experimental results and simulations.

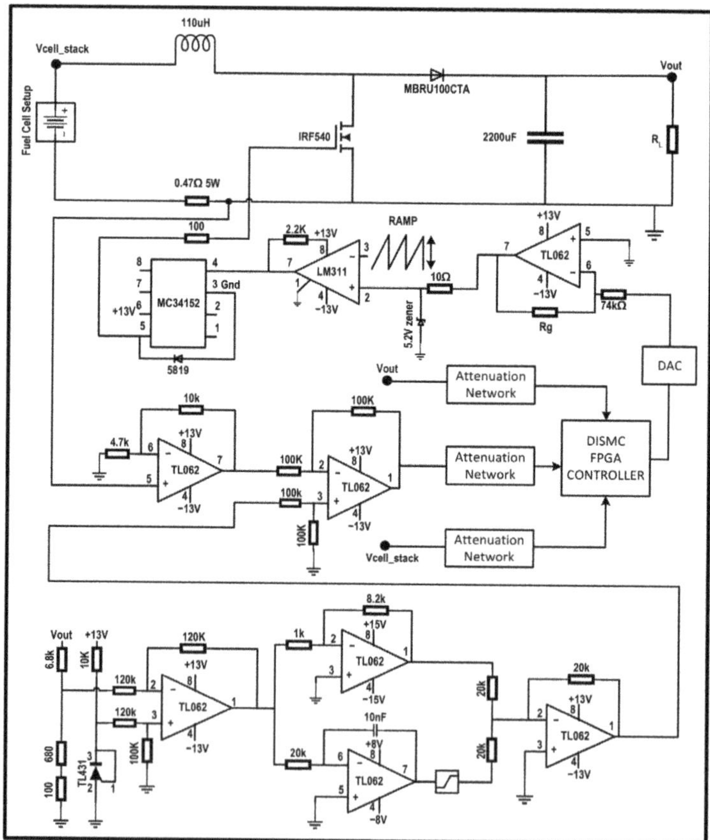


**Figure 4.** Schematic of the system.

### 5.1. Steady-State Error and Transient Response

For the sake of experimental comparison between the proposed technique and conventional controllers, a PI controller is implemented, and the results are shown in Figure 5.

Experiments show that the conventional system has a rise time, $t_r$, of 45.2 ms and settles, $t_s$, in 83.82 ms. Computer simulations of the same are also performed to validate the mathematical model. Simulations show that the PI-based system has a rise and settling time of 44.1 ms and 82.5 ms respectively. The experimental results of the proposed DISMC-based PEMFC system are shown in Figure 6. The system shows superior performance with a rise time, $t_r$, of 29.2 ms and a settling time, $t_s$, of 54.32 ms, whereas the simulations show 28.3 ms and 52.69 ms of the rise and settling time, respectively. There is a very close match between the simulations and the experimental findings; however, there is a small difference between the simulation and practical results due to the unmodeled dynamics and measuring inaccuracies. Both of the controllers are regulated at 32 V output and exhibit no steady-state error. The system shows an improvement of 35.4% in rise time and 34.0% in settling time, which is a direct result of the fact that the discontinuous function is implemented directly, and no indirect method is used to calculate the equivalent control.

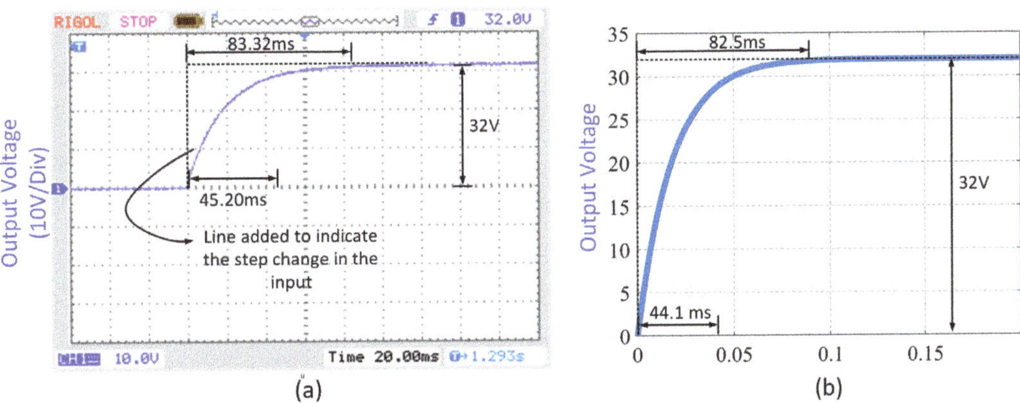


**Figure 5.** Results using conventional PI controller with fuel cell. (**a**) Hardware results. (**b**) Simulation results.

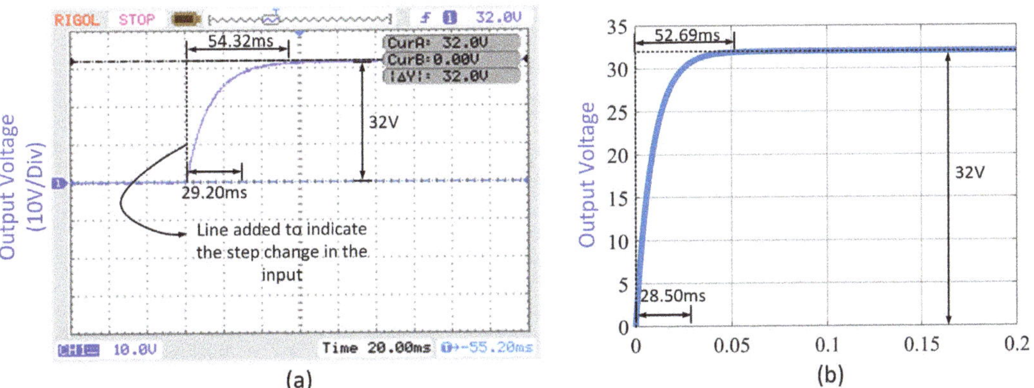


**Figure 6.** Results showing performance of the proposed DISMC with PEMFC. (**a**) Hardware results. (**b**) Simulation results.

The figure exhibiting the control effort obtained through simulation is shown in Figure 7.

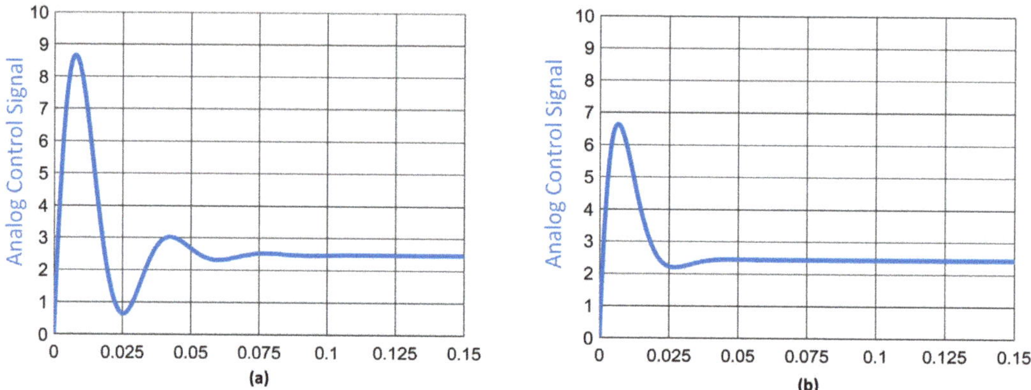

**Figure 7.** Control effort obtained via simulation. (**a**) PID controller. (**b**) DISMC controller.

*5.2. Robustness to Load Variations*

In order to validate the robustness of the system, an experimental setup using a Rigol function generator with a switching frequency of 10 kHz is used. The function generator's output is connected to the base of a NPN transistor C1383. This drives a PNP power transistor TIP147, which connects and disconnects an additional load of 47 $\Omega$ from the nominal load of 82 $\Omega$. Thus, experiments are conducted with the load resistance abruptly increased to assess the robustness of the proposed DISMC. Figure 8 shows that the conventional PI system exhibits an undershoot of 3.20 V, while DISMC demonstrates an improvement of 50% by decreasing the voltage dip to 1.6 V. PI recovers in 257 µs, while DISMC recovers in 85 µs. This shows a 67% improvement in recovery time.

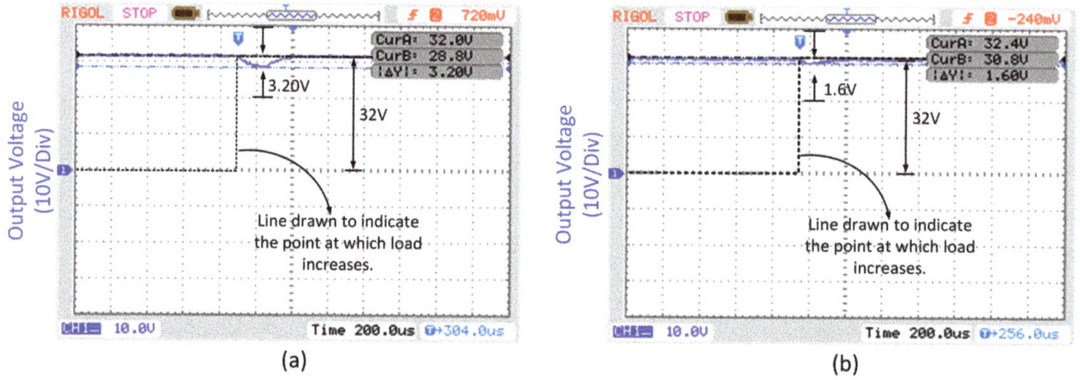


(a)                    (b)

**Figure 8.** Results showing robustness of the PI and proposed DISMC with PEMFC. (**a**) PI controller. (**b**) DISMC controller.

## 6. Conclusions

With the rising $CO_2$ pollution, researchers are focusing for the development of different clean energy solutions, and the fuel cell is among those electrochemical sources which are eco-friendly with zero emissions. Different types of SMC techniques have been reported in the literature to operate fuel cells under optimal conditions. However, there is a significant gap in the integration of power electronic techniques with the fuel cell to experimentally evaluate the performance. This research reports a significant improvement in the performance of the overall system using DISMC. In addition to this, a mathematical proof of the system's stability is provided. The results display 35.4%, 34% and 50% improvements in the rise time, settling time and robustness, respectively. Moreover, simulations are performed that confirm a close resemblance with the experimental results. Future work could be extended in the direction to incorporate system time delays and temperature variations.

**Author Contributions:** A.Y. and A.R.Y. conceptualized the research and acted as the project administrators. M.B.S., S.Z., M.R., R.N., R.A.E.A. and S.B. acted as the investigators. All authors have read and agreed to the published version of the manuscript.

**Funding:** This work was supported through the Annual Funding track by the Deanship of Scientific Research (DSR), Vice Presidency for Graduate Studies and Scientific Research, King Faisal University, Saudi Arabia, Project Number AN000498 transformed to Project Number GRANT1504.

**Data Availability Statement:** Not applicable.

**Acknowledgments:** The authors would like to thank Aamer Iqbal Bhatti for his excellent guidance. We would also like to thank Karen Stanforth for her valuable inputs.

**Conflicts of Interest:** The authors declare no conflict of interest.

## References

1. Sharma, A.; Khan, R.A.; Sharma, A.; Kashyap, D.; Rajput, S. A Novel Opposition-Based Arithmetic Optimization Algorithm for Parameter Extraction of PEM Fuel Cell. *Electronics* **2021**, *10*, 2834. [CrossRef]
2. Chen, W.M.; Kim, H. Energy, economic, and social impacts of a clean energy economic policy: Fuel cells deployment in Delaware. *Energy Policy* **2020**, *144*, 111617. [CrossRef]
3. Xing, L.; Bai, X.; Gao, Y.; Cao, Z. Improving clean electrical power generation: A theoretical modelling analysis of a molten sodium hydroxide direct carbon fuel cell with low pollution. *J. Clean. Prod.* **2021**, *281*, 124623. [CrossRef]
4. Godula-Jopek, D.I.h.A.; Westenberger, A.F. Fuel Cell Types: PEMFC/DMFC/AFC/PAFC//MCFC/SOFC/. In *Encyclopedia of Energy Storage*; Cabeza, L.F., Ed.; Elsevier: Oxford, UK, 2022; pp. 250–265. [CrossRef]
5. Hosseini, S.E. 4.12-Hydrogen and Fuel Cells in Transport Road, Rail, Air, and Sea. In *Comprehensive Renewable Energy*, 2nd ed.; Letcher, T.M., Ed.; Elsevier: Oxford, UK, 2022; pp. 317–342. [CrossRef]
6. Yang, B.; Li, J.; Li, Y.; Guo, Z.; Zeng, K.; Shu, H.; Cao, P.; Ren, Y. A critical survey of proton exchange membrane fuel cell system control: Summaries, advances, and perspectives. *Int. J. Hydrogen Energy* **2022**, *47*, 9986–10020. [CrossRef]
7. Lin-Kwong-Chon, C.; Grondin-Pérez, B.; Kadjo, J.J.A.; Damour, C.; Benne, M. A review of adaptive neural control applied to proton exchange membrane fuel cell systems. *Annu. Rev. Control* **2019**, *47*, 133–154. [CrossRef]
8. Yuan, H.; Dai, H.; Wei, X.; Ming, P. Model-based observers for internal states estimation and control of proton exchange membrane fuel cell system: A review. *J. Power Sources* **2020**, *468*, 228376. [CrossRef]
9. Yuan, X.Z.; Nayoze-Coynel, C.; Shaigan, N.; Fisher, D.; Zhao, N.; Zamel, N.; Gazdzicki, P.; Ulsh, M.; Friedrich, K.A.; Girard, F.; et al. A review of functions, attributes, properties and measurements for the quality control of proton exchange membrane fuel cell components. *J. Power Sources* **2021**, *491*, 229540. [CrossRef]
10. Hou, J.; Yang, M.; Ke, C.; Zhang, J. Control logics and strategies for air supply in PEM fuel cell engines. *Appl. Energy* **2020**, *269*, 115059. [CrossRef]
11. Oliveira, T.R.; Krstić, M.; Tsubakino, D. Extremum Seeking for Static Maps With Delays. *IEEE Trans. Autom. Control* **2017**, *62*, 1911–1926. [CrossRef]
12. He, H.; Quan, S.; Sun, F.; Wang, Y.X. Model Predictive Control With Lifetime Constraints Based Energy Management Strategy for Proton Exchange Membrane Fuel Cell Hybrid Power Systems. *IEEE Trans. Ind. Electron.* **2020**, *67*, 9012–9023. [CrossRef]
13. Chatrattanawet, N.; Hakhen, T.; Kheawhom, S.; Arpornwichanop, A. Control structure design and robust model predictive control for controlling a proton exchange membrane fuel cell. *J. Clean. Prod.* **2017**, *148*, 934–947. [CrossRef]
14. Chen, Q.; Gao, L.; Dougal, R.A.; Quan, S. Multiple model predictive control for a hybrid proton exchange membrane fuel cell system. *J. Power Sources* **2009**, *191*, 473–482. [CrossRef]
15. Li, G.; Fu, H.; Madonski, R.; Czeczot, J.; Nowak, P.; Lakomy, K.; Sun, L. Feed-forward offset-free model predictive temperature control for proton exchange membrane fuel cell: An experimental study. *ISA Trans.* 2021, *in press*. [CrossRef]
16. Ekaputri, C.; Syaichu-Rohman, A. Model predictive control (MPC) design and implementation using algorithm-3 on board SPARTAN 6 FPGA SP605 evaluation kit. In Proceedings of the 2013 3rd International Conference on Instrumentation Control and Automation (ICA), Bali, Indonesia, 28–30 August 2013; pp. 115–120. [CrossRef]
17. Liu, Z.; Chen, H.; Peng, L.; Ye, X.; Xu, S.; Zhang, T. Feedforward-decoupled closed-loop fuzzy proportion-integral-derivative control of air supply system of proton exchange membrane fuel cell. *Energy* **2022**, *240*, 122490. [CrossRef]
18. Mahali, A.P.; Babu Gara, U.B.; Mullapudi, S. Fuzzy Logic based Control Strategies for Proton Exchange Membrane Fuel Cell System. *IFAC-PapersOnLine* **2022**, *55*, 703–708. [CrossRef]

19. Chen, X.; Xu, J.; Fang, Y.; Li, W.; Ding, Y.; Wan, Z.; Wang, X.; Tu, Z. Temperature and humidity management of PEM fuel cell power system using multi-input and multi-output fuzzy method. *Appl. Therm. Eng.* **2022**, *203*, 117865. [CrossRef]
20. AbouOmar, M.S.; Su, Y.; Zhang, H.; Shi, B.; Wan, L. Observer-based interval type-2 fuzzy PID controller for PEMFC air feeding system using novel hybrid neural network algorithm-differential evolution optimizer. *Alex. Eng. J.* **2022**, *61*, 7353–7375. [CrossRef]
21. Behrooz, F.; Mariun, N.; Marhaban, M.; Mohd Radzi, M.A.; Ramli, A. Review of Control Techniques for HVAC Systems—Nonlinearity Approaches Based on Fuzzy Cognitive Maps. *Energies* **2018**, *11*, 495. [CrossRef]
22. Souissi, A. Adaptive sliding mode control of a PEM fuel cell system based on the super twisting algorithm. *Energy Rep.* **2021**, *7*, 3390–3399. [CrossRef]
23. Derbeli, M.; Barambones, O.; Farhat, M.; Ramos-Hernanz, J.A.; Sbita, L. Robust high order sliding mode control for performance improvement of PEM fuel cell power systems. *Int. J. Hydrogen Energy* **2020**, *45*, 29222–29234. [CrossRef]
24. Javaid, U.; Mehmood, A.; Arshad, A.; Imtiaz, F.; Iqbal, J. Operational Efficiency Improvement of PEM Fuel Cell—A Sliding Mode Based Modern Control Approach. *IEEE Access* **2020**, *8*, 95823–95831. [CrossRef]
25. Javaid, U.; Iqbal, J.; Mehmood, A.; Uppal, A.A. Performance improvement in polymer electrolytic membrane fuel cell based on nonlinear control strategies—A comprehensive study. *PLoS ONE* **2022**, *17*, e0264205. [CrossRef] [PubMed]
26. Choe, S.Y.; Lee, J.G.; Ahn, J.W.; Baek, S.H. Integrated modeling and control of a PEM fuel cell power system with a PWM DC/DC converter. *J. Power Sources* **2007**, *164*, 614–623. [CrossRef]
27. Chen, K.; Hou, Y.; Jiang, C.; Pan, X.; Hao, D. Experimental investigation on statistical characteristics of cell voltage distribution for a PEMFC stack under dynamic driving cycle. *Int. J. Hydrogen Energy* **2021**, *46*, 38469–38481. [CrossRef]
28. Chen, X.; Fang, Y.; Liu, Q.; He, L.; Zhao, Y.; Huang, T.; Wan, Z.; Wang, X. Temperature and voltage dynamic control of PEMFC Stack using MPC method. *Energy Rep.* **2022**, *8*, 798–808. [CrossRef]
29. Cao, Y.; Li, Y.; Zhang, G.; Jermsittiparsert, K.; Nasseri, M. An efficient terminal voltage control for PEMFC based on an improved version of whale optimization algorithm. *Energy Rep.* **2020**, *6*, 530–542. [CrossRef]
30. Roux Oliveira, T.; Revoredo, T. Monitoring Function based Extremum Seeking for Fuel Cell Energy Applications. In Proceedings of the International Congress of Mechanical Engineering (COBEM), Rio de Janeiro, Brazil, 6–11 December 2015.
31. Awais, M.; Yasin, A.R.; Riaz, M.; Saqib, B.; Zia, S.; Yasin, A. Robust Sliding Mode Control of a Unipolar Power Inverter. *Energies* **2021**, *14*, 5405. [CrossRef]
32. Sankar, K.; Saravanakumar, G.; Jana, A.K. Nonlinear multivariable control of an integrated PEM fuel cell system with a DC-DC boost converter. *Chem. Eng. Res. Des.* **2021**, *167*, 141–156. [CrossRef]
33. Kuo, J.K.; Wang, C.F. An integrated simulation model for PEM fuel cell power systems with a buck DC–DC converter. *Int. J. Hydrogen Energy* **2011**, *36*, 11846–11855. [CrossRef]
34. Sabanovic, A. Buck converter regulator operating in the sliding mode. In Proceedings of the PCI, Orlando, FL, USA, 19–21 April 1983.
35. Venkataramanan, R. *Sliding Mode Control of Power Converters*; California Institute of Technology: Pasadena, CA, USA, 1986.
36. Carpita, M.; Marchesoni, M. Experimental study of a power conditioning system using sliding mode control. *IEEE Trans. Power Electron.* **1996**, *11*, 731–742. ISBN 0885-8993. [CrossRef]
37. Hizarci, H.; Pekperlak, U.; Arifoglu, U. Conducted emission suppression using an EMI filter for grid-tied three-phase/level T-type solar inverter. *IEEE Access* **2021**, *9*, 67417–67431. ISBN 2169-3536. [CrossRef]
38. Qi, W.; Li, S.; Tan, S.C.; Hui, S.R. Parabolic-modulated sliding-mode voltage control of a buck converter. *IEEE Trans. Ind. Electron.* **2017**, *65*, 844–854. ISBN 0278-0046. [CrossRef]
39. Repecho, V.; Biel, D.; Ramos-Lara, R.; Vega, P.G. Fixed-switching frequency interleaved sliding mode eight-phase synchronous buck converter. *IEEE Trans. Power Electron.* **2017**, *33*, 676–688. ISBN 0885-8993. [CrossRef]
40. Al-Wesabi, I.; Fang, Z.; Wei, Z.; Dong, H. Direct sliding mode control for dynamic instabilities in DC-link voltage of standalone photovoltaic systems with a small capacitor. *Electronics* **2022**, *11*, 133. [CrossRef]
41. Yao, S.; Gao, G.; Gao, Z. On multi-axis motion synchronization: The cascade control structure and integrated smc–adrc design. *ISA Trans.* **2021**, *109*, 259–268. ISBN 0019-0578. [CrossRef]
42. Yasin, A.R.; Ashraf, M.; Bhatti, A.I. A Novel Filter Extracted Equivalent Control Based Fixed Frequency Sliding Mode Approach for Power Electronic Converters. *Energies* **2019**, *12*, 853. [CrossRef]
43. Gao, M.; Wang, D.; Li, Y.; Yuan, T. Fixed frequency pulse-width modulation based integrated sliding mode controller for phase-shifted full-bridge converters. *IEEE Access* **2017**, *6*, 2181–2192. ISBN 2169-3536. [CrossRef]
44. Repecho, V.; Biel, D.; Arias, A. Fixed switching period discrete-time sliding mode current control of a PMSM. *IEEE Trans. Ind. Electron.* **2017**, *65*, 2039–2048. ISBN 0278-0046. [CrossRef]
45. Chincholkar, S.H.; Jiang, W.; Chan, C.Y. An improved PWM-based sliding-mode controller for a DC–DC cascade boost converter. *IEEE Trans. Circuits Syst. II Express Briefs* **2017**, *65*, 1639–1643. ISBN 1549-7747. [CrossRef]
46. Li, J.; Liu, Z.; Su, Q. Improved adaptive backstepping sliding mode control for a three-phase PWM AC–DC converter. *IET Control Theory Appl.* **2019**, *13*, 854–860. ISBN 1751-8644. [CrossRef]
47. Repecho, V.; Biel, D.; Olm, J.M.; Colet, E.F. Switching frequency regulation in sliding mode control by a hysteresis band controller. *IEEE Trans. Power Electron.* **2016**, *32*, 1557–1569. ISBN 0885-8993. [CrossRef]

48. Ortiz-Castrillón, J.R.; Mejía-Ruiz, G.E.; Muñoz-Galeano, N.; López-Lezama, J.M.; Cano-Quintero, J.B. A sliding surface for controlling a semi-bridgeless boost converter with power factor correction and adaptive hysteresis band. *Appl. Sci.* **2021**, *11*, 1873. ISBN 2076-3417. [CrossRef]

49. Hoyos, F.E.; Candelo-Becerra, J.E.; Hoyos Velasco, C.I. Application of zero average dynamics and fixed point induction control techniques to control the speed of a DC motor with a buck converter. *Appl. Sci.* **2020**, *10*, 1807. ISBN 2076-3417. [CrossRef]

50. Muñoz, J.G.; Angulo, F.; Angulo-Garcia, D. Zero average surface controlled boost-flyback converter. *Energies* **2020**, *14*, 57. ISBN 1996-1073. [CrossRef]

51. Yasin, A.R.; Ashraf, M.; Bhatti, A.I.; Uppal, A.A. Fixed frequency sliding mode control of renewable energy resources in DC micro grid. *Asian J. Control* **2019**, *21*, 2074–2086. ISBN 1561-8625. [CrossRef]

52. Yasin, A.R.; Ashraf, M.; Bhatti, A.I. Fixed frequency sliding mode control of power converters for improved dynamic response in DC micro-grids. *Energies* **2018**, *11*, 2799. ISBN 1996-1073. [CrossRef]

53. Li, Z.; Wang, X.; Kong, M.; Chen, X. Bidirectional harmonic current control of brushless doubly fed motor drive system based on a fractional unidirectional converter under a weak grid. *IEEE Access* **2021**, *9*, 19926–19938. ISBN 2169-3536. [CrossRef]

54. Tan, S.C.; Lai, Y.M.; Tse, C.K.; Cheung, M.K. A fixed-frequency pulsewidth modulation based quasi-sliding-mode controller for buck converters. *IEEE Trans. Power Electron.* **2005**, *20*, 1379–1392. ISBN 0885-8993. [CrossRef]

55. Khan, I.; Bhatti, A.I.; Arshad, A.; Khan, Q. Robustness and Performance Parameterization of Smooth Second Order Sliding Mode Control. *Int. J. Control Autom. Syst.* **2016**, *14*, 681–690. [CrossRef]

56. Abrishamifar, A.; Ahmad, A.; Mohamadian, M. Fixed switching frequency sliding mode control for single-phase unipolar inverters. *IEEE Trans. Power Electron.* **2011**, *27*, 2507–2514. ISBN 0885-8993. [CrossRef]

57. Ye, J.; Malysz, P.; Emadi, A. A fixed-switching-frequency integral sliding mode current controller for switched reluctance motor drives. *IEEE J. Emerg. Sel. Top. Power Electron.* **2014**, *3*, 381–394. ISBN 2168-6777. [CrossRef]

58. Huerta, S.C.; Alou, P.; Garcia, O.; Oliver, J.A.; Prieto, R.; Cobos, J. Hysteretic mixed-signal controller for high-frequency DC–DC converters operating at constant switching frequency. *IEEE Trans. Power Electron.* **2012**, *27*, 2690–2696. ISBN 0885-8993. [CrossRef]

59. Vidal-Idiarte, E.; Marcos-Pastor, A.; Garcia, G.; Cid-Pastor, A.; Martinez-Salamero, L. Discrete-time sliding-mode-based digital pulse width modulation control of a boost converter. *IET Power Electron.* **2015**, *8*, 708–714. ISBN 1755-4535. [CrossRef]

60. Yan, W.T.; Ho, C.N.M.; Chung, H.S.H.; Au, K.T. Fixed-frequency boundary control of buck converter with second-order switching surface. *IEEE Trans. Power Electron.* **2009**, *24*, 2193–2201. ISBN 0885-8993. [CrossRef]

61. Venkateswaran, R.; Yesudhas, A.A.; Lee, S.R.; Joo, Y.H. Integral sliding mode control for extracting stable output power and regulating DC-link voltage in PMVG-based wind turbine system. *Int. J. Electr. Power Energy Syst.* **2023**, *144*, 108482. [CrossRef]

62. Armghan, H.; Yang, M.; Armghan, A.; Ali, N. Double integral action based sliding mode controller design for the back-to-back converters in grid-connected hybrid wind-PV system. *Int. J. Electr. Power Energy Syst.* **2021**, *127*, 106655. [CrossRef]

63. Riaz, M.; Yasin, A.R.; Arshad Uppal, A.; Yasin, A. A novel dynamic integral sliding mode control for power electronic converters. *Sci. Prog.* **2021**, *104*, 00368504211044848. ISBN 0036-8504. [CrossRef]

64. Khan, Q.; Bhatti, A.I.; Iqbal, S.; Iqbal, M. Dynamic integral sliding mode for MIMO uncertain nonlinear systems. *Int. J. Control Autom. Syst.* **2011**, *9*, 151–160. [CrossRef]

65. Omran, A.; Lucchesi, A.; Smith, D.; Alaswad, A.; Amiri, A.; Wilberforce, T.; Sodré, J.R.; Olabi, A.G. Mathematical model of a proton-exchange membrane (PEM) fuel cell. *Int. J. Thermofluids* **2021**, *11*, 100110. [CrossRef]

66. Baschuk, J.J.; Li, X. A general formulation for a mathematical PEM fuel cell model. *J. Power Sources* **2005**, *142*, 134–153. [CrossRef]

67. Xue, X.D.; Cheng, K.W.E.; Sutanto, D. Unified mathematical modelling of steady-state and dynamic voltage–current characteristics for PEM fuel cells. *Electrochim. Acta* **2006**, *52*, 1135–1144. ISBN 0013-4686. [CrossRef]

 *electronics*

*Article*

# Design of Sensorless Speed Control System for Permanent Magnet Linear Synchronous Motor Based on Fuzzy Super-Twisted Sliding Mode Observer

Zheng Li *, Jinsong Wang, Shaohua Wang, Shengdi Feng, Yiding Zhu and Hexu Sun *

School of Electrical Engineering, Hebei University of Science and Technology, Shijiazhuang 050018, China; wangjingsong@stu.hebust.edu.cn (J.W.); w18262637195@163.com (S.W.); fsdlyt901412@163.com (S.F.); zhuyiding@stu.hebust.edu.cn (Y.Z.)
* Correspondence: lizheng@hebust.edu.cn (Z.L.); hxsun@hebust.edu.cn (H.S.); Tel.: +86-311-81668722 (Z.L.)

**Abstract:** To improve the tracking capability and sensorless estimation accuracy of a permanent magnet linear synchronous motor (PMLSM) control system, a sensorless control system based on a continuous terminal sliding mode controller (CT-SMC) and fuzzy super-twisted sliding mode observer (F-ST-SMO) was designed. Compared with a conventional slide mode control, CT-SMC can reach the equilibrium point in limited time to ensure the continuity of control and achieve fast tracking of reference speed. Based on the PMLSM design of F-ST-SMO, a super-twisted sliding mode algorithm is used to replace the traditional first order sliding mode algorithm. Meanwhile, fuzzy rules are introduced to adjust the sliding mode gain adaptively, which replaces the fixed gain of traditional SMO and reduces chattering of the system. Finally, the effectiveness and superiority of the designed control system are proven by simulation and experiment.

**Keywords:** permanent magnet synchronous linear motor; continuous terminal sliding mode; sliding mode control; fuzzy controller; super-twisted sliding mode observer

**Citation:** Li, Z.; Wang, J.; Wang, S.; Feng, S.; Zhu, Y.; Sun, H. Design of Sensorless Speed Control System for Permanent Magnet Linear Synchronous Motor Based on Fuzzy Super-Twisted Sliding Mode Observer. *Electronics* **2022**, *11*, 1394. https://doi.org/10.3390/electronics11091394

Academic Editor: Hamid Reza Karimi

Received: 26 March 2022
Accepted: 24 April 2022
Published: 27 April 2022

**Publisher's Note:** MDPI stays neutral with regard to jurisdictional claims in published maps and institutional affiliations.



## 1. Introduction

PMLSM has been widely used in modern industries such as robotics and military due to its easy structure, high reliability and high efficiency [1,2]. To solve the limitations of a conventional PI control, many new control algorithms such as active disturbance rejection control, adaptive control, neural network control, and SMC have been proposed. Among them, SMC is a kind of nonlinear control algorithm with strong robustness. This method is insensitive to perturbation and has the advantages of quick response. It is applied to PMLSM to enhance the system control capability. However, the robustness of the SMC is achieved by using a large switching control gain, which often produces chattering. References [3,4] added an active disturbance rejection speed control in fast terminal sliding mode, which improved the robustness of the system and the ability to track a given speed, but the speed overshoot became larger, and the control input would have a singular problem. To reduce system jitter, Ref. [5] designed an adaptive law to dynamically adjust the switching gain of the system. While satisfying the control accuracy of the system, the system jitter was reduced. The authors of [6,7] combined a sliding mode control with direct torque theory, which improved the dynamic response capability of the system, but it increased the switching gain of the system and strengthened the system chattering. According to [8,9], in order to reduce the buffeting issue, the saturation function took the place of the sign function, the system switching was more stable, and the chattering of the system was reduced, but the sliding mode switching was difficult to achieve, and the anti-interference ability was weakened.

Meanwhile, in most high-performance servo control systems, the position and speed of the mover are critical parameters. In the PMLSM vector control, it is usually necessary to install a mechanical sensor on the motor to feedback the speed of the motor mover in real time, forming a closed-loop operation of the motor control system. While improving the detection accuracy of the system, the mechanical sensor also increases the burden of the system. To reduce the burden and cost of the system, a sensorless control strategy is necessary for its application in PMLSM [10]. Commonly used sensorless algorithms include Kalman filter algorithm, model reference adaptive algorithm, Sliding Mode Observer algorithm, etc. Among them, the Kalman filter can estimate the system state online and then realize the real-time monitoring of the system, but the amount of calculation is too large, and the adjustment time of the system becomes longer [11]. The model reference adaptive algorithm realizes the identification of the motor parameters through the appropriate adaptive law. Although it improves the robustness of the system, the design parameters are too many, the system becomes complex, and the dynamic response capability becomes weak [12].

Because SMO is independent of motor parameters, it has strong robustness to system disturbances. Compared with other speed sensorless control methods such as the Kalman filter means and model reference adaptive control algorithm, this method is simple in design and requires less computation and more dynamic responsiveness, with better advantages. The authors in [13,14] used an improved index convergence law to weaken the system buffeting caused by the sliding mode observer. Although it can achieve purpose to a certain extent, when the error change is zero, the sliding mode control rate is also zero, which is disadvantageous to the control of the system. The authors of [15] proposed an adaptive SMO, which has good robustness. The authors of [16,17] designed an extended Kalman filter to obtain continuous back EMF and to improve observation accuracy, but the amount of calculation is too big, which increases the control cost of the system. The authors of [18–20] proposed a cascade sliding mode observer, which improved the robustness of the system, but the cascade made the tracking ability of the system worse, resulting in a certain phase difference.

Therefore, based on the theory of PMLSM, a control system combining CT-SMC and F-ST-SMO is designed to optimize the overall control system. The CT-SMC solves the singularity problem of the terminal sliding mode and makes the approach process of the algorithm have a time limit through the stability condition constraint, standardizes the approach trajectory, and makes the system reach the sliding mode surface in a certain time. F-ST-SMO using higher-order features keeps the continuity of the output, so as to weaken the hf switching chattering of sliding mode. Thus, the introduction of fuzzy rules can solve the upper bound of the boundary function in a selected control algorithm in the actual problem, in which is difficult to dynamically adjust the sliding mode gain coefficient of the fuzzy rules to reduce chattering near the sliding mode surface. The simulation and experiment prove that the designed system simplifies the mechanical structure of the control system and improves the dynamic performance and control precision. The main innovations of this paper are as follows:

(1) Based on the theoretical basis of SMC and taking PMLSM as the control object, a control system combining CT-SMC and F-ST-SMO is designed.

(2) To solve the problem of the large error of traditional SMO observations, a hyperdistortion algorithm is introduced to maintain the continuity of output to weaken chattering caused by high-frequency switching in sliding mode. Fuzzy rules are introduced to dynamically adjust the sliding mode gain coefficient to reduce chattering near the sliding mode surface.

(3) System verification. The dynamic performance of the designed control system is verified by comparing the CTSMC control system with a traditional SMC and PI control system. The observation performance of F-ST-SMO is compared with that of traditional SMO, and the error analysis is made with the data of the mechanical sensor.

## 2. Design of Terminal Sliding Mode Speed Controller

### 2.1. PMLSM Mathematical Model

The mathematical model for establishing the PMLSM in the synchronous rotating coordinate system $d - q$ is as follows:

$$\begin{cases} \frac{di_d}{dt} = -\frac{R_s}{L_d}i_d + \frac{\pi L_q}{\tau L_d}vi_q + \frac{u_d}{L_d} \\ \frac{di_q}{dt} = -\frac{R_s}{L_q}i_q - \frac{\pi L_d}{\tau L_q}vi_d + \frac{u_q}{L_q} - \frac{\pi\psi_f}{\tau L_q}v \end{cases} \tag{1}$$

where $R_s$ is the stator resistance; $\tau$ is the PMLSM pole pitch; $v$ is the speed of the linear motor; $u_d$ and $u_q$ are the voltage of the $d - q$ axis; $i_d$ and $i_q$ are the current of the $d - q$ axis; $L_d$ and $L_q$ are the inductance components of the $d - q$ axis; $\psi_f$ is the permanent magnet flux linkage.

Since $L_d = L_q$ in the PMLSM used, the thrust equation can be written as

$$F_{em} = p_n\frac{3\pi}{2\tau}\psi_f i_q \tag{2}$$

where $F_{em}$ is the thrust of the linear motor, and $p_n$ is the number of pole pairs of the linear motor.

The mechanical motion equation of PMLSM is as follows:

$$m\frac{dv}{dt} = F_{em} - f - Bv \tag{3}$$

where $m$ is the quality of the mover, $B$ is the viscous friction factor, and $f$ is the system disturbance. Simultaneously, formulas (2) and (3) can be obtained

$$\frac{dv}{dt} = a_M i_q + b_M v + c_M f \tag{4}$$

where $a_M = \frac{1}{m}p_n\frac{3\pi}{2\tau}\psi_f$; $b_M = -\frac{B}{m}$; $c_M = -\frac{1}{m}$.

### 2.2. Design of CT-SMC

The velocity loop controller is designed using a first-order CT-SMC algorithm. From the linear motor equation of motion, the first-order velocity differential equation can be written as

$$\begin{aligned} \dot{v} &= a_M i_q + b_M v + c_M f \\ &= a_M i_q{}^* + b_M v + c_M f - a_M(i_q{}^* - i_q) \\ &= bi_q{}^* + d \end{aligned} \tag{5}$$

where

$$\begin{cases} b = a_M = \frac{1}{m}p_n\frac{3\pi}{2\tau}\psi_f \\ d = b_M v + c_M f - a_M(i_q^* - i_q) \end{cases} \tag{6}$$

$d$ can be regarded as the lumped disturbance of the system. The derivative of $d$ in the system is bounded and satisfies the following conditions:

$$\left|\ddot{d}\right| \leq k_{ed} \tag{7}$$

where $k_{ed}$ is an ordinary constant and $k_{ed} > 0$.

Let $v_r$ be the specified speed of the system, then the speed tracking error is defined as:

$$e = v_r - v \tag{8}$$

Calculate the first derivative of the velocity tracking error, and substitute Equation (5), it can be deduced that

$$\dot{e} = \dot{v}_r - \dot{v} = \dot{v}_r - bi_q^* - d \tag{9}$$

Terminal sliding surface for design is as follows:

$$s = \dot{e} + c\,\mathrm{sgn}(e)|e|^{p/q} \tag{10}$$

where $c > 0, 0 < p/q < 1$.

The speed control law based on CT-SMC is designed as follows:

$$\begin{cases} i_q^* = b^{-1}(u_{eq} + u_v) \\ u_{eq} = \dot{v}_r + c\,\mathrm{sgn}(e)|e|^{\alpha} \\ u_v = k_v \int_0^t \mathrm{sgn}(s)dt \end{cases} \tag{11}$$

where $k_v$ is an ordinary constant and $k_v > 0$, $t$ is the running time of the system $\alpha \in (0,1)$; $\mathrm{sgn}(?)$ is a symbolic function; $b = a_M = \frac{1}{m}p_n\frac{3\pi}{2\tau}\psi_f$ is the control gain. $u_{eq}$ and $u_v$ are equivalent control laws.

### 2.3. Stability Proof of CT-SMC

If system (5) satisfies $k_v > k_{ed}$, under the action of the control law (10), the velocity error of the system will converge to the equilibrium point within a finite time.

**Proof.** According to Equation (9), the sliding surface (10) can be written in the following form:

$$s = \dot{e} + c\,\mathrm{sgn}(e)|e|^{p/q} = \dot{v}_r - bi_q^* - d + c\,\mathrm{sgn}(e)|e|^{p/q} \tag{12}$$

Substituting formula (11) into (12), we can obtain:

$$\begin{aligned} s\ &= \dot{v}_r - bi_q^* - d + c\,\mathrm{sgn}(e)|e|^{p/q} \\ &= c\,\mathrm{sgn}(e)|e|^{p/q} + \dot{v}_r - (u_{eq} + u_v) - d \\ &= c\,\mathrm{sgn}(e)|e|^{p/q} + \dot{v}_r - (\dot{v}_r + c\,\mathrm{sgn}(e)|e|^{p/q} + u_v) - d \\ &= -u_v - d \end{aligned} \tag{13}$$

The Lyapunov function is used to judge the stability of CT-SMC, and the constructed function is

$$V = \frac{1}{2}s^2 \tag{14}$$

Taking the first derivative of it and substituting it into Equation (13), we can obtain

$$\dot{V} = s\dot{s} = s[-\dot{u}_v - \dot{d}] \tag{15}$$

Substituting Equation (11) into the above equation, we can obtain

$$\begin{aligned} \dot{V} &= s(-\dot{u}_v - \dot{d}) = s(-k_v\mathrm{sgn}(s) - \dot{d}) \\ &= -k_v|s| - \dot{d}s \end{aligned} \tag{16}$$

Formula (16) can be written in the following form:

$$\begin{aligned} \dot{V} &= -k_v|s| - \dot{d}s \le -k_v|s| + \left|\dot{d}\right||s| \\ &= -[k_v - \left|\dot{d}\right|]|s| \le -[k_v - k_{cd}]|s| \\ &= -\sqrt{2}(k_v - k_{ed})V^{\frac{1}{2}} \end{aligned} \tag{17}$$

According to Equation (17), if $k_v > k_{ed}$ it is true. Then, $\dot{V} < 0$ is established. It shows that the velocity error will reach the sliding surface $s = 0$ in a limited time. In this way, there can be a first-order nonlinear differential equation with finite time convergence: $s = \dot{e} + c\,\mathrm{sgn}(e)|e|^{\alpha} = 0$.

When $c > 0$, the velocity deviation will converge to the equilibrium state in a finite time along the sliding mode surface $s = 0$.

When the fractional power is reduced to integer power 1, the sliding surface in this paper is reduced to:

$$s = \dot{e} + ce \tag{18}$$

The controller degenerates to:

$$\begin{cases} i_q^* = b^{-1}(u_{eq} + u_v) \\ u_{eq} = \dot{v}_r + ce \\ u_v = k_v \int_0^t \text{sgn}(s)dt \end{cases} \tag{19}$$

According to the above design, the control structure based on the CT-SMC method can be drawn as shown in Figure 1. The red PMLSM module represents the current closed-loop structure including the permanent magnet synchronous linear motor and other components in Figure 1. $\square$

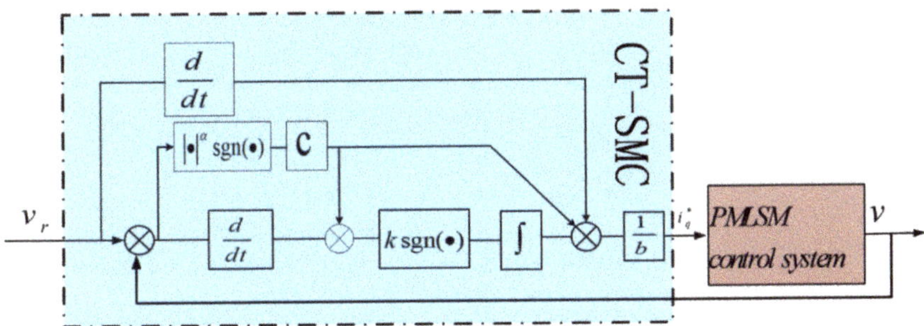


**Figure 1.** System block diagram of PMLSM based on CT-SMC.

## 3. Design of F-ST-SMO

### 3.1. Traditional SMO

The electrical equation of PMLSM in the $\alpha\beta$ coordinate system is as follows:

$$\begin{cases} \frac{di_\alpha}{dt} = \frac{1}{L}(-Ri_\alpha + u_\alpha - E_\alpha) \\ \frac{di_\beta}{dt} = \frac{1}{L}(-Ri_\beta + u_\beta - E_\beta) \end{cases} \tag{20}$$

where $E_\alpha = -\frac{\pi}{\tau}v\psi_f \sin\theta$, $E_\beta = -\frac{\pi}{\tau}v\psi_f \cos\theta$ can be regarded as $\alpha\beta$ induced electromotive force in coordinate system. To acquire the extended back EMF of the estimated value, the SMO can be designed as

$$\begin{cases} \frac{d\hat{i}_\alpha}{dt} = \frac{1}{L}(-R\hat{i}_\alpha + u_\alpha - z_\alpha) \\ \frac{d\hat{i}_\beta}{dt} = \frac{1}{L}(-R\hat{i}_\beta + u_\beta - z_\beta) \end{cases} \tag{21}$$

$$\begin{cases} z_\alpha = k\text{sgn}(\hat{i}_\alpha - i_\alpha) \\ z_\beta = k\text{sgn}(\hat{i}_\beta - i_\beta) \end{cases} \tag{22}$$

where $\hat{i}_\alpha, \hat{i}_\beta$ is the observed value of current; $k$ is the traditional SMO gain.

Subtracting from Equations (20) and (21), the current error state equation is

$$\begin{cases} \frac{d\tilde{i}_\alpha}{dt} = \frac{1}{L}(-R\tilde{i}_\alpha - z_\alpha + E_\alpha) \\ \frac{d\tilde{i}_\beta}{dt} = \frac{1}{L}(-R\tilde{i}_\beta - z_\beta + E_\beta) \end{cases} \tag{23}$$

where $\tilde{i}_\alpha = \hat{i}_\alpha - i_\alpha, \tilde{i}_\beta = \hat{i}_\beta - i_\beta$ is the current observation deviation.

SMO is used to estimate the current, and its sliding mode surface function is as follows

$$\tilde{i} = [\tilde{i}_a \tilde{i}_\beta]^T = 0 \tag{24}$$

SMO enters sliding mode when the following conditions are met

$$\tilde{i}^T \dot{\tilde{i}} < 0 \tag{25}$$

When the sliding mode gain satisfies Equation (25), then

$$\tilde{i} = \dot{\tilde{i}} = 0 \tag{26}$$

Substituting Equation (26) into Equation (23), we can obtain

$$E = [k\mathrm{sgn}(\hat{i}_a - i_a) k\mathrm{sgn}(\hat{i}_\beta - i_\beta)]^T \tag{27}$$

According to Equation (27), it is known that the back EMF estimation is associated with the high frequency switching signal. However, the estimation method based on the arc tangent function directly substitutes the high-frequency switching signal into the division operation of the arc tangent function, resulting in the phenomenon of high-frequency chattering.

### 3.2. Super-Twisted Control Algorithm

For the traditional first-order sliding mode observer, the estimated back EMF has a chattering phenomenon due to the influence of the sign function, which leads to a tracking error in the estimated mover velocity. Although the chattering can be reduced by low-pass filtering, it will cause a phase delay. To this end, an ST algorithm is introduced to solve the chattering phenomenon caused by the traditional SMO.

The ST algorithm can be shown in formula (28):

$$\begin{cases} \frac{dx}{dt} = k^* |\tilde{x}|^{\frac{1}{2}} \mathrm{sgn}(\tilde{x}) + x_m \\ \frac{dx_m}{dt} = k^* \mathrm{sgn}(\tilde{x}) \end{cases} \tag{28}$$

where $x$ is the state variable, $\tilde{x}$ is the deviation between the approximated value and the true value, $x_m$ is a custom intermediate variable, $k^*$ is the ST sliding mode gain.

The state variable $x$ in the ST algorithm is replaced by the current signal $\hat{i}_a$ and $\hat{i}_\beta$ estimated by PMLSM, and the SMO of PMLSM based on the ST algorithm is obtained, as shown in Equation (29).

$$\begin{cases} \frac{d\hat{i}_a}{dt} = \frac{1}{L}(-R\hat{i}_a + u_a - k_f |\tilde{i}_a|^{\frac{1}{2}} \mathrm{sgn}(\tilde{i}_a) - \int k_s \cdot k_f \mathrm{sgn}(\tilde{i}_a) dt) \\ \frac{d\hat{i}_\beta}{dt} = \frac{1}{L}(-R\hat{i}_\beta + u_\beta - k_f |\tilde{i}_\beta|^{\frac{1}{2}} \mathrm{sgn}(\tilde{i}_\beta) - \int k_s \cdot k_f \mathrm{sgn}(\tilde{i}_\beta) dt) \end{cases} \tag{29}$$

where $k_f$ is the boundary function, and $k_s$ is a small constant, which is used to reduce the influence caused by the error differential of fuzzy control input.

The current deviation equation acquired by subtracting Equation (20) from Equation (29) is:

$$\begin{cases} \frac{d\tilde{i}_a}{dt} = \frac{1}{L}(-R\tilde{i}_a - k_f |\tilde{i}_a|^{\frac{1}{2}} \mathrm{sgn}(\tilde{i}_a) - \int k_s \cdot k_f \mathrm{sgn}(\tilde{i}_a) dt + E_a) \\ \frac{d\tilde{i}_\beta}{dt} = \frac{1}{L}(-R\tilde{i}_\beta - k_f |\tilde{i}_\beta|^{\frac{1}{2}} \mathrm{sgn}(\tilde{i}_\beta) - \int k_s \cdot k_f \mathrm{sgn}(\tilde{i}_\beta) dt + E_\beta) \end{cases} \tag{30}$$

where $\tilde{i}_a = \hat{i}_a - i_a, \tilde{i}_\beta = \hat{i}_\beta - i_\beta$ is the deviation of current observation.

When ST-SMO is stable, the estimated value is equal to the actual value, that is, the PMLSM current estimation error and change rate are approximately zero $(\widetilde{i}_{\alpha\beta} = \dot{\widetilde{i}}_{\alpha\beta} = 0)$. At this point, the back EMF of the PMLSM estimated by the ST-SMO can be obtained:

$$\begin{cases} E_a = k_f \left|\widetilde{i}_\alpha\right|^{\frac{1}{2}} \mathrm{sgn}(\widetilde{i}_\alpha) + \int k_s \cdot k_f \mathrm{sgn}(\widetilde{i}_\alpha) dt \\ E_\beta = k_f \left|\widetilde{i}_\beta\right|^{\frac{1}{2}} \mathrm{sgn}(\widetilde{i}_\beta) + \int k_s \cdot k_f \mathrm{sgn}(\widetilde{i}_\beta) dt \end{cases} \tag{31}$$

In order to improve the chattering problem in SMC, the ST control algorithm is selected to ensure the continuity of the output. The algorithm can weaken the chattering caused by the first-order sliding mode control and improve the dynamic response and anti-interference ability of the system. However, the ST control algorithm relies too much on the upper bound of the boundary function, which is hard to obtain in practice. In addition to the serious influence of the sign function on the buffeting phenomenon of a sliding mode observer, there is also sliding mode gain. In a general sliding mode system, in order to maintain the rapid response ability of the system, a larger sliding mode gain is usually selected, but this will also produce a larger buffeting. When the moving point is close to the sliding mode surface, a small gain is enough. Therefore, if the sliding mode gain can be adjusted adaptively according to the position of the moving point relative to the sliding mode surface, the buffeting of the system can be effectively suppressed. In order to realize the adaptive adjustment of the sliding mode gain, a fuzzy control rule is introduced.

### 3.3. Design of Fuzzy Controller

Fuzzy control has strong adaptability to external disturbances because it does not completely rely on mathematical models, but on fuzzy rules. It is especially suitable for nonlinear control systems.

The boundary function $k_f$ is estimated according to the sliding mode reachable condition. The input variables of the fuzzy control system are $\widetilde{i}_{a\beta}$ and $\dot{\widetilde{i}}_{a\beta}$, and the output variable is $k_f$. If $\widetilde{i}_{a\beta}$ is large, it indicates that there is a large difference between $\hat{i}_{a\beta}$ and $i_{a\beta}$, and the moving point is far from the sliding mode surface, and then $k_f$ should increase. If $\widetilde{i}_{a\beta}$ is small, it means that the motion point is close to the sliding surface, and then $k_f$ should decrease. If $\dot{\widetilde{i}}_{a\beta}$ is large, it indicates that the moving point is rapidly moving away from or close to the sliding mode surface, and the gain needs to be increased to strengthen the control effect; otherwise, the gain needs to be reduced.

The input variables are defined in the domain of {−0.002 0.002}, and the output variables in the domain of {1000 1400}. The input of fuzzy language is {NB (negative big), NS (negative small), Z(zero), PS (positive small), PB (positive big)}, and the fuzzy language value of the output is {PS (positive small), S(small), M(medium), B(big), PB (positive big)}. Control rules obtained by designing $k_f$ are shown in Table 1.

**Table 1.** Control rules.

| $\widetilde{i}_{a\beta}$ | $\dot{i}_{\alpha\beta}$ | | | | |
|---|---|---|---|---|---|
| | **NB** | **NS** | **ZO** | **PS** | **PB** |
| NB | PB | PB | B | B | M |
| NS | PB | B | B | M | M |
| ZO | B | M | M | S | S |
| PS | S | M | M | B | B |
| PB | M | B | B | PB | PB |

### 3.4. Construction of F-ST-SMO Model

Based on the current model of the PMLSM in the $\alpha\beta$ coordinate system, such as formula (32):

$$\dot{i}_{\alpha\beta} = \frac{1}{L}(-Ri_{\alpha\beta} + u_{\alpha\beta} - E_{\alpha\beta}) \tag{32}$$

where $i_{\alpha\beta} = [i_\alpha \ i_\beta]^T$ is the stator $\alpha\beta$ axis current; $R$ is stator resistance; $L$ is the stator inductance; $u_{\alpha\beta} = [u_\alpha \ u_\beta]^T$ is the voltage of stator $\alpha\beta$ axis; $E_{\alpha\beta} = [E_\alpha \ E_\beta]^T$ is the inverse electromotive force of the linear electric motor. PMLSM current equation based on F-ST-SMO can be restated as:

$$\dot{\hat{i}}_{\alpha\beta} = \frac{1}{L}(-R\hat{i}_{\alpha\beta} + u_{\alpha\beta} - z_{\alpha\beta}) \tag{33}$$

where $\hat{i}_{\alpha\beta} = [\hat{i}_\alpha \ \hat{i}_\beta]^T$ is the estimation of stator $\alpha\beta$ axis current; $z_{\alpha\beta} = k_f\left|\tilde{i}_{\alpha\beta}\right|^{\frac{1}{2}}\mathrm{sgn}(\tilde{i}_{\alpha\beta}) + \int k_s \cdot k_f\mathrm{sgn}(\tilde{i}_{\alpha\beta})dt$ is the control law. Where $k_f$ is the boundary function, $k_f$ is derived from fuzzy control rules. The main task of the observer is to select a suitable control law to minimize the estimation error and finally estimate the motor stator $\alpha\beta$ axis back EMF. When the sliding mode switching function approaches the sliding mode switching surface, $\tilde{i}_{\alpha\beta} = \dot{\tilde{i}}_{\alpha\beta} = 0$ is obtained, and Equation (31) is obtained. To obtain a continuous estimate of the back EMF, a low-pass filter is required, as shown in the equation:

$$\hat{E}_{\alpha\beta} = \frac{\omega_c}{s + \omega_c}z_{\alpha\beta} \tag{34}$$

where $w_c = 2\pi f_c$, $f_c$ indicates the cut-off frequency of the low-pass filter.

To obtain the velocity information of the mover, the estimated value of velocity is expressed as:

$$\hat{\omega}_e = \frac{\sqrt{E_\alpha{}^2 + E_\beta{}^2}}{\psi_f} \tag{35}$$

where $\hat{\omega}_e = \frac{\pi}{\tau}\hat{v}$, and we can finally obtain an estimate of the velocity of the actuator.

To sum up, the control principle of F-ST-SMO is to first use the input voltage signal through the sliding mode observer designed by the PMLSM current state equation to obtain the current estimated value of the SMO, and then compare it with the actual PMLSM current signal. The error and rate of change of the current signal are obtained by making the difference, the control gain of the sliding mode observer is dynamically regulated through the fuzzy control rules, the estimated back EMF is obtained through the ST algorithm, and finally the mover speed is obtained. The principle of the F-ST-SMO is shown in Figure 2.

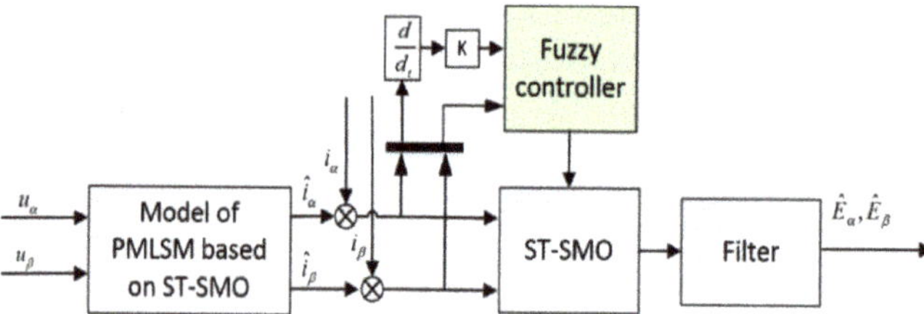


**Figure 2.** F-ST-SMO algorithm implementation diagram.

## 4. Simulation and Experimental Results

### 4.1. Comparison of Speed Controllers

The block diagram of the PMLSM control system designed in this paper is set up in MATLAB/SIMULINK, as shown in Figure 3. The motor driving parameters used are shown in Table 2, and the improved SMC is compared with the traditional PI and traditional SMC control system.

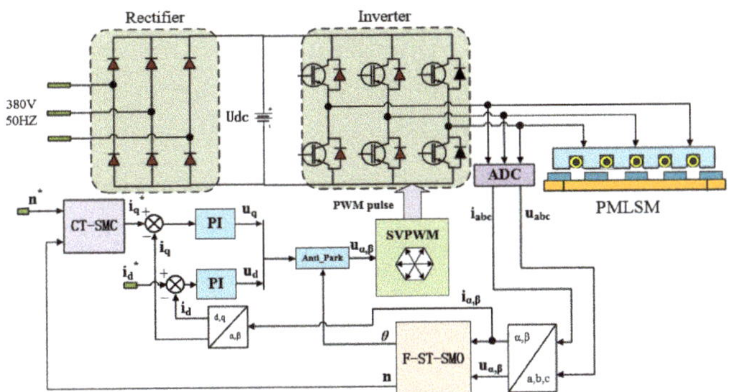

**Figure 3.** Control system block diagram.

**Table 2.** Main parameters of straight line.

| Parameter | Value |
|---|---|
| stator resistance Rs/$\Omega$ | 4.0 |
| d-q axis inductance Ldq/mH | 8.2 |
| Mover mass m/kg | 1.425 |
| Viscous friction coefficient B/N/m·s | 44 |
| Polar distance $\tau$/m | 0.016 |
| DC Bus Voltage U/V | 48 |

### 4.1.1. Constant Load, Varying Speed

To analyze the dynamic response performance of the designed control system, the traditional PI control and traditional SMC are set up as a comparison. The speed change of the given system is ($1\text{m/s} \rightarrow 2\text{m/s} \rightarrow 3\text{m/s}$) to observe the tracking effect of PMLSM for different speeds and the dynamic response performance when the speed changes. Figure 4 shows the comparison of the operation speed of the CT-SMC system, traditional PI control system and SMC system.

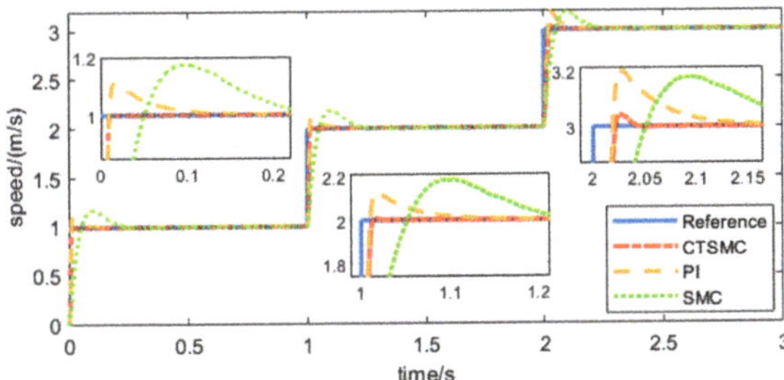

**Figure 4.** Speed comparison.

To analyze the speed contrast diagram, the CT-SMC system shown in this paper has almost no overshoot in the process from 0 to 1 m/s and from 1 to 2 m/s. Although there is overshoot in the process from 2 to 3 m/s, the overshoot is much smaller than that in the traditional PI control and traditional SMC system. In these three speed increases, the process from 2 to 3 m/s is the worst dynamic response time of CT-SMC designed in this paper. The CT-SMC system still shows superior control performance versus the traditional control. In the process of speed change, the adjustment time of CT-SMC is less than 0.05 s, the adjustment time of PI is more than 0.1 s, while the adjustment time of conventional SMC is more than 0.2 s.

Figure 5 shows the thrust variation comparison of the three control systems. Among the three control modes, it can be seen that the SMO system shows strong thrust fluctuation due to its inherent characteristics. The CT-SMC system designed in this paper has a similar trend of thrust change with the traditional PI control system, but when the speed changes, the thrust changes significantly, resulting in a long overshoot time.

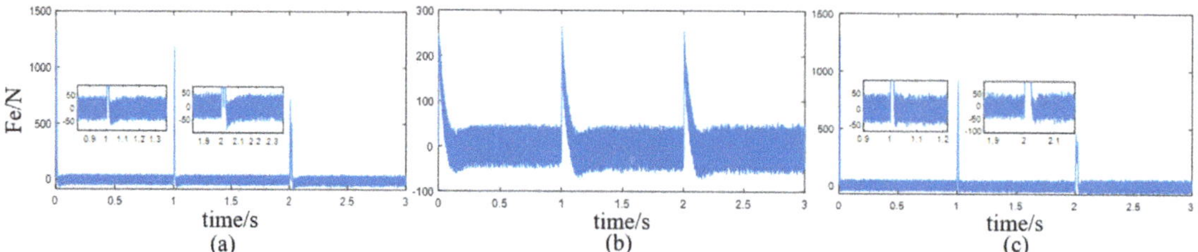

**Figure 5.** The thrust comparison of the three control system: (**a**) thrust change of the CT-SMC system; (**b**) thrust change of the PI control system; (**c**) thrust change of the SMC system.

### 4.1.2. Constant Speed, Varying Load

Given that the system speed is 1.5 m/s, and the running time is set as 1 s, random load is added into the control system as shown in Figure 6. Figure 7 shows the comparison of the running speed of the three systems. The overshoot and overshoot time of the CT-SMC control system are much less than those of the other two control systems.

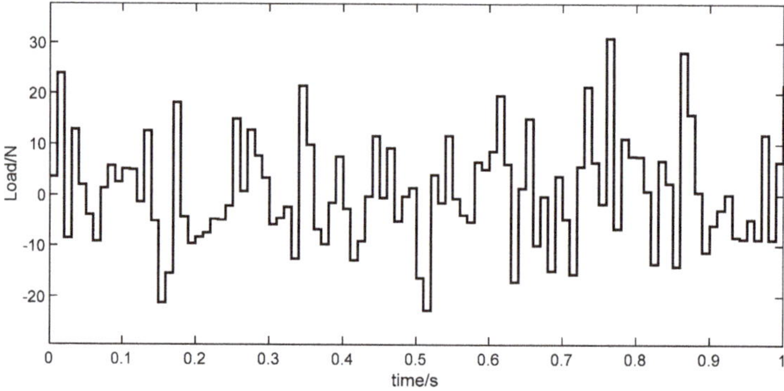

**Figure 6.** Random load added to the system.

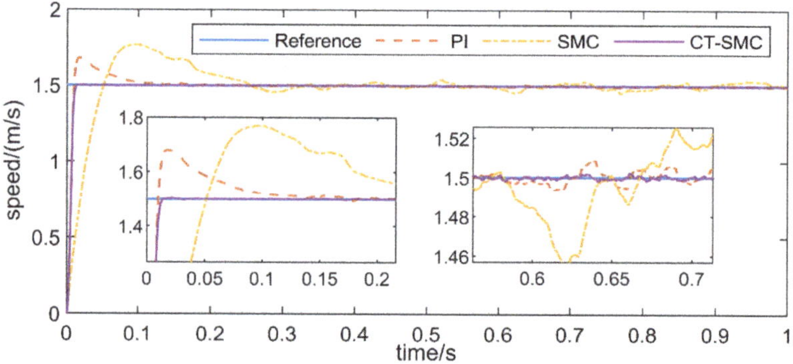


**Figure 7.** Comparison of running speed.

### 4.1.3. Observer Comparison

Through the simulation verification, the superior control performance of the CT-SMC system designed in this paper is proven. On the basis of the CT-SMC system, the traditional

SMO is added to compare with the designed F-ST-SMO. The speed of the given system is 1.5 m/s, and the speed rises to 2 m/s at 1 s, so as to analyze the tracking situation and dynamic response ability of the observer at different speeds.

It can be seen from Figure 8 that the traditional SMO observation speed will have large chattering, and the observer directly gives the error caused by chattering to the speed controller of the control system, resulting in a decrease in the overall accuracy of the control system. In the simulation diagram, the medium tracking speed fluctuates around the reference speed. In the F-ST-SMO system designed in this paper, due to the action of the fuzzy controller, the output changes to the sliding mode gain, which validly reduces the buffeting of the traditional SMO and has higher accuracy.

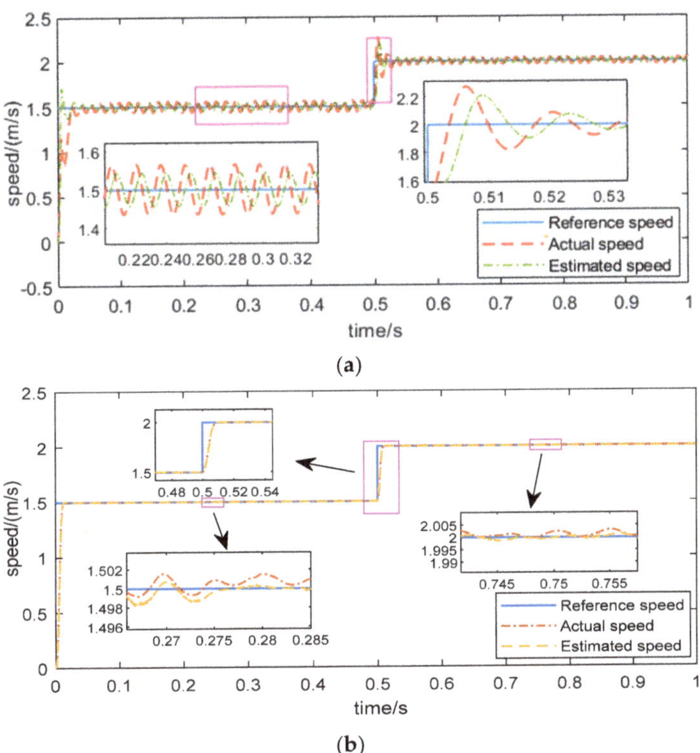

**Figure 8.** Comparison of observation effects of observers: (**a**) waveform of observed velocity and actual velocity of conventional SMO system; (**b**) waveform of observed velocity and actual velocity of THE F-ST-SMO system.

*4.2. Experiment*

Compared with the simulation environment, the experimental environment has more external disturbances, and there is a heating phenomenon in the operation of the electrical generator. Building an experimental platform for verification is conducive to further improving the control system, such that the designed algorithm can be better and faster applied in practical applications. The experimental platform as shown in Figure 9 is built in this paper, which is mainly composed of upper computer, servo driver, control card, grating sensor and precision linear motor. The on–off controlled electromagnetic weight is used as the load of the linear motor. During the operation of the linear motor, the sensor transmits the position signal and current signal to the control board for closed-loop control operation. The encoder is used to compare with the observer designed in this paper. The motor parameters selected in this paper are shown in Table 2.

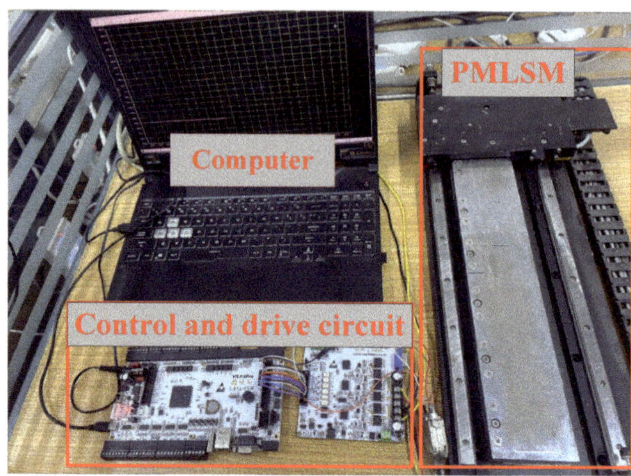


**Figure 9.** Experimental platform.

First, the proposed CT-SMC control system is built and compared with the traditional PI control system, and then the traditional SMC control system, the dynamic performance, and anti-disturbance performance of the controller are analyzed. The given speed of the system was set as 0.8–1.5 m/s. Figure 10 shows the dynamic effect under three control strategies. Figure 10a–c shows the comparison diagrams of the running speed and reference speed of the PI control system, SMC control system and CT-SMC control system, respectively. Figure 10d–f shows the comparison of overshoot, regulation time and error of the three control systems. It can be concluded from the analysis that the traditional SMC control system has more room for improvement compared with the current more mature PI control system. It has a larger overshoot and longer adjustment time. Through the improved CT-SMC, the overshoot and adjustment time are greatly reduced, and the error is also smaller.

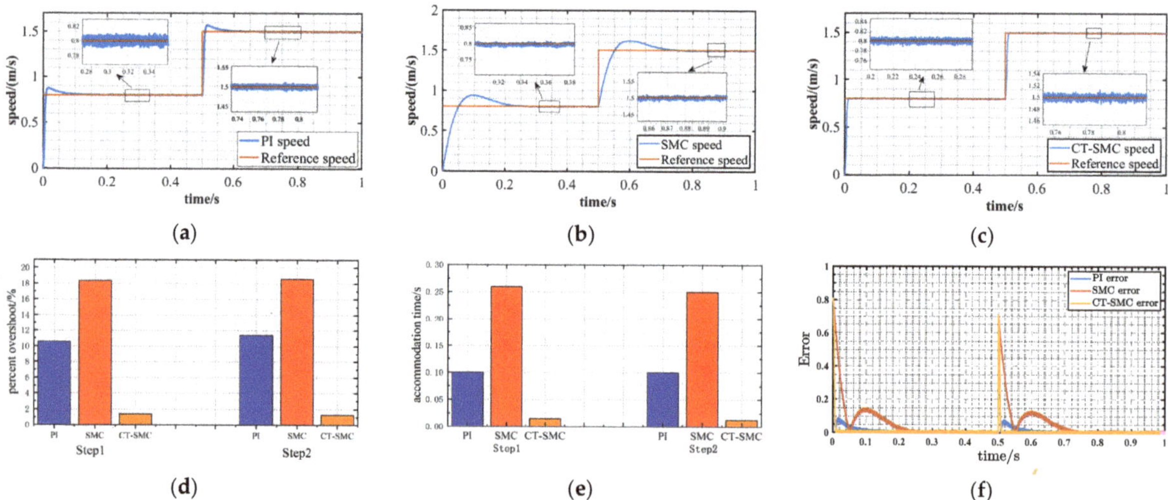

**Figure 10.** Dynamic performance comparison of the three control effects: (**a**) speed of PI control system; (**b**) speed of SMC control system; (**c**) speed of CT-SMC control system; (**d**) overshoot comparison of the three control systems; (**e**) adjustment time comparison of the three control systems; (**f**) error comparison of the three control systems.

Figure 11 shows the operation of the three control systems at a given speed of 1.2 m/s and with a 50 N load added at 0.4 s. Figure 11a–c shows the operating speed waveforms of the three control systems, and Figure 11d–f shows the speed errors of the three control systems after loading. According to the analysis, when the load was added, the fluctuation of the SMC control system decreased slightly compared with PI, but there was still a large fluctuation. The improved CT-SMC control system not only has good dynamic performance, but also has good anti-disturbance performance.

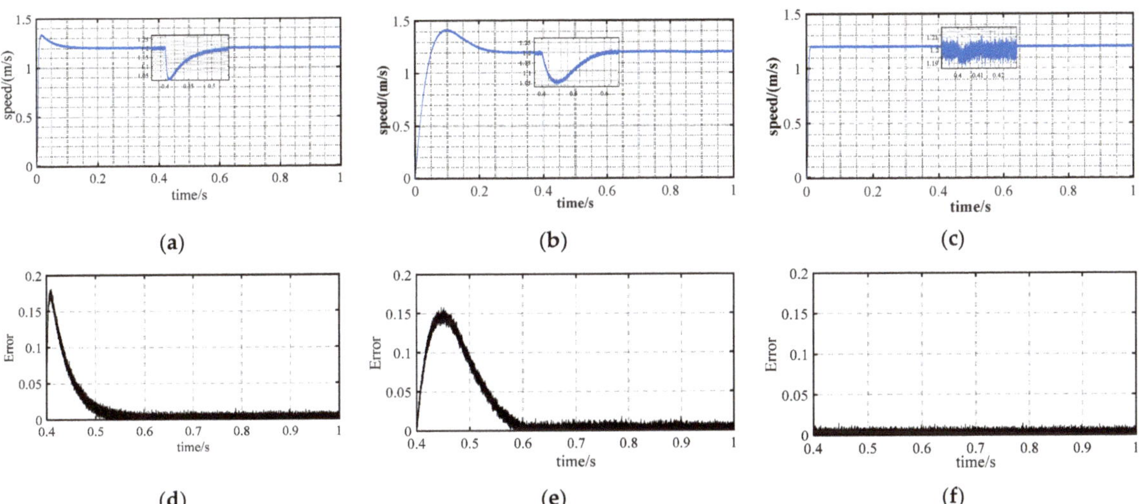

**Figure 11.** Under the three control strategies, the operation condition with load added at 0.4 s: (**a**) speed of PI control system; (**b**) speed of SMC control system; (**c**) speed of CT-SMC control system; (**d**) speed error of PI control system after 0.4 s; (**e**) speed error of SMC control system after 0.4 s; (**f**) speed error of CT-SMC control system after 0.4 s.

CT-SMC can have a good control effect in dynamic response and anti-disturbance performance. On the basis of selecting CT-SMC for the controller, traditional SMO and F-ST-SMO are added for speed observation. Figure 12 is the speed operation when the given system speed is 1.5 m, and the observed speed of the observer is fed back to the speed controller. Mechanical sensors are used for comparison with estimated speed. The traditional SMO has a large jitter and a large distortion in the speed-up stage, and the speed also has a large jitter after the speed reaches the reference speed. In the speed observation of F-ST-SMO, the observed speed is close to the real speed in both thethe speed rising stage and the speed stable operation stage. Compared with the traditional SMO, the F-ST-SMO is much smaller than the traditional SMO.

The experimental results show that the F-ST-SMO sensorless control system based on CT-SMO performs well in dynamic performance and anti-disturbance performance. The sensorless control system simplifies the mechanical structure of the control system and greatly improves the control accuracy compared with the traditional SMO.

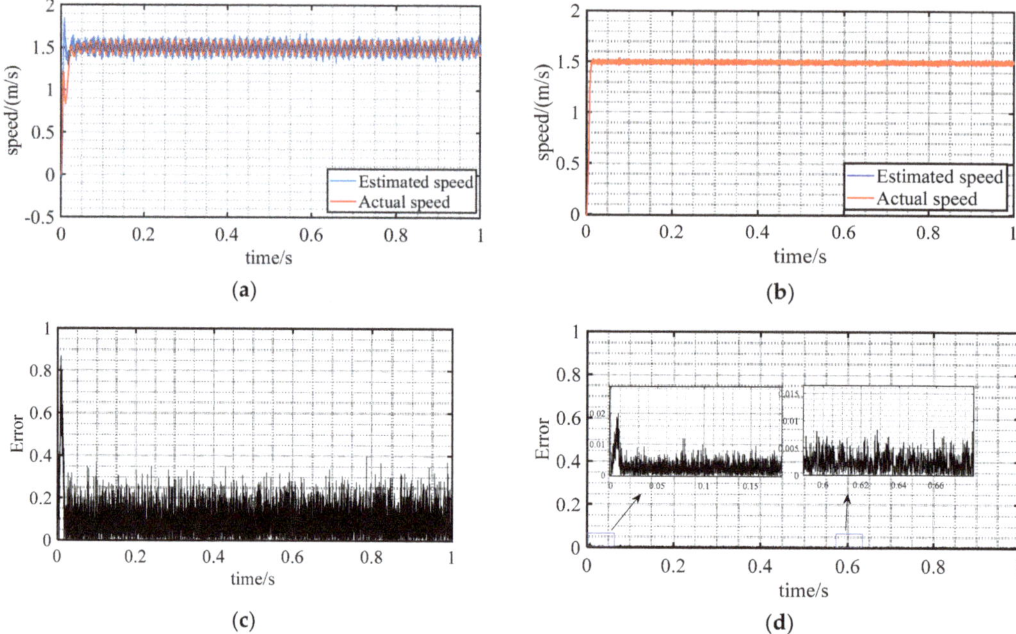

**Figure 12.** Comparison of the operation of the two observers; (**a**) comparison between the observation speed and the actual speed of conventional SMO; (**b**) comparison between the observation speed and the actual speed of F-ST-SMO; (**c**) observation error of conventional SMO; (**d**) observation error of F-ST-SMO.

## 5. Conclusions

In this paper, a CT-SMC and F-ST-SMO control system based on the PMLSM mathematical model is proposed. This method provides a solution for high-precision operation of a PMLSM sensorless system. The mechanical motion equations of PMLSM are analyzed, and a speed closed-loop controller is designed using a continuous terminal sliding mode control algorithm. The voltage equation of PMLSM was rewritten in the static coordinate system, and F-ST-SMO was designed to replace the traditional mechanical sensor. The simulation and experimental verification of the designed control system under variable speed and load conditions show that the dynamic performance of the designed control system is better, and the position tracking error is greatly reduced.

**Author Contributions:** Writing—review and editing, supervision, project administration, funding acquisition, Z.L.; writing—review and editing, supervision, project administration, funding acquisition, H.S. (Z.L. and H.S. contributed equally to this work as corresponding authors); methodology, software, writing—original draft preparation, J.W. and S.W.; validation, S.F. and Y.Z. All authors have read and agreed to the published version of the manuscript.

**Funding:** This work is supported by the National Natural Science Foundation of China (no. 51877070, U20A20198, 51577048), the Natural Science Foundation of Hebei Province of China (no. E2021208008), the Talent Engineering Training Support Project of Hebei Province (A201905008), the National Engineering Laboratory of Energy-saving Motor and Control Technique, Anhui University (no. KFKT201901).

**Conflicts of Interest:** The authors declare no conflict of interest.

## References


1. Liu, J.-L.; Xiao, F.; Shen, Y.; Mai, Z.-Q.; Li, C.-R. Position-sensorless control technology of permanent-magnet synchronous motor—A review. *Trans. China Electrotech. Soc.* **2017**, *32*, 76–88.
2. Yang, C.; Ma, T.; Che, Z.; Zhou, L. An adaptive-gain sliding mode observer for sensorless control of permanent magnet linear synchronous motors. *IEEE Access* **2018**, *6*, 3469–3478. [CrossRef]
3. Ding, B.; Xu, D.; Jiang, B.; Shi, P.; Yang, W. Disturbance-observer-based terminal sliding mode control for linear traction system with prescribed performance. *IEEE Trans. Transp. Electrif.* **2021**, *7*, 649–658. [CrossRef]
4. Zhao, X.; Fu, D. Adaptive neural network nonsingular fast terminal sliding mode control for permanent magnet linear synchronous motor. *IEEE Access* **2019**, *7*, 180361–180372. [CrossRef]
5. Roy, S.; Baldi, S.; Fridman, L.M. On adaptive sliding mode control without a priori bounded uncertainty. *Automatica* **2020**, *111*, 108650. [CrossRef]
6. Cheema, M.A.M.; Fletcher, J.E.; Farshadnia, M.; Rahman, M.F. Sliding mode based combined speed and direct thrust force control of linear permanent magnet synchronous motors with first-order plus integral sliding condition. *IEEE Trans. Power Electron.* **2019**, *34*, 2526–2538. [CrossRef]
7. Wang, W.; Lin, H.; Yang, H.; Liu, W.; Lyu, S. Second-order sliding mode-based direct torque control of variable-flux memory machine. *IEEE Access* **2020**, *8*, 34981–34992. [CrossRef]
8. Sun, X.; Cao, J.; Lei, G.; Guo, Y.; Zhu, J. A composite sliding mode control for SPMSM drives based on a new hybrid reaching law with disturbance compensation. *IEEE Trans. Transp. Electrif.* **2021**, *7*, 1427–1436. [CrossRef]
9. Jin, H.; Zhao, X. Approach angle-based saturation function of modified complementary sliding mode control for PMLSM. *IEEE Access* **2019**, *7*, 126014–126024. [CrossRef]
10. Cheema, M.A.M.; Fletcher, J.E.; Farshadnia, M.; Xiao, D.; Rahman, M.F. Combined speed and direct thrust force control of linear permanent-magnet synchronous motors with sensorless speed estimation using a sliding-mode control with integral action. *IEEE Trans. Ind. Electron.* **2017**, *64*, 3489–3501. [CrossRef]
11. Yang, R.; Li, L.-Y.; Wang, M.-Y.; Zhang, C.-M.; Zeng-Gu, Y.-M. Force ripple estimation and compensation of PMLSM with incremental extended state modeling-based kalman filter: A practical tuning method. *IEEE Access* **2019**, *7*, 108331–108342. [CrossRef]
12. Wen, T.; Wang, Z.; Xiang, B.; Han, B.; Li, H. Sensorless Control of Segmented PMLSM for Long-Distance Auto-Transportation System Based on Parameter Calibration. *IEEE Access* **2020**, *8*, 102467–102476. [CrossRef]
13. Wang, A.; Jia, X.; Dong, S. A new exponential reaching law of sliding mode control to improve performance of permanent magnet synchronous motor. *IEEE Trans. Magn.* **2013**, *49*, 2409–2412. [CrossRef]
14. Wang, Y.; Feng, Y.; Zhang, X.; Liang, J. A new reaching law for antidisturbance sliding-mode control of PMSM speed regulation system. *IEEE Trans. Power Electron.* **2020**, *35*, 4117–4126. [CrossRef]
15. Roy, S.; Baldi, S.; Ioannou, P.A. An Adaptive Control Framework for Underactuated Switched Euler-Lagrange Systems. *IEEE Trans. Autom. Control* **2021**. [CrossRef]
16. Xu, Z.; Rahman, M.F. Comparison of a sliding observer and a kalman filter for direct-torque-controlled IPM synchronous motor drives. *IEEE Trans. Ind. Electron.* **2012**, *59*, 4179–4188. [CrossRef]
17. Cao, R.; Jiang, N.; Lu, M. Sensorless control of linear flux-switching permanent magnet motor based on extended kalman filter. *IEEE Trans. Ind. Electron.* **2020**, *67*, 5971–5979. [CrossRef]
18. Zhou, Z.; Xia, C.; Yan, Y.; Wang, Z.; Shi, T. Disturbances attenuation of permanent magnet synchronous motor drives using cascaded predictive-integral-resonant controllers. *IEEE Trans. Power Electron.* **2018**, *33*, 1514–1527. [CrossRef]
19. Wen, T.; Liu, Y.; Wu, Q.H.; Qiu, L. Cascaded sliding-mode observer and its applications in output feedback control part I: Observer design and stability analysis. *CSEE J. Power Energy Syst.* **2021**, *7*, 295–306.
20. Rsetam, K.; Cao, Z.; Man, Z. Cascaded-extended-state-observer-based sliding-mode control for underactuated flexible joint robot. *IEEE Trans. Ind. Electron.* **2020**, *67*, 10822–10832. [CrossRef]

MDPI
St. Alban-Anlage 66
4052 Basel
Switzerland
Tel. +41 61 683 77 34
Fax +41 61 302 89 18
www.mdpi.com

*Electronics* Editorial Office
E-mail: electronics@mdpi.com
www.mdpi.com/journal/electronics